Recent Developments in Food Gels

Recent Developments in Food Gels

Editors

Baskaran Stephen Inbaraj
Kandi Sridhar
Minaxi Sharma

MDPI

Basel • Beijing • Wuhan • Barcelona • Belgrade • Novi Sad • Cluj • Manchester

Editors

Baskaran Stephen Inbaraj
Fu Jen Catholic University
New Taipei City, Taiwan

Kandi Sridhar
Karpagam Academy of
Higher Education
(Deemed to be University)
Coimbatore, India

Minaxi Sharma
CARAH
Ath, Belgium

Editorial Office
MDPI
St. Alban-Anlage 66
4052 Basel, Switzerland

This is a reprint of articles from the Special Issue published online in the open access journal *Gels* (ISSN 2310-2861) (available at: https://www.mdpi.com/journal/gels/special_issues/developments_food_gels).

For citation purposes, cite each article independently as indicated on the article page online and as indicated below:

Lastname, A.A.; Lastname, B.B. Article Title. *Journal Name* **Year**, *Volume Number*, Page Range.

ISBN 978-3-0365-9698-3 (Hbk)
ISBN 978-3-0365-9699-0 (PDF)
doi.org/10.3390/books978-3-0365-9699-0

Contents

Preface

Food gels have played a pivotal role in the food industry, contributing to the enhancement of texture, visual appeal, and stability of various food products. Recently, studies on food gels have endured substantial advancements and opened up novel opportunities and potential advantages for the food industry. Although significant progress has been made in synthetic polymer chemistry to overcome some of the challenges in gel formulation, there are persisting challenges, such as sustainability, renewability, and cost-effectiveness, which require further attention and innovation. Therefore, there is a growing interest towards developing gels sourced from natural plant-based substances with minimal environmental impact. Food biopolymers are promising, including proteins and polysaccharides possessing a versatile range of functionalities and physical gelation characteristics. These contribute significantly to prolonging shelf-life, reducing fat content, enhancing satiety, and enabling the 3D printing of complex food shapes. Yet, additional research is imperative to comprehend the fabrication techniques, gelling mechanisms, and the structural/mechanical attributes of food gels. In this regard, to enhance our scientific knowledge and comprehension of natural products derived from food gels, we organized the Special Issue below, dealing with but not limited to:

- Food gel fabrication with novel processing methods;
- Polymerization/crosslinking methods;
- Elucidation of molecular mechanisms;
- Innovative analytical approaches to characterization, the molecular structure–functionality relationship, and food gel–body interaction.

This Special Issue presents a collection of research articles, reviews, and perspectives that highlight some of the recent advances in this rapidly evolving field. Under this Special Issue, we invited globally renowned researchers to contribute to this topic, resulting in a total of 16 submissions, out of which 13 articles were accepted after rigorous peer review for final publication.

Baskaran Stephen Inbaraj, Kandi Sridhar, and Minaxi Sharma
Editors

gels

Editorial

Editorial on Special Issue "Recent Developments in Food Gels"

Kandi Sridhar [1], Minaxi Sharma [2,*] and Baskaran Stephen Inbaraj [3,*]

1 Department of Food Technology, Karpagam Academy of Higher Education (Deemed to Be University), Coimbatore 641021, India
2 CARAH ASBL, Rue Paul Pastur, 11-7800 Ath, Belgium
3 Department of Food Science, Fu Jen Catholic University, New Taipei City 242062, Taiwan
* Correspondence: m.sharma@carah.be (M.S.); 138547@mail.fju.edu.tw (B.S.I.)

Food gels have been a crucial component in the food industry for many years. They have been used to improve the texture, appearance, and stability of food products [1]. However, with recent advancements in technology and research, food gels have taken a giant leap forward, offering new possibilities and potential benefits for the global food industry [2]. While significant progress has been made in synthetic polymer chemistry to overcome some of the challenges in gel design, emerging challenges, such as sustainability, renewability, and cost-effectiveness, remain to be overcome. To address these challenges, there is a growing interest in preparing gels from natural sources with low environmental impact. Food biopolymers, such as proteins and polysaccharides are particularly promising due to their affordable, edible, biocompatible, biodegradable, and renewable properties, with a diverse range of functionalities and physical gelation characteristics [3]. Compared to synthetic gels, food gels play a critical role in modern food design by imparting desired sensory, rheological, textural, and functional properties, increasing shelf-life, reducing fat, enhancing satiety, and enabling 3D printing of complex food shapes [4]. However, further research is needed to understand the fabrication methods, gelling mechanisms, and structural/mechanical properties of food gels. It is also important to investigate how these design principles impact the rheological and tribological properties of foods to improve food quality and modify nutrient delivery, without affecting the sensory properties or altering drug/bioactive targeting within the gastrointestinal tract. Therefore, a team of researchers with expertise in food gels, novel food product development, functional foods, and extraction of bioactive compounds have organized a Special Issue entitled "Recent Developments in Food Gels" to be published in Gels (ISSN 2310-2861), an international open access journal on physical (supramolecular) and chemical gel-based materials, under the section "Gel Applications". In this Special Issue, we delved into the latest research and innovations in the world of food gels and explored the potential they hold by collecting research and review articles dealing with but not limited to:

- Food gel fabrication with novel processing methods;
- Polymerization/crosslinking methods;
- Elucidation of molecular mechanisms;
- Innovative analytical approaches to characterization, molecular structure–functionality relationships, and food gel–body interaction.

This Special Issue presents a collection of research articles, reviews, and perspectives that highlight some of the recent advances in this rapidly evolving field. The articles in this issue cover a broad range of topics, including the use of novel gelation agents, the characterization of the structure and rheological properties of food gels, and the application of food gels in novel food products. For this Special Issue, we had invited globally renowned researchers to contribute to this topic, resulting in a total of 16 submissions, out of which 13 articles were accepted after rigorous peer-review for final publication.

One of the key themes in this issue is the addition of novel ingredients to gels and studying their characteristics. For example, Popov et al. investigated the effects of incorporating hogweed pectin into an apple pectin hydrogel. The researchers found that the

Citation: Sridhar, K.; Sharma, M.; Stephen Inbaraj, B. Editorial on Special Issue "Recent Developments in Food Gels". *Gels* **2023**, *9*, 899. https://doi.org/10.3390/gels9110899

Received: 1 November 2023
Accepted: 7 November 2023
Published: 13 November 2023

addition of hogweed pectin led to an increase in the gelation temperature and gel strength of the apple pectin hydrogel. Additionally, the hogweed pectin was found to improve the sensory properties of the hydrogel, specifically its texture and overall acceptability. The study also found that the concentration of hogweed pectin had an impact on the properties of the hydrogel, with higher concentrations resulting in a stronger and more elastic gel. However, excessively high concentrations of hogweed pectin led to a decrease in the sensory properties of the hydrogel, likely due to a more pronounced earthy flavor. Similarly, another study by Arai et al. used wet-ground okra to improve the mechanical properties and intermolecular forces of soybean protein isolate (SPI) gels. The results showed that the incorporation of wet grinder-treated okra into SPI gels improved their mechanical properties, including hardness and elasticity, compared to control gels without okra. Likewise, a study investigated the effect of oxidation modification induced by malondialdehyde on the physicochemical properties and gel characteristics of duck myofibrillar proteins. The results showed that the oxidation modification caused significant changes in the physicochemical properties of the proteins, such as decreased solubility and increased surface hydrophobicity. Furthermore, the gels made from the modified proteins had lower a gel strength, water-holding capacity, and thermal stability compared to the non-modified proteins. All these studies suggested that the addition and/or modification of different ingredients can significantly affect the properties of the resulting gels, which could have implications in the food industry for the development of products with desirable textures and functional properties.

Several articles in this issue describe the enrichment of food gels with plant-based sources. For instance, Belova et al. explored the potential of using callus tissue of narrow-leaved lupin (*Lupinus angustifolius*) as a bioactive filler in 3D-printed k-carrageenan food gels. The researchers prepared the callus tissue using a tissue culture technique and characterized its chemical composition and antioxidant activity. They then incorporated the callus tissue into the k-carrageenan gel and evaluated the effects on the physical, mechanical, and rheological properties of the gel. The results showed that the incorporation of the callus tissue improved the antioxidant activity of the gel and did not significantly affect its physical and mechanical properties. With a similar approach, a study by Carvajal-Mena et al. optimized the 3D printing parameters for salmon gelatin gels using artificial neural networks and the response surface methodology. The study considered the interactions between different independent variables and used artificial neural networks to train and learn from the experimental data of the response variables. The authors performed a regression analysis of training, validation, and testing for all the experiments of apparent viscosity, hardness, and dimensional stability. The optimal printing conditions were found to be 24 mm/s for the extrusion speed, 0.70 mm for the nozzle diameter, and 0.50 mm for the nozzle height to achieve 380.85 Pa·s for viscosity, 7.75 N for hardness, and 98.87% for dimensional stability. These predicted values were not significantly different from the experimental data under the optimal printing conditions. Another study by Popov, Smirnov, Paderin, Khramova. et al. used the antioxidant pectin obtained from fireweed (*Chamaenerion angustifolium*) to improve the mechanical and rheological properties of an agar gel. This study further evaluated the effect of adding different concentrations of pectin on the mechanical and rheological properties of the agar gel, as well as its simulated digestibility and oral processing. The results showed that the addition of pectin increased the gel strength, elasticity, and viscosity, and improved its antioxidant properties. The simulated digestion study revealed that the pectin-enriched gel had a slower rate of disintegration, indicating improved stability and the potential for controlled release of bioactive compounds. The sensory analysis showed that the gel with the highest concentration of pectin had the highest overall acceptance. Therefore, these studies highlighted the potential of incorporating plant-based materials for the development of functional gels with improved mechanical, rheological, and antioxidant properties.

Food gels can play an important role in coating applications to provide a smooth and uniform coating, which can help to prevent moisture loss and oxidation, and protect the food product from contamination. With the aim of developing a composite edible coating, Heristika et al. used gelatin–pectin incorporating garlic essential oil on the physicochemical characteristics of red chili (*Capsicum annuum* L.). The study found that the edible coating improved the shelf life of the red chili peppers by reducing the rate of moisture loss and weight loss. Additionally, the coating was found to have antimicrobial properties, which inhibited the growth of bacteria on the surface of the peppers. Another study by Trodtfeld et al. developed a composite gel made from biodegradable compounds, including prolamin, d-mannose, and citric acid, as a coating to increase the oxygen barrier of food packaging materials. The gels were physically cross-linked with particles synthesized from tetraethyl orthosilicate and tetramethyl orthosilicate precursors to improve stability and the mechanical properties. The article concluded that the composite gel holds promise for oxygen-barrier food packaging and is safe for consumer contact, but further research is needed to optimize the stability of the oxygen barrier in humid environments and investigate the potential sensitizing effects of biodegradable materials on consumers. These findings have implications for the development of new food preservation technologies that incorporate natural ingredients such as garlic essential oil.

Similarly, Fauzan et al. developed an eco-friendly, biodegradable, and sustainable active packaging material using fish gelatin-based edible film incorporated with *Ficus carica* L. leaf extract. The results showed that adding *Ficus carica* L. leaf extract to gelatin films significantly affected their tensile strength, elongation at break, transmittance and transparency, solubility, water vapor permeability, antioxidant activities, and antibacterial activity. The most promising result was obtained in the edible film with 10% *Ficus carica* L. leaf extract among all the samples. The study's overall findings showed that the fish gelatin-based films incorporating *Ficus carica* L. leaf extract have a good potential as an eco-friendly, biodegradable, and sustainable active packaging.

Finally, this issue contains four systematic review articles that explore the principles and formation mechanisms of hydrogels, and their current status and various applications in the food industry. Nath et al. summarize the various applications of food hydrogels, such as in food texture modification, nutrient delivery, and 3D printing, thereby highlighted the potential of hydrogels as a versatile tool for designing innovative food products. Another review by Kaur et al. provides an overview of the current state of research on milk protein-based nanohydrogels. The authors discuss the various methods for synthesizing milk protein-based nanohydrogels, their unique properties such as biocompatibility, stability, ability to encapsulate bioactive compounds, and potential applications of milk protein-based nanohydrogels in the food industry. Similarly, another article discusses the formulation, processing, and potential applications of food emulsion gels made from plant-based ingredients. The review concludes that the emulsion gels were semi-solid systems with a gel network structure that integrate the characteristics of emulsions and gels. These are used in the food industry to create texture, deliver functional food ingredients, and reduce the fat content of products. The article also discusses that plant-based emulsion gels have promising prospects as a delivery system for functional ingredients and as a healthy alternative to traditional fats. A review by Said et al. discusses the extraction and characterization of pectin from various sources, as well as the preparation of pectin hydrogels using different methods. The review focuses on the various crosslinking methods used to form hydrogels, with a focus on physical, chemical, and interpenetrating polymer network approaches. The article also highlights the potential applications of pectin hydrogels in the food industry, such as encapsulating bioactive substances, improving stability, and controlled release.

In conclusion, recent advances in food gels have been nothing short of revolutionary for the food industry. From the development of plant-based gels to the increased use of natural gelling agents, these innovations are changing the way we think about food and the role gels play in preserving and improving the quality of food products. These developments are not only improving the sustainability of the food industry, but also providing new opportunities for food product innovation and the creation of healthier, more functional food products. The future of food gels is exciting and holds even more potential for the food industry. We encourage researchers and companies to continue exploring the possibilities of food gels and to push the boundaries of what is possible. We believe that with continued research and innovation, food gels will play an even greater role in shaping the food industry in the years to come. We hope that this issue will provide valuable insights and inspiration for those in the food industry and spark further discussion and research in this field.

List of Contributions

1. Popov, S.; Smirnov, V.; Khramova, D.; Paderin, N.; Chistiakova, E.; Ptashkin, D.; Vityazev, F. Effect of hogweed pectin on rheological, mechanical, and sensory properties of apple pectin hydrogel. *Gels* **2023**, *9*, 225.
2. Arai, Y.; Nishinari, K.; Nagano, T. Wet grinder-treated okara improved both mechanical properties and intermolecular forces of soybean protein isolate gels. *Gels* **2022**, *8*, 616.
3. Zhu, X.; Ma, Z.; Zhang, X.; Huang, X.; Liu, J.; Zhuang, X. Effect of malondialdehyde-induced oxidation modification on physicochemical changes and gel characteristics of duck myofibrillar proteins. *Gels* **2022**, *8*, 633.
4. Belova, K.; Dushina, E.; Popov, S.; Zlobin, A.; Martinson, E.; Vityazev, F.; Litvinets, S. Enrichment of 3D-printed k-carrageenan food gel with callus tissue of narrow-leaved lupin *Lupinus angustifolius*. *Gels* **2023**, *9*, 45.
5. Carvajal-Mena, N.; Tabilo-Munizaga, G.; Saldaña, M.D.A.; Pérez-Won, M.; Herrera-Lavados, C.; Lemus-Mondaca, R.; Moreno-Osorio, L. Three-dimensional printing parameter optimization for salmon gelatin gels using artificial neural networks and response surface methodology: Influence on physicochemical and digestibility properties. *Gels* **2023**, *9*, 766.
6. Popov, S.; Smirnov, V.; Paderin, N.; Khramova, D.; Chistiakova, E.; Vityazev, F.; Golovchenko, V. Enrichment of agar gel with antioxidant pectin from fireweed: Mechanical and rheological properties, simulated digestibility, and oral processing. *Gels* **2022**, *8*, 708.
7. Heristika, W.; Ningrum, A.; Supriyadi; Munawaroh, H.S.H.; Show, P.L. Development of composite edible coating from gelatin-pectin incorporated garlic essential oil on physicochemical characteristics of red chili *(Capsicum annnum* L.). *Gels* **2023**, *9*, 49.
8. Trodtfeld, F.; Tölke, T.; Wiegand, C. Developing a prolamin-based gel for food packaging: In-vitro assessment of cytocompatibility. *Gels* **2023**, *9*, 740.
9. Fauzan, H.R.; Ningrum, A.; Supriyadi, S. Evaluation of a fish gelatin-based edible film incorporated with *Ficus carica* L. leaves extract as active packing. *Gels* **2023**, *9*, 1–13.
10. Nath, P.C.; Debnath, S.; Sridhar, K.; Inbaraj, B.S.; Nayak, P.K.; Sharma, M. A comprehensive review of food hydrogels: Principles, formation mechanisms, microstructure, and its applications. *Gels* **2023**, *9*, 1.
11. Kaur, M.; Bains, A.; Chawla, P.; Yadav, R.; Kumar, A.; Inbaraj, B.S.; Sridhar, K.; Sharma, M. Milk protein-based nanohydrogels: Current status and applications. *Gels* **2022**, *8*, 432.
12. Yiu, C.C.-Y.; Liang, S.W.; Mukhtar, K.; Kim, W.; Wang, Y.; Selomulya, C. Food emulsion gels from plant-based ingredients: Formulation, processing, and potential applications. *Gels* **2023**, *9*, 366.
13. Said, N.S.; Olawuyi, I.F.; Lee, W.Y. Pectin hydrogels: Gel-Forming behaviors, mechanisms, and food applications. *Gels* **2023**, *9*, 732.

Author Contributions: Conceptualization, K.S., B.S.I. and M.S.; methodology, K.S. and M.S.; software, M.S.; validation, B.S.I. and M.S.; formal analysis, K.S., B.S.I. and M.S.; investigation, K.S., B.S.I. and M.S.; resources, M.S. and K.S.; data curation, K.S., B.S.I. and M.S.; writing—original draft preparation, K.S. and M.S.; writing—review and editing, K.S., B.S.I. and M.S.; visualization, B.S.I. and M.S.; supervision, M.S.; project administration, B.S.I. and M.S.; funding acquisition, M.S. and B.S.I. All authors have read and agreed to the published version of the manuscript.

Funding: This research received no external funding.

Acknowledgments: We would like to thank all of the authors and contributors for their contributions to this Special Issue on recent advances in food gels.

Conflicts of Interest: The authors declare no conflict of interest.

References

1. Banerjee, S.; Bhattacharya, S. Food Gels: Gelling Process and New Applications. *Crit. Rev. Food Sci. Nutr.* **2012**, *52*, 334–346. [CrossRef] [PubMed]
2. Cao, Y.; Mezzenga, R. Design principles of food gels. *Nat. Food* **2020**, *1*, 106–118. [CrossRef] [PubMed]
3. Baranwal, J.; Barse, B.; Fais, A.; Delogu, G.L.; Kumar, A. Biopolymer: A sustainable material for food and medical applications. *Polymers* **2022**, *14*, 983. [CrossRef] [PubMed]
4. Liu, B.; Yang, H.; Zhu, C.; Xiao, J.; Cao, H.; Simal-Gandara, J.; Li, Y.; Fan, D.; Deng, J. A comprehensive review of food gels: Formation mechanisms, functions, applications, and challenges. *Crit. Rev. Food Sci. Nutr.* **2022**, 1–23. [CrossRef] [PubMed]

gels

Article

Effect of Hogweed Pectin on Rheological, Mechanical, and Sensory Properties of Apple Pectin Hydrogel

Sergey Popov *, Vasily Smirnov, Daria Khramova, Nikita Paderin, Elizaveta Chistiakova, Dmitry Ptashkin and Fedor Vityazev

Institute of Physiology of Federal Research Centre "Komi Science Centre of the Urals Branch of the Russian Academy of Sciences", 50 Pervomaiskaya Str., 167982 Syktyvkar, Russia
* Correspondence: s.v.popov@inbox.ru; Tel./Fax: +78-21-224-1001

Abstract: This study aims to develop hydrogels from apple pectin (AP) and hogweed pectin (HP) in multiple ratios (4:0; 3:1; 2:2; 1:3; and 0:4) using ionotropic gelling with calcium gluconate. Rheological and textural analyses, electromyography, a sensory analysis, and the digestibility of the hydrogels were determined. Increasing the HP content in the mixed hydrogel increased its strength. The Young's modulus and tangent after flow point values were higher for mixed hydrogels than for pure AP and HP hydrogels, suggesting a synergistic effect. The HP hydrogel increased the chewing duration, number of chews, and masticatory muscle activity. Pectin hydrogels received the same likeness scores and differed only in regard to perceived hardness and brittleness. The galacturonic acid was found predominantly in the incubation medium after the digestion of the pure AP hydrogel in simulated intestinal (SIF) and colonic (SCF) fluids. Galacturonic acid was slightly released from HP-containing hydrogels during chewing and treatment with simulated gastric fluid (SGF) and SIF, as well as in significant amounts during SCF treatment. Thus, new food hydrogels with new rheological, textural, and sensory properties can be obtained from a mixture of two low-methyl-esterified pectins (LMPs) with different structures.

Keywords: low-methyl-esterified pectin; hydrogel; ionotropic gelation; texture; rheology; electromyography; sensory analysis; simulated digestion

Citation: Popov, S.; Smirnov, V.; Khramova, D.; Paderin, N.; Chistiakova, E.; Ptashkin, D.; Vityazev, F. Effect of Hogweed Pectin on Rheological, Mechanical, and Sensory Properties of Apple Pectin Hydrogel. *Gels* **2023**, *9*, 225. https://doi.org/10.3390/gels9030225

Academic Editors: Baskaran Stephen Inbaraj, Kandi Sridhar and Minaxi Sharma

Received: 17 February 2023
Revised: 9 March 2023
Accepted: 13 March 2023
Published: 14 March 2023

1. Introduction

Hydrogels are of increasing interest in food applications, as they make it possible to obtain food that meets modern consumer requirements. Currently, there is an intensive development of new food gels for the creation of healthy food products with high sensory scores. Customizable hydrogel structure design can improve the nutritional value, health content, taste, and texture properties of food-grade hydrogels [1]. Plant-derived polysaccharides have a high potential for the creation of food hydrogels due to their safety and inexpensive production. [2]. Pectin is an acidic polysaccharide that is a component that is located in the spaces between walls of cells of flowering plants. [3]. Pectin-based gelled foods are becoming increasingly popular thanks to the gelling properties and biological activities of pectin [4–6].

Pectin is not broken down in the upper gastrointestinal tract and is fermented by colon bacteria [5]. Therefore, pectin has gained a lot of importance in its use as a food matrix for the targeted release of bioactivities to the colon [7]. Nevertheless, the usage of hydrogels made of commercial pectins as colon delivery systems is restricted due to intense swelling, declining mechanical properties, and quick degradation in the small intestine [4]. Citrus and apple pectins are the most commonly used for hydrogel preparation [7]. Generally, the macromolecule of pectin consists of nonbranched and branched areas. The main structural domain of pectin is homogalacturonan, which is made of 1,4-α-D-galacturonic acid residues. The homogalacturonic part largely determines the gel-forming and emulsifying properties of pectin [8]. Therefore, traditional ways of extracting pectin from plant raw material based on long treatment with acids at high temperatures destroy less stable ramified regions so that commercial pectins often consist primarily of linear homogalacturonan [9]. However, many natural pectins additionally contain such structural domains as rhamnogalacturonan II, xylogalacturonan, and others. Moreover, an important branched domain of pectins is rhamnogalacturonan type I in which arabinose and galactose are attached to rhamnose residues embedded in galacturonan, forming side chains [10]. Carboxyls of galacturonic acids are partially methyl-esterified, separating pectins into high-methyl-esterified pectins (HMPs) and LMPs. HMPs form gels at low pH, and gelling LMPs are provided by multivalent cations [11]. Gelling of LMPs involves free carboxyls interacting with cations to create "egg-box" formations, polysaccharide chain dimerization, and the aggregation of dimers into a gel network [11,12].

Rhamnogalacturonan-I-containing pectin from hogweed H. sosnówskyi (HP) was earlier isolated and characterized [13]. Hogweed pectin was found to be an LMP and formed a stronger calcium hydrogel than commercial apple pectin (AP) [14]. A hydrogel made from HP was found to be more stable than a hydrogel made from AP in saline with pH 3.0–8.0 [15]. However, the degradation of HP hydrogels during digestion has not been investigated yet. We assume here that HP hydrogels are more stable in the gastrointestinal environment than AP hydrogels and that the addition of HP changes the properties of AP hydrogels. It was previously shown that the addition of an HMP to an LMP produced a stronger mixed gel due to a synergistic effect [16], whereas the gel properties of a mixture of two LMPs have not been determined earlier.

This investigation aims to estimate the influence of HP (degree of methyl esterification ca. 21%) on the rheological and mechanical properties of AP (degree of methyl esterification ca. 43%) hydrogels. Chewing, bolus formation, sensory scores, and the simulated digestibility of pure and mixed hydrogels made from AP and HP are also compared.

2. Results and Discussion

2.1. Preparation of Pectin Hydrogels

Ionotropic hydrogels were created by immersing a dialysis bag containing a pectin solution (50 mL) in a calcium gluconate solution (200 mL). After 48 h, a solid hydrogel was formed in the dialysis bag in the form of a cylinder with a length of 7–8 cm, a diameter of 2.4–2.8 cm, and a weight of 44–48 g. The cylindrical hydrogel was then cut into puck-shaped samples at a predetermined height for subsequent analyses.

Five types of pectin hydrogels (PG1–PG5) were obtained using AP and HP in different ratios (4:0; 3:1; 2:2; 1:3, and 0:4). The PGs had the same density (1.07–1.13 mg/mm^3) and water content, measured as the weight loss (WL) during drying (80.7–82.7%), regardless of the pectin composition. Increasing the HP content in the hydrogels slightly increased the pH. The pH values for PG1, PG3, and PG5 were 4.42 $\pm$ 0.05, 4.55 $\pm$ 0.04, and 4.70 $\pm$ 0.08, respectively.

An external gelation method used in functional food applications of LMPs is typically based on heating and cooling calcium solutions [5,16]. To encapsulate thermally unstable bioactive substances, pectin gel beads are usually obtained at room temperature by dropping a pectin solution into a solution of calcium ions [17–19]. We received pectin hydrogels at room temperature in order to keep the possibility of loading them with functional ingredients open. At the same time, larger sizes of the obtained hydrogel samples than gel beads of a typical diameter of 1–2 mm were more convenient for studying the textural properties of hydrogel. Calcium gluconate rather than calcium chloride was used to improve consumer acceptance of the hydrogels, as calcium chloride has a strong, bitter taste. Sucrose was shown to promote the formation of pectin gels because its hydroxyl groups could stabilize Ca–pectin crosslinks and form hydrogen bonds with water molecules, immobilizing free water and concentrating the polymer environment for gelation [16]. Similar values of hydrogel density and water content among the obtained PGs were expected. First, the hydrogels contained an equal amount of pectin (4%), and second, an equal concentration of sucrose (10%) provided a similar water-keeping capacity.

The pH shift caused by the addition of the HP seemed to be associated with a lower methyl esterification of HP than AP. The greater number of free carboxyl groups in the HP provided more sites of interaction with calcium ions, which apparently led to a higher content of calcium ions in the HP hydrogel than in the AP hydrogel. Therefore, there were fewer dissociated carboxyl groups and fewer free hydrogen ions in HP hydrogel homogenates prepared for pH measurement than in AP ones. As a result, the pH of the HP hydrogel was higher since samples made from HP hydrogel represented the calcium pectate of a weaker acid and stronger base than samples made from AP hydrogel.

2.2. Rheological Properties of Pectin Hydrogels

The rheological data were analyzed in the linear (LVE) and nonlinear (n-LVE) viscoelastic regions. The storage modulus G' of all the PG samples was greater than the loss modulus G'' throughout the LVE region. The G'_{LVE} and G''_{LVE} values increased with increasing HP content in the hydrogels, indicating an increase in network rigidity and total structural strength (Table 1). The G'_{LVE} and G''_{LVE} values of PG5 were 1.9 and 2.3 times higher than those of PG1, respectively. The loss factor, tan $[\delta]_{LVE}$, represented a material's physical behavior. The tan $[\delta]_{LVE}$ was 0.20–0.27 for PG1–PG5, indicating that pectin hydrogels had a solid-like behavior. The limiting strain (γL) at which G' sharply diminished with the increase in strain in PG4 and PG5 was lower than that of PG1. The fracture strain (γFR) measured at the fracture point decreased as the hydrogels' HP content increased. The γFR value of PG5 was 2.5 times lower than that of PG1 (Table 1). The fracture stress (τFP), where $G' = G''$, and the complex modulus (G*FP) at τFP did not differ between PG1 and PG5. The slope of the loss tangent in the nonlinear region (tan [δ] AF) described the transition of the gels from the elastic state to the viscous state since it showed the ratios of the viscous and elastic constituents of materials at large amounts of deformation. PG1, PG2, PG4, and PG5 showed similar tan $[\delta]_{AF}$ values without any significant differences among them. PG3 showed a higher tan $[\delta]_{AF}$ value than PG1, PG2, PG4, and PG5, which indicated the highest spreadability of the hydrogel containing equal amounts of AP and HP (Table 1).

Table 1. Strain sweep for pectin hydrogels PG1–PG5 in the amplitude sweep test (1 Hz. 20 °C).

Parameters	PG1	PG2	PG3	PG4	PG5
G'_{LVE} (Pa)	51,636 ± 4741 ^{a}	61,209 ± 7204 ^{b}	74,486 ± 10,993 ^{c}	88,753 ± 5467 ^{d}	98,533 ± 14,595 ^{e}
G''_{LVE} (Pa)	10,966 ± 2033 ^{a}	12,997 ± 3039 ^{a}	16,562 ± 4640 ^{b}	17,428 ± 2411 ^{c}	25,357 ± 4587 ^{d}
G^*_{LVE}	53,643 ± 5536 ^{a}	63,046 ± 5936 ^{b}	77,100 ± 9036 ^{c}	89,529 ± 6592 ^{d}	101,313 ± 15,488 ^{e}
Tan $[\delta]_{LVE}$	0.21 ± 0.04 ^{a}	0.22 ± 0.09 ^{a}	0.24 ± 0.12 ^{a}	0.20 ± 0.04 ^{a}	0.27 ± 0.10 ^{a}
γL (%)	0.35 ± 0.08 ^{a}	0.23 ± 0.02 ^{a,b}	0.20 ± 0.01 ^{a,b}	0.19 ± 0.02 ^{b}	0.15 ± 0.04 ^{b}
τFr (Pa)	251 ± 151 ^{a}	155 ± 48 ^{a}	170 ± 18 ^{a}	216 ± 42 ^{a}	114 ± 24 ^{a}
γFr	0.75 ± 0.10 ^{a}	0.54 ± 0.03 ^{b}	0.48 ± 0.05 ^{b,c}	0.44 ± 0.01 ^{c}	0.30 ± 0.06 ^{d}
G^*_{FP} (Pa)	24,252 ± 9158 ^{a}	29,605 ± 8375 ^{a}	38,404 ± 5752 ^{a}	47,546 ± 9452 ^{a}	53,702 ± 39,039 ^{a}
Tan $[\delta]_{AF}$	0.34 ± 0.13 ^{a,c,d}	0.36 ± 0.12 ^{a,b,c,d}	0.66 ± 0.06 ^{b}	0.39 ± 0.06 ^{a,c}	0.19 ± 0.04 ^{a,d}
G^*max/G^*_{LVE}	1.25 ± 0.21 ^{a}	1.26 ± 0.17 ^{a}	1.20 ± 0.02 ^{a}	1.05 ± 0.08 ^{a}	1.53 ± 0.18 ^{a}

Means with standard deviations are given ($n = 3$). Differences are significant ($p < 0.05$) between means labeled with different lowercase letters (a, b, c, and d). Storage modulus (G'_{LVE}), loss modulus (G''_{LVE}), complex modulus (G^*_{LVE}), limiting value of strain (γL), loss tangent (TAN $[\delta]_{LVE}$), fracture stress (τFr), corresponding complex modulus (G^*_{FP}) with the stress at flow point, and slope of the loss tangent after flow point (tan [δ]AF).

The rheological parameters of the frequency sweep tests in the LVE range are shown in Table 2. G' and G'' frequency dependencies are described by the power law equation with high correlation ($R^2 = 0.91$–0.99). The G' values increased as the hydrogels' HP content increased. The G' values of PG5 were 1.4 times higher than those of PG1. The frequency dependences of the elastic (k'), loss (k''), and complex (A) moduli increased with increase in the HP content in the hydrogels. However, the overall loss tangent (k''/k') and complex viscosity slope (η*S) did not differ among PG1–PG5. All PGs seemed to show elastic behavior, as revealed by the overall loss tangent (k''/k') values in the range 0.13–0.19, as well as the low slope of both moduli indicated by $0.08 \leq n' \leq 0.10$ and $0.04 \leq n'' \leq 0.10$ (Table 2).

Table 2. Frequency dependence of viscoelastic parameters for pectin hydrogels PG1–PG5, viscosity, and frequency ($0.3 < \omega < 70.0$ or at 0.54/1.11/10.50/30.60 Hz).

	Parameters		PG1	PG2	PG3	PG4	PG5
	k′ (Pa*s)		71,741 ± 20,277 ^{a}	86,851 ± 9022 ^{a}	88,590 ± 3574 ^{a}	83,490 ± 32,656 ^{a,b}	127,855 ± 15,415 ^{b}
	k″ (Pa*s)		10,307 ± 2719 ^{a}	11,295 ± 2339 ^{a}	13,227 ± 92 ^{a,b}	15,140 ± 4187 ^{a,b}	19,114 ± 5748 ^{b}
	A		75,313 ± 16,622 ^{a}	83,298 ± 5382 ^{a}	84,316 ± 4866	101,487 ± 37,040 ^{a,b}	135,162 ± 25,722 ^{b}
	k″/k′		0.14 ± 0.00 ^{a}	0.13 ± 0.02 ^{a}	0.15 ± 0.01 ^{a,b}	0.19 ± 0.03 ^{b}	0.17 ± 0.02 ^{a,b}
	η^*_s		12,039 ± 2919 ^{a}	12,701 ± 2685 ^{a}	10,096 ± 4678 ^{a}	12,346 ± 4791 ^{a}	18,773 ± 6072 ^{a}
	n'		0.08 ± 0.02 ^{a}	0.09 ± 0.02 ^{a}	0.10 ± 0.02 ^{a}	0.08 ± 0.05 ^{a}	0.09 ± 0.01 ^{a}
	n''		0.10 ± 0.02 ^{a,c}	0.10 ± 0.01 ^{a}	0.08 ± 0.01 ^{a,c}	0.04 ± 0.01 ^{b}	0.08 ± 0.05 ^{a,b,c}
	z		14.2 ± 3.7 ^{a}	10.8 ± 3.0 ^{a}	9.6 ± 2.4 ^{a}	9.3 ± 1.2 ^{a}	9.7 ± 1.3 ^{a}
Frequency (Hz)	0.54	G′ (kPa)	72.8 ± 18.1 ^{a}	74.4 ± 15.7 ^{a,b}	62.3 ± 29.5 ^{ab}	78.7 ± 32.3 ^{a,b}	118.9 ± 25.7 ^{b}
		G″ (kPa)	11.3 ± 2.4 ^{a}	12.9 ± 3.7 ^{a,b}	12.9 ± 4.2 ^{b}	19.6 ± 2.9 ^{b}	31.8 ± 6.3 ^{c}
	1.11	G′ (kPa)	72.4 ± 20.7 ^{a}	81.0 ± 17.0 ^{a}	69.1 ± 28.0 ^{a}	89.2 ± 32.7 ^{a}	132.5 ± 27.6 ^{b}
		G″ (kPa)	10.4 ± 2.9 ^{a}	11.1 ± 2.1 ^{a,b}	10.7 ± 4.0 ^{ab}	13.7 ± 3.7 ^{a,b}	19.2 ± 6.2 ^{b}
	10.50	G′ (kPa)	87.1 ± 26.7 ^{a}	96.9 ± 19.4 ^{a}	84.0 ± 34.8 ^{a}	100.4 ± 34.8 ^{a,b}	174.2 ± 36.6 ^{b}
		G″ (kPa)	13.0 ± 4.1 ^{a}	14.1 ± 2.9 ^{a}	12.1 ± 4.8 ^{a}	16.8 ± 5.1 ^{a,b}	26.9 ± 5.6 ^{b}
	30.60	G′ (kPa)	94.7 ± 31.6 ^{a}	108.3 ± 22.7 ^{a}	93.0 ± 38.9 ^{a}	100.9 ± 37.7 ^{a}	192.0 ± 36.4 ^{b}
		G″ (kPa)	15.7 ± 5.9 ^{a}	16.0 ± 3.5 ^{a}	13.7 ± 5.6 ^{a}	17.8 ± 5.8 ^{a,b}	30.7 ± 8.8 ^{b}

Means with standard deviations are given ($n = 3$). Differences are significant ($p < 0.05$) between means labeled with different lowercase letters (a, b, and c). The frequency dependences of the elastic (k′ and n'), loss (k″ and n''), and complex (A and z) moduli; overall loss tangent (k′/k″); and the slope of complex viscosity (η*S).

The viscosity values of the PGs are given in Table 3.

Table 3. Summary of power law parameters for the relationship between storage modulus or viscosity and frequency ($0.03 < \omega < 70.00$ Hz or at 10 and 50 Hz) of pectin hydrogels PG1–PG5.

Hydrogel	Viscosity				
	K (Pa*s)	R^2	n	η_{app} 10 (Hz)	η_{app} 50 (Hz)
PG1	11,942	0.999	−0.946	1333.10 ± 409.6	273.9 ± 109.4
PG2	12,677	0.999	−0.904	1482.5 ± 297.0	343.6 ± 79.6
PG3	10,780	0.999	−0.899	1285.1 ± 531.2	291.4 ± 125.1
PG4	13,310	0.997	−0.926	1541.4 ± 532.7	303.2 ± 111.2
PG5	18,847	0.999	−0.908	2162.8 ± 872.7	502.0 ± 198.3

Using an agglomerative hierarchical cluster analysis, Alghooneh et al. [20] divided the rheological properties of hydrocolloids into properties related to the strength, number, and distance of linkages, as well as the timescale of the junction zone. The rheological parameters of the first group, such as G'_{LVE}, k′, k″, and A, increased with increasing HP content in the hydrogels, indicating an increase in the strength of the linkage in the HP-containing hydrogels. The rheological parameters of the group "number of linkages" (G^*max/G^*_{LVE}, τFr, n', and z), "timescale of junction zone" (Tan $[\delta]_{LVE}$), k″/k′, and η*s), and "distance of linkage" (n'') did not differ among PG1–PG5, indicating the same number of linkages, the same time required for the transformation of hydrogel network chains to the thermodynamically ideal state, and the same average linear distance between two adjacent crosslinks in the gel networks.

2.3. Mechanical Properties of Pectin Hydrogels

The PGs showed force–distance curves of very different sizes in the puncture test (Figure 1). Table 4 shows the results of the puncture test of the PGs.

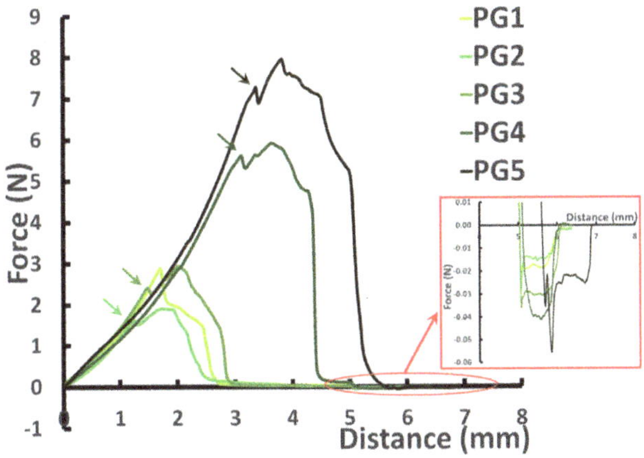

Figure 1. Representative force–distance curves of pectin hydrogels PG1–PG5. The arrows show the first significant break. The insertion window demonstrates the curve area, which corresponded to adhesiveness.

Table 4. Mechanical properties of pectin hydrogels PG1–PG5 in the puncture test.

Hydrogel	Hardness (N)	Fracturability (N)	Consistency (mJ)	Adhesiveness (mN)	Brittleness (mm)	Young's Modulus (kPa)
PG1	3.09 ± 0.41 [a]	3.09 ± 0.41 [a]	3.90 ± 0.51 [a]	30.1 ± 3.4 [a]	1.9 ± 0.2 [a]	608 ± 52 [a]
PG2	2.63 ± 0.49 [a]	2.56 ± 0.57 [a]	3.50 ± 0.44 [a]	26.7 ± 5.7 [a]	1.7 ± 0.1 [b]	521 ± 103 [b]
PG3	4.09 ± 0.98 [b]	3.93 ± 0.97 [b]	6.23 ± 1.36 [b]	39.5 ± 5.1 [b]	2.1 ± 0.3 [a]	782 ± 91 [c]
PG4	6.16 ± 0.70 [c]	6.04 ± 0.71 [c]	13.49 ± 2.69 [c]	50.7 ± 16.1 [b,c]	3.5 ± 0.2 [c]	488 ± 161 [a,b]
PG5	7.37 ± 1.82 [c]	7.03 ± 1.71 [c]	18.06 ± 5.29 [d]	52.5 ± 11.4 [c]	4.1 ± 0.6 [d]	474 ± 188 [a,b]

Means with standard deviations are given ($n = 8$). Differences are significant ($p < 0.05$) between means labeled with different lowercase letters (a, b, c, and d).

Hardness, measured as the maximum peak force [21], increased with increasing HP content in the hydrogels, indicating better stability of the HP gel networks under large deformation. The hardness value of PG5 was 2.4 times higher than that of PG1. The curve of PG1 did not have a fracture until the maximum peak was reached, resulting in the same hardness and fracturability values for PG1. Fracturability increased as the hydrogels' HP content increased. The curves of PG3–PG5 exhibited a significant break in the curve before reaching the maximum peak (shown by arrows in Figure 1). The fracturability value of PG5 was 2.3 times higher than that of PG1. The consistency represented the required work for deforming the material [22]. Similar to hardness and fracturability, PG5 and PG1 demonstrated the highest and lowest consistencies, respectively. Further, PG5 showed the highest values of adhesiveness, determined as the maximum negative peak below the baseline during withdrawal. A synergistic effect of AP and HP on the Young's modulus of mixed hydrogels was discovered. The Young's modulus [23] was the highest for PG3, which contained equal amounts of AP and HP. The Young's modulus value of PG3 exceeded those of PG1 and PG5 by 29 and 65%, respectively. Interestingly, the Young's modulus was positively correlated ($R = 0.88$, $p < 0.05$) with tan [δ]AF.

PG1–PG3 had a close serum release in the range of 0.62–0.71%. The serum release values of PG4 and PG5 were 0.48 ± 0.08 and 0.49 ± 0.03% ($p < 0.05$ vs. PG1–PG3), respectively.

The changes in mechanical parameters (hardness and consistency) were consistent with the changes in rheological parameters and confirmed a stronger linkage in HP hydrogels compared to AP hydrogels [24]. There are conflicting data for the effect of textural hardness on liking and acceptability evaluations of hydrogel foods. In a number of studies, increases in the hardness of hydrogels have led to decreases in their liking scores [25–28]. However, studies [29,30] have reported an increase in overall liking of foods with higher hardness, whereas others [31–33] have found no effect of hardness on acceptability. Increasing food hardness may be effective at slowing energy intake [34], and therefore, HP hydrogels may be useful for enhancing satiety and combating obesity.

Hydrogels containing HP demonstrated higher fracturability and adhesiveness than AP hydrogels, similar to hardness and consistency. Fractures can occur when all the bonds in a gel network are broken [20], so the increase in fracturability indicated higher numbers of linkages in the HP hydrogels. It should be noted, however, that G*max/G*LVE, τFr, n', and z values, which were related to the number of linkages, did not differ among PG1–PG5. Increased adhesiveness indicated an increase in the linkage distance, which contradicted the rheological analysis, which revealed equal n'' values in all the PGs. Different degrees of deformation in mechanical and rheological tests could explain the discrepancy in results.

Differences in the chemical structures of HP and AP apparently determined the different properties of pectin hydrogels PG1–PG5, which were made of different contents of HP and AP. It was assumed that Ca^{2+} ions bound to free COO groups in the homogalacturonic region. Then, dimerization of polysaccharide chains and dimer–dimer interactions occurred due to hydrogen bonds [11,12]. The lower degree of methyl esterification seemed to provide a higher density of ionic crosslinks in the HP compared to the AP hydrogel network (Figure 2). In addition, HP macromolecules had a higher content of arabinose residues (3.3 vs. 0.8 mol%), which indicated a more pronounced rhamnogalacturonan I region than that of AP. The arabinose-modulated gelling of pectin because of hydrogen bonding interactions provided additional entanglements [35,36]. In a study [35], it was shown that a polysaccharide with a halved content of arabinose yielded a less strong gel. Zheng et al. [36] reported increased gel strength because of the limitation of network chain mobility provided by arabinose residues. Therefore, pectin hydrogel enhancement by HP may also be due to arabinose of the side chains, which added to the effect of more free carboxyl groups in the backbone (Figure 2).

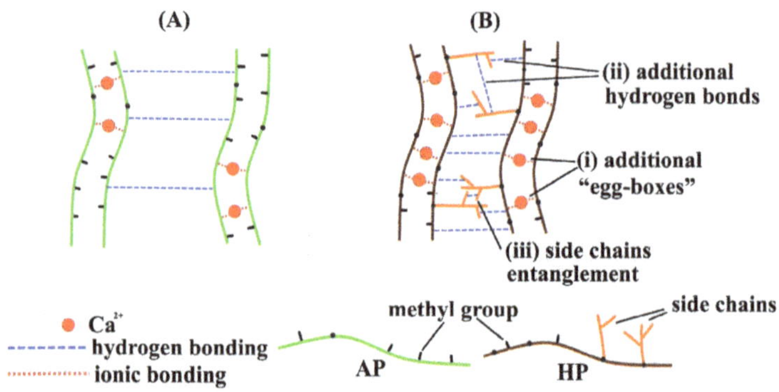

Figure 2. Scheme of a proposed chain interaction in AP (**A**) and HP (**B**) hydrogels. Suggested reasons for strengthening the HP hydrogel: (i) an increase in the number of Ca^{2+} crosslinks due to a lower degree of esterification; (ii) the formation of additional hydrogen bonds due to more sugar side chains; and (iii) the entanglement of sugar side chains.

2.4. Oral Processing and Sensory Evaluation of Pectin Hydrogels

Electromyography (EMG) parameters did not differ during the chewing of PG1 and PG3 (Figure 3). Chewing PG5 required 19% more time (Figure 3A) and 20% more chews (Figure 3B) than PG1 and PG3. Masseter and temporalis activities were 30% higher when chewing PG5 than when chewing PG1 and PG3 (Figure 3C).

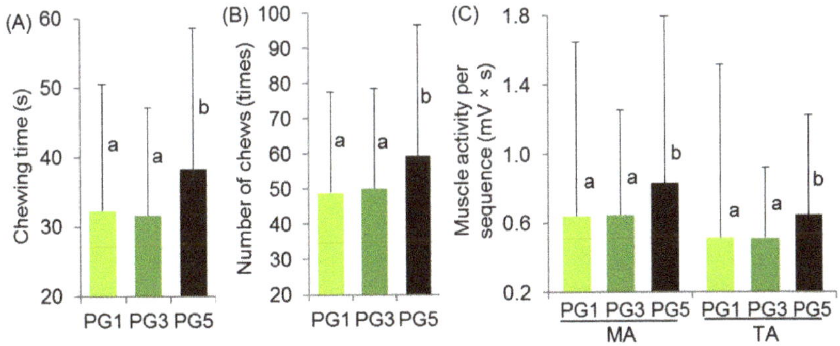

Figure 3. Electromyography data on PG chewing time (A), number (B), and activity of the masseter (MA) and temporal (TA) muscles (C). Means with standard deviations are given. Differences are significant ($p < 0.05$) between means labeled with different lowercase letters (a and b) ($n = 16$).

As known, the incorporation of saliva into a bolus while chewing transforms the initial value of moisture content in solid or semisolid foods to prepare a ready-to-swallow bolus in the mouth [37]. A gravimetrical analysis of saliva content in the bolus formed from the PGs revealed that saliva uptake varied between 22.6 and 29.1 wt% in pectin hydrogels (Table 5).

Table 5. Salivary parameters and viscosity of the bolus from PG1, PG3, and PG5.

Hydrogel	Saliva Parameters		Viscosity Parameters				
	SU * (%)	SIR ** (g/min)	K (Pa × s)	R^2	n	η_{app} 10 (s^{-1})	η_{app} 50 (s^{-1})
PG1	22.6 ± 14.0 [a]	3.04 ± 2.6 [a]	194.49	0.972	−0.788	50.6 ± 22.1 [a]	2.8 ± 1.1 [a]
PG3	24.0 ± 12.7 [a]	3.42 ± 2.7 [a]	181.39	0.962	−0.778	47.2 ± 19.7 [a]	2.9 ± 0.9 [a]
PG5	29.1 ± 13.4 [b]	3.38 ± 2.2 [a]	170.65	0.974	−0.811	43.5 ± 25.7 [a]	2.1 ± 0.7 [a]

Means with standard deviations are given. * SU—saliva uptake; ** SIR—saliva incorporation rate. Differences are significant ($p < 0.05$) between means labeled with different lowercase letters (a and b) ($n = 15$).

Saliva uptake has been found to vary between 3 and 10 wt% in different gels [38,39]. PG3 failed to affect salivation compared to PG1 (Table 5). A 29% increase in saliva in the PG5 bolus was comparable to that of PG1. Because the compositions of the PGs were similar, serum release appeared to be a critical initial characteristic of pectin hydrogels that could affect saliva secretion during oral processing and subsequent saliva uptake into the bolus. Saliva not only hydrates the bolus but also lubricates it by lowering friction due to its watery character and the presence of salivary mucins, thereby making swallowing the bolus safe [37,40]. Therefore, PG5, with its low serum release (0.49%), needed more saliva incorporation to enhance lubrication than PG1 (serum release 0.66%). Similar to our findings, low serum release agar-gelatin gels incorporated more saliva into the bolus [38]. The highest increase in saliva uptake was observed for the bolus of the hardest PG5 samples. It might seem intuitively obvious that the textural differences between PG1 and PG5 determined salivary flow while chewing hydrogels. However, the saliva incorporation rate (SIR) into the bolus was similar for all the PGs (Table 5). Obviously, the hardest PG5 samples needed more chews to swallow, likely due to the more complex microstructure of the PG5 mesh, which required more mechanical mixing and saliva incorporation to form a cohesive bolus. This result is consistent with a previous study on carrageenan gels [41], which reported that prolonged chewing time increased saliva incorporation. The boluses demonstrated shear-thinning behavior (Figure 4), which is in agreement with previous results [42].

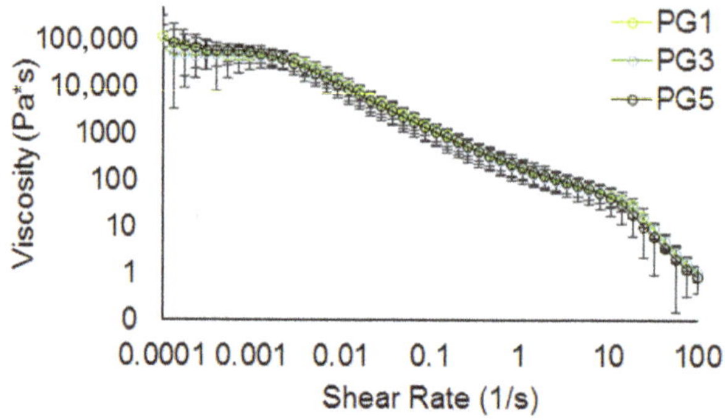

Figure 4. Flow curves of boluses from PGs.

A resulting power law index η and consistency parameter K for the bolus are shown in Table 5. All volunteers presented a safe and efficacious swallow of the bolus from all the PGs. The viscosity of the bolus dramatically decreased under mastication and was significantly lower than the initial viscosity of the pectin hydrogel samples at orally relevant shear rates (50 s^{-1}) (Table 3), although the viscosity values were higher than those of thickened fluids for dysphagia management [43]. The viscosity values were not significantly different among three PGs at the swallowing point, despite the differences in the initial viscosity, when volunteers were given PG1, PG3, and PG5 with viscosities of 273.9, 291.4, and 502.0 Pa*s at shear rates of 50 s^{-1}, respectively (Table 3). In practice, subjects adjust the duration and intensity of chewing to obtain a bolus that is easy to swallow. [39,44,45]. Since samples of PG5 were the hardest, volunteers adapted their chewing to obtain boluses with similar rheological characteristics, regardless of hydrogel composition. The increase in saliva uptake in the PG5 bolus indicates that more saliva was incorporated into the bolus to give it the viscosity that was needed for safe and efficient swallowing. Thus, a long duration of chewing, more chewing movements, and higher muscle activity while chewing PG5 appeared to be required to achieve comfortable swallowing characteristics of the bolus.

Sixteen nontrained subjects rated the overall and consistency liking using a nine-point hedonic scale and the perceived intensity of texture attributes using a 100 mm visual analog scale for the sensory evaluation of pectin hydrogels. PG1, PG3, and PG5 received the same likeness scores, which lay between slightly disliking (score $\simeq$ 4) and neither liking nor disliking (score $\simeq$5). Figure 5 shows the mean values of the perceived texture scores during the early (hardness, brittleness, springiness, and moisture), middle (adhesiveness and chewiness), and late (easiness to swallow) chewing of the PGs.

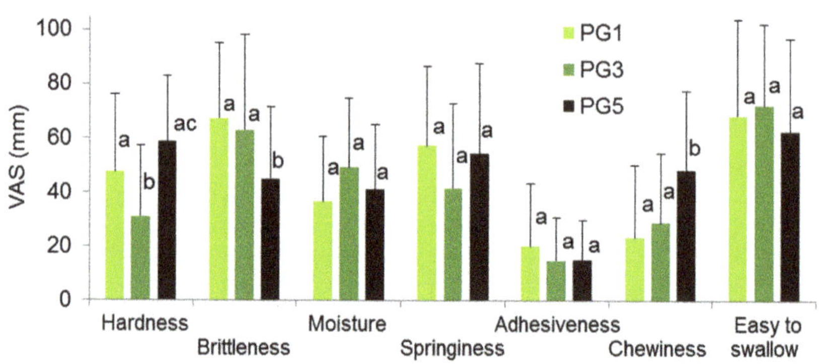

Figure 5. Sensory attributes of PGs. Means with standard deviations are given. Differences are significant ($p < 0.05$) between means labeled with different lowercase letters (a, b, and c) ($n = 16$).

The PGs differed only in regard to perceived hardness and brittleness. The perceived hardness of PG1 and PG5 did not differ, while the perceived hardness of PG3 was lower by 35 and 48% than that of PG1 and PG5, respectively. The perceived brittleness of PG1 and PG3 was similar, whereas the perceived brittleness of PG5 was lower by 33 and 29% than that of PG1 and PG3, respectively. An unexpected result was that, in our study, the sensory assessments of hardness and brittleness did not coincide with the data from the instrumental analysis.

2.5. Correlation Analysis of Mechanical, Rheological, and Sensory Properties of Pectin Hydrogels

The interrelationships between the sensory attributes and the mechanical and rheological properties of PGs were investigated using Pearson's coefficient (Table 6). Perceived and instrumental hardness were not correlated, while perceived and instrumental brittleness were inversely correlated. Consistent with [42,46], perceived hardness was positively correlated with the viscosity of the pectin hydrogels. A positive relationship among the perceived hardness, elasticity, and adhesiveness of the PGs was found. Young's modulus and consistency were inversely related to perceived hardness. Fracturability was inversely correlated with almost all the rheological parameters, while chewiness was positively correlated with almost all the rheological parameters, which is consistent with a previous study [42].

Table 6. Correlations between volunteers' evaluations of texture and the mechanical and rheological properties of PGs.

	Sensory Attributes [1]						
	1	2	3	4	5	6	7
			General properties				
SR	−0.26	0.31 *	−0.01	−0.03	0.09	−0.38 *	0.10
WL	−0.29 *	−0.01	0.21	−0.22	−0.11	0.01	0.07
			Rheological properties				
G′LVE (Pa)	0.17	−0.30 *	0.07	−0.03	−0.11	0.36 *	−0.07
G″LVE (Pa)	0.21	−0.31 *	0.05	−0.01	−0.11	0.37 *	−0.08
γFR	−0.12	0.28	−0.10	0.06	0.12	−0.34 *	0.06
k′ (Pa*s)	0.21	−0.31 *	0.05	−0.01	−0.11	0.37 *	−0.08
k″ (Pa*s)	0.23	−0.31 *	0.04	0.01	−0.10	0.37 *	−0.09
η*s	0.37 *	−0.29 *	−0.08	0.12	−0.04	0.34 *	−0.12
A	0.29 *	−0.31 *	0.00	0.05	−0.08	0.37 *	−0.10
			Viscosity				
η10	0.34 *	−0.31 *	−0.04	0.09	−0.06	0.36 *	−0.11
η55	0.31 *	−0.31 *	−0.02	0.06	−0.07	0.37 *	−0.11
			Mechanical properties				
Hardness	0.26	−0.31 *	0.01	0.03	−0.09	0.38 *	−0.10
Fracturability	0.27	−0.31 *	0.01	0.03	−0.09	0.38 *	−0.10
Brittleness	0.31 *	−0.31 *	−0.02	0.06	−0.07	0.37 *	−0.11
Adhesiveness	0.40 *	−0.23	−0.14	0.18	0.01	0.27	−0.12
E	−0.40 *	0.22	0.15	−0.18	−0.01	−0.26	0.12
Consistency	−0.29 *	−0.31 *	0.00	0.04	−0.08	0.37 *	−0.10

SR—serum release; WL—water content measured as weight loss (WL) during drying; E—Young's modulus. [1] Sensory attributes: 1—hardness; 2—brittleness; 3—moisture; 4—springiness; 5—adhesiveness; 6—chewiness; 7—easy to swallow. * $p < 0.05$.

Correlations among the liking and sensory-textural attributes are presented in Table 7. Overall liking was shown to positively correlate with consistency liking, taste, and ease to swallow. In turn, liking consistency was positively related to perceived brittleness, moisture, and swallowing ease. In agreement with the EMG data, perceived hardness and chewiness were positively correlated. An inverse correlation among perceived hardness, moisture, and ease of swallowing was revealed. Perceived brittleness was positively correlated with moisture (Table 7).

Table 7. Pearson's correlations among liking and sensory attributes of PGs.

	Liking and Sensory Attributes										
	1	2	3	4	5	6	7	8	9	10	11
Overall liking	-										
Consistency liking	0.65 *	-									
Aroma	0.03	0.04	-								
Taste	0.36 *	0.22	0.35 *	-							
Hardness	−0.07	−0.22	0.08	0.08	-						
Brittleness	0.22	0.39 *	0.41 *	0.50 *	−0.12	-					
Moisture	0.05	0.37 *	−0.14	0.17	−0.32 *	0.31 *	-				
Springiness	−0.16	0.12	0.43 *	0.07	0.27	0.23	0.19	-			
Adhesiveness	0.14	−0.03	0.10	0.17	−0.12	0.18	0.18	−0.01	-		
Chewiness	−0.13	−0.20	−0.06	−0.06	0.62 *	−0.20	−0.27	0.05	−0.10	-	
Ease to swallow	0.32 *	0.43 *	0.18	0.24	−0.42 *	0.28	0.33 *	−0.01	0.04	−0.36 *	-

* $p < 0.05$.

Several studies have previously demonstrated the interrelationships between the liking and sensory textural attributes of food gels. A study [47] discovered a positive relationship between the hardness and chewability of k-carrageenan hydrogels and k-carrageenan mixed with sodium alginate hydrogels. A sensory analysis of gels made from three hydrocolloids showed a positive correlation between firmness and elasticity, chewiness, and cohesiveness [42].

The correlation analysis of sensory attributes and oral-processing parameters revealed a positive relationship between perceived springiness and temporal muscle activity, as well as between perceived adhesiveness and bolus viscosity (Table 8). Chewiness was positively correlated with masseter muscle activity and saliva uptake, but it was inversely correlated with bolus viscosity. Easy swallowing was positively correlated with masseter and temporal muscle activity.

Table 8. Pearson's correlations between sensory attributes and oral-processing parameters of PGs.

	Sensory Attributes [1]						
	1	2	3	4	5	6	7
Duration	−0.02	−0.17	−0.28	−0.02	−0.05	0.07	−0.12
Number of chews	−0.02	−0.11	−0.27	−0.05	0.05	0.12	−0.13
MA (mV)	0.23	−0.04	−0.17	0.27	−0.05	0.31 *	−0.46 *
TA (mV)	0.23	−0.13	−0.19	0.29 *	−0.13	0.17	−0.55 *
MA (mV × s)	0.12	−0.10	−0.23	0.12	−0.06	0.23	−0.38 *
TA (mV × s)	0.12	−0.17	−0.28	0.13	−0.13	0.20	−0.42 *
Saliva uptake	0.23	−0.07	−0.16	0.03	0.02	0.36 *	0.08
Saliva incorporation rate	0.22	−0.02	−0.02	0.02	0.01	0.26	−0.04
Viscosity of bolus:							
$\eta 10$	−0.19	0.01	0.26	−0.06	0.14	−0.30 *	0.12
$\eta 55$	−0.10	−0.01	0.14	−0.10	0.34 *	−0.20	−0.06

MA and TA—masseter and temporalis muscle activity, respectively. [1] Sensory attributes: 1—hardness; 2—brittleness; 3—moisture; 4—springiness; 5—adhesiveness; 6—chewiness; 7—easy to swallow. * $p < 0.05$.

2.6. Simulated Digestibility of Pectin Hydrogels

The content of total sugars gradually decreased during the digestion of PG1, PG3, and PG5 (Figure 6).

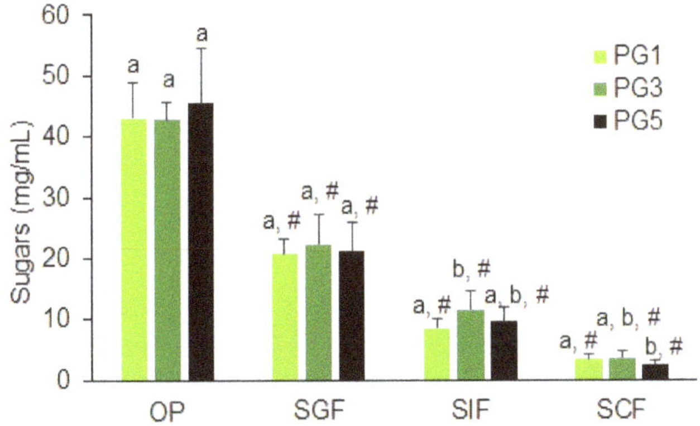

Figure 6. The amount of total sugars measured using phenol-sulfur method released from PGs during successive oral in vivo (OP) and gastrointestinal in vitro phases of digestion. SGF, SIF, and SCF—simulated gastric, intestinal, and colonic fluids, respectively. Means with standard deviations are given. Differences are significant ($p < 0.05$) between means labeled with different lowercase letters (a and b); #—indicates significant differences vs. the previous phase ($n = 6$, $p < 0.05$).

Galacturonic acid was released predominantly after the digestion of PG1 in SIF and SCF (Figure 7A). Galacturonic acid was slightly released from PG5 during the initial stages of digestion (OP, SGF, and SIF) and in significant amounts during the SCF phase. The release of galacturonic acid from PG3 had an intermediate intensity compared to PG1 and PG5 in SIF and did not increase in SCF compared to SIF. During digestion in OP and SGF, calcium was intensively released from the PGs regardless of pectin composition, probably due to the strong destruction of the hydrogels' structure during chewing in vivo (Figure 7B). During digestion in SIF, calcium was released from PG3 and PG5 6.3 and 4.8 times less than from PG1, respectively.

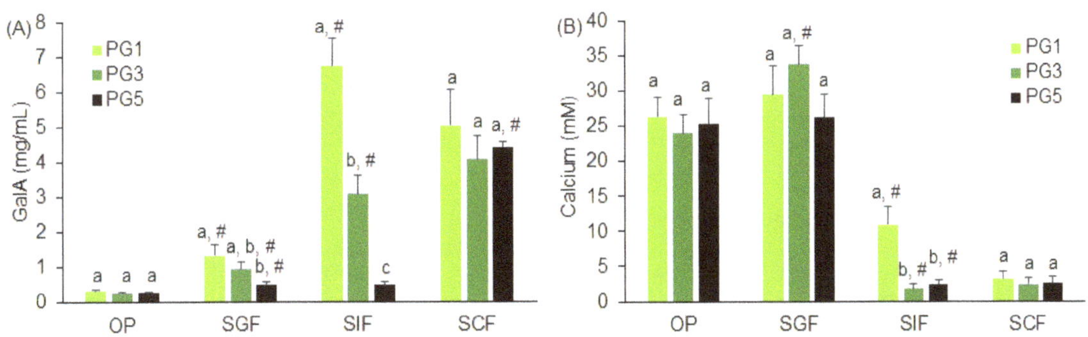

Figure 7. The amounts of galacturonic acid (GalA) (**A**) and calcium (**B**) released from PGs during successive oral in vivo (OP) and gastrointestinal in vitro phases of digestion. SGF, SIF, and SCF—simulated gastric, intestinal, and colonic fluids, respectively. Means with standard deviations are given. Differences are significant ($p < 0.05$) between means labeled with different lowercase letters (a, b, and c); #—indicates significant differences vs. the previous phase ($n = 6$, $p < 0.05$).

The release of calcium during in vivo chewing could be attributed to the release of free calcium contained in the gel matrix rather than crosslinked calcium. The concentration of crosslinked calcium for PG preparation was excessive and corresponded to a stoichiometric ratio ($R = 2(Ca^{2+})/(COO^{-})$) in the range of 4.1–5.7. The strongest gel was formed at R = 2–3 [48]. It was also suggested that high calcium concentrations may hinder the formation of gel networks when R >> 1 [48].

Similarly, the large amounts of total sugars in the first phases of digestion were likely due to the release of sucrose and not to the destruction of gel network polysaccharides. According to galacturonic acid release data, the HP gel network was more resistant to digestion than the AP gel network. The more stable HP hydrogel in the upper gastrointestinal tract appeared to be a better fit than the AP hydrogel as a vehicle for the delivery of biologically active substances to the colon.

3. Conclusions

The prospects for the use of HP as a hydrocolloid for the production of food gels were evaluated. The advantages of HP food hydrogels compared with AP food hydrogels were demonstrated regarding stronger structure and higher digestion resistance. Rheological and textural measurements showed that the strength and number of linkages in the HP hydrogels were higher than in the AP hydrogels, so increasing the HP content in the mixed hydrogel increased its strength. The synergistic gelation between pectins occurred because the Young's modulus and tangent after flow point values were higher for mixed hydrogels that contained equal amounts of AP and HP than for pure AP and HP hydrogels. Stronger HP hydrogels increased chewing duration and intensity. More intense and longer oral processing of HP hydrogels provided a ready-to-swallow bolus with the same viscosity and salivary wetness as an AP hydrogel bolus. Untrained volunteers gave pectin hydrogels high marks for overall and texture liking, indicating that they had a high potential for consumer acceptance. Among the sensory textural attributes, HP was characterized by reduced brittleness, which had a an inverse correlation with almost all the rheological and mechanical properties. The main functional advantage of HP-containing hydrogels was their resistance to digestion in the small intestine. Therefore, HP hydrogels could be promising as a food matrix carrier of biologically active substances delivered to the large intestine. Thus, new food hydrogels with new rheological, textural, and sensory properties could be obtained from a mixture of low-methyl-esterified pectins with different structures.

4. Materials and Methods

4.1. Materials

The isolation and chemical properties of HP were previously described [14]. Apple pectin AU701 (Herbstreith & Fox GmbH, Neuemberg, Germany) was used as AP. Table 9 shows the characteristics of AP and HP. Calcium gluconate was supplied by Zhejiang Tianyi Food Additives (Tongxiang, Zhejiang, China); sucrose was purchased from a local supermarket.

Table 9. Chemical characteristics of AP and HP.

Pectin	Monosaccharides (mol%) [a]						Rha/GalA	RG-I% [b]	(Ara + Gal)/Rha [c]	DM [d]	M_w, kDa	M_w/M_n
	UA	Gal	Xyl	Glc	Rha	Ara						
HP	90.6 ± 0.7	3.2 ± 0.2	1.0 ± 0.3	0.5 ± 0.2	2.1 ± 0.1	3.3 ± 0.1	0.02	10.61	3.13	21	538	4.1
AP	89.5 ± 0.7	2.6 ± 2.6	3.8 ± 0.1	1.7 ± 0.1	1.6 ± 0.1	0.8 ± 0.5	0.02	6.61	2.14	43	401	5.2

[a] Data are calculated as molar percentages. [b] Rhamnogalacturonan I = 2Rhamnose% + Arabinose% + Galactose%. [c] Average length of rhamnogalacturonan I side chains. [d] Degree of methyl esterification. UA: uronic acids; GalA: galacturonic acid; Gal: galactose; Xyl: xylose; Glc: glucose; Rha: rhamnose; Ara: arabinose.

4.2. Preparation of Pectin Hydrogels

AP, HP (2 g), three different blend combinations of AP (1.5 g) and HP (0.5 g), AP (1 g) and HP (1 g), or AP (0.5 g) and HP (1.5 g) were dissolved in deionized water (50 mL) with the addition of sucrose (5 g). The solutions were heated (60 °C) under continuous magnetic stirring (200 rpm) for 60 min for better dissolution and then cooled to room temperature. Dialysis tubes with a pore size of 14 kPa (Sigma-Aldrich Co, St. Louis, UO, USA) were filled with the solutions obtained (50 mL) and were held in a solution of 0.3 M calcium gluconate and 10% sucrose (200 mL) for 48 h at 25 °C for pectin gelling (Figure 8A).

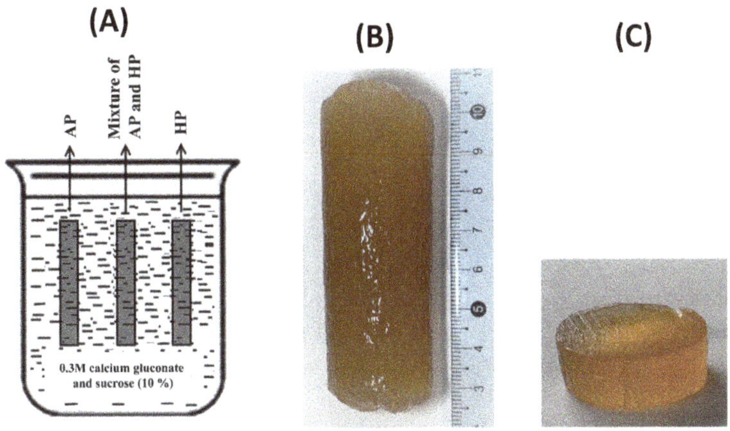

Figure 8. Preparation of hydrogels. General scheme (**A**), a representative image of PG3 hydrogel formed in a dialysis bag (**B**), and puck-shaped PG3 sample used for analyses (**C**).

Cylinder-shaped (7–8 cm high) hydrogels were formed (Figure 8B). After being removed from the dialysis tubes, the hydrogels were washed with distilled water and cut into single-serving puck-shaped pieces at a predetermined height for subsequent analyses (Figure 8C). Five types of pectin hydrogels were obtained depending on the content of AP and HP in the pectin mixture: PG1—4% AP; PG2—3% AP mixed with 1% HP; PG3—2% AP mixed with 2% HP; PG4—1% AP mixed with 3% HP; and PG5—4% HP (Table 10).

Table 10. Final compositions of the five pectin gel samples.

Hydrogel	AP (*w*/*v* %)	HP (*w*/*v* %)	Sucrose (*w*/*v* %)
PG1	4	-	10
PG2	3	1	10
PG3	2	2	10
PG4	1	3	10
PG5	-	4	10

4.3. Characterization of Pectin Hydrogels

4.3.1. General Characterization

The pH was determined for hydrogel aqueous homogenates (1:10 (*w*/*v*)) using an S20 SevenEasy™ pH meter (Mettler-Toledo AG, Schwerzenbach, Switzerland). The weight of 1 cm square hydrogel cubes (n = 8) was measured (AG245, Mettler Toledo International, Greifensee, Switzerland) to determine the density as weight/volume. Weight loss (WL) was determined using a gravimetric method [15].

4.3.2. Rheological Characterization of Pectin Hydrogels

A rotational-type rheometer (Anton Paar, Physica MCR 302, Graz, Austria) equipped with a parallel plate geometry (diameter 25 mm; gap 4.0 mm) was used for the strain and frequency sweep measurements.

Strain sweep evaluation was performed from 0.01 to 100% strain amplitudes using a controlled shear rate mode at 20 °C at a constant frequency and stress of 1 Hz and 9.0 Pa, respectively. The storage modulus (G'_{LVE}), loss modulus (G''_{LVE}), loss tangent (tan δ) in LVE, complex modulus (G^*_{LVE}), limiting value of strain (the strain at which biopolymers enter from linear viscoelastic region to nonlinear viscoelastic region, γL), and stress at flow point (τFP) with the corresponding complex modulus (G^*_{FP}), fracture stress (τFr), and slope of the loss tangent after flow point (tan δ_{AF}) were determined [20]. The shear strain dependence of G′ and G′ was determined using the power law equation:

$$G' = A' \times \omega^{n'}, \quad (1)$$

$$G'' = A'' \times \omega^{n''}, \quad (2)$$

where ω is the angular shear strain (%); A′ (Pa*s) and A″ (Pa*s) are intercepts; and n′ and n'' are the slopes of G′ and G″ frequency dependence, respectively. A′ = A″ is a measure of the contribution of the viscous component in relation to the elastic component and represents the overall loss tangent of the material.

For the frequency sweep experiments, the obtained mechanical spectra were characterized by the values of G′ and G″ (Pa) as a function of frequency in the range of 0.3–70.0 Hz at 20 °C and a constant stress of 9.0 Pa. The loss factor tan δ was calculated as the ratio of G″ and G′ [49]. The power law function [50] was expressed as follows:

$$\eta = K_c \times y^n, \quad (3)$$

where η is the steady viscosity, K_c is the consistency constant, y is the shear rate, and n is the power law index or flow behavior index.

The degrees of frequency dependence for G′ and G″ were determined by the power law parameters [51], which are expressed as follows:

$$G' = k' \times \omega^{n'}, \quad (4)$$

$$G'' = k'' \times \omega^{n''}, \quad (5)$$

where G′ and G″ are the storage and loss moduli, respectively; ω is the oscillation frequency (Hz); and k′ and k″ are constants. In addition, the complex dynamic viscosity frequency dependence η*s was determined.

The strength of the network (A, Pa*$s^{1/z}$) and the network extension parameter (z) were evaluated as follows according to [52]:

$$G' = G^* \times (\omega) = A \times \omega^{1/z}, \quad (6)$$

where G* (Pa) is the complex modulus.

4.3.3. Instrumental Texture Characterization of Pectin Hydrogels

A puncture test (probe P/5 mm, depth 4 mm) for the PG1, PG2, PG3, PG4, and PG5 samples (0.4 cm high) was carried out using a TA-XT Plus (Texture Technologies Corp., Stable Micro Systems, Godalming, UK) instrument at room temperature.

Serum release was determined as the weight ratio of the released serum to the initial weight of a hydrogel after compression to 80% of its original height at room temperature [53].

4.3.4. Characterization of Oral Processing of Pectin Hydrogels

Sixteen volunteers of both sexes without masticatory or swallowing dysfunctions were involved equally in the research. Three test sessions were performed by each subject: (1) EMG recording under unilateral chewing; (2) sensory score assessment; and (3) bolus collection after free chewing. Puck-shaped pieces (~6 g, 1 cm high) of the hydrogel samples (PG1, PG3, and PG5) were presented to each participant on a plastic spoon.

EMG activity from the superficial masseter and anterior temporalis was monitored using bipolar electrodes (11 × 25 mm) that were separated from each other by approximately 20 mm. Before EMG recording, participants were instructed to chew freely until it was easy to swallow [53].

Then, participants evaluated nine sensory attributes using a 100 mm visual analog scale and acceptability using a 9-point hedonic scale [54].

In the third session, participants were asked to chew a fixed, preweighed quantity (6 g, 1 cm high) of PG1, PG3, or PG5 samples in a single mouthful and spit out the bolus just before swallowing it. The weight of the wet bolus was calculated as the combined weight of the expectorated material minus the weight of rinsing water [55]. Salivary uptake was calculated using the following equation:

$$SU = (WB - WH)/WH \times 100, \tag{7}$$

where SU is saliva uptake (%), and the WB and WH are weights of the wet bolus (g) and wet hydrogel sample (g), respectively. The rate of saliva incorporation (SIR) was calculated using the following equation:

$$SIR = (WB - WH)/TC, \tag{8}$$

where SIR is saliva incorporation (g/min), TC is the time of chewing (min), and WB and WH are weights of the wet bolus (g) and wet hydrogel sample (g), respectively.

The apparent viscosities (flow curves) of the bolus were obtained at shear rates ranging from 0.0001 to 100 s^{-1} at 37 °C using a rotational-type rheometer [42].

4.3.5. In Vivo Oral Phase (OP) and Static In Vitro Gastrointestinal Digestion

Six healthy volunteers chewed each type of hydrogel (4 g) 20 times and spat the bolus into a beaker [56]. Immediately thereafter, 4.0 mL of water was added to the beaker, it was stirred, and all the fluid was separated for analysis. The gel fragments were transferred to a 20 mL sheathed glass container for further in vitro digestion. In vitro gastrointestinal digestion was approached by a method [53] using SGF (pH 1.5, 0.08 M HCl, and 0.03 M NaCl), SIF (pH 6.8, 0.05 M KH_2PO_4, and 0.02 M NaOH), and SCF (0.01 M KH_2PO_4, 0.05 M $NaHPO_4$, and pectinase: 1.7 mg/mL). The contents of galacturonic acid and calcium ions were determined in the fluid after each phase of digestion. For this, aliquots (1–2 mL) of incubation medium were taken and centrifuged, and the resulting supernatant was precipitated with a fourfold volume of 96% ethanol. The precipitate was washed twice with 96% ethanol and dissolved with 3 mL of H_2O. The resulting solution was used to determine the content of galacturonic acid by the reaction of the sample with 3,5-dimethylphenol in the presence of concentrated H_2SO_4 [57]. The alcohol supernatant was used to determine the total amount of sugars using the phenol-sulfur method. The concentration of calcium was determined using a Calcium-Agat kit (Agat-Med, Moscow, Russia).

4.4. *Statistical Analysis*

All statistical analyses were performed using Statistica 10.0 (StatSoft, Inc., Tulsa, OK, USA). Results are presented as means ± standard deviations (SDs). The differences among the means in serum release, the rheological and mechanical parameters, and the digestion studies were estimated with one-way ANOVA and Tukey's HSD test. A one-way repeated ANOVA and Fisher's LSD post hoc test were applied to determine differences in the EMG, sensory, saliva, and bolus variables for different hydrogels. Pearson's correlations were calculated to study the relationships among the rheological, mechanical, and sensory properties of the hydrogels.

Author Contributions: Conceptualization, S.P.; methodology, F.V.; validation, S.P. and V.S.; formal analysis, V.S.; investigation, V.S., N.P., D.K., D.P., E.C. and F.V.; resources, S.P.; data curation, V.S.; writing—original draft preparation, V.S.; writing—review and editing, S.P.; visualization, F.V.; supervision, S.P.; project administration, S.P.; funding acquisition, S.P. All authors have read and agreed to the published version of the manuscript.

Funding: This research did not receive any specific grant from funding agencies in the public, commercial, or not-for-profit sectors.

Institutional Review Board Statement: This research was carried out within the framework of the research topic of the Institute of Physiology of Komi Science Center of the Urals Branch of the Russian Academy of Sciences FUUU-2022-0066 (1021051201895-9). The study was conducted according to the guidelines of the Declaration of Helsinki and approved by the Ethical Committee of the Komi Science Center of the Russian Academy of Sciences (date of approval: 10 March 2022).

Informed Consent Statement: Informed consent was obtained from all subjects involved in the study.

Data Availability Statement: The data that support the findings of this study are available from the corresponding author upon reasonable request.

Conflicts of Interest: The authors declare no conflict of interest.

References

1. Nath, P.C.; Debnath, S.; Sridhar, K.; Inbaraj, B.S.; Nayak, P.K.; Sharma, M. A Comprehensive Review of Food Hydrogels: Principles, Formation Mechanisms, Microstructure, and Its Applications. *Gels* **2023**, *9*, 1. [CrossRef]
2. Manzoor, M.; Singh, J.; Bandral, J.D.; Gani, A.; Shams, R. Food hydrocolloids: Functional, nutraceutical and novel applications for delivery of bioactive compounds. *Int. J. Biol. Macromol.* **2020**, *165*, 554–567. [CrossRef]
3. Yang, Y.; Anderson, C.T. Biosynthesis, localisation, and function of pectins in plants. In *Pectin: Technological and Physiological Properties*; Springer: Cham, Switzerland, 2020; pp. 1–15.
4. Padma Ishwarya, S.; Sandhya, R.; Nisha, P. Advances and prospects in the food applications of pectin hydrogels. *Crit. Rev. Food Sci. Nutr.* **2021**, *62*, 4393–4417. [CrossRef] [PubMed]
5. Naqash, F.; Masoodi, F.A.; Rather, S.A.; Wani, S.M.; Gani, A. Emerging concepts in the nutraceutical and functional properties of pectin—A review. *Carbohydr. Polym.* **2017**, *168*, 227–239. [CrossRef] [PubMed]
6. Minzanova, S.T.; Mironov, V.F.; Arkhipova, D.M.; Khabibullina, A.V.; Mironova, L.G.; Zakirova, Y.M.; Milyukov, V.A. Biological activity and pharmacological application of pectic polysaccharides: A review. *Polymers* **2018**, *10*, 1407. [CrossRef]
7. Lara-Espinoza, C.; Carvajal-Millán, E.; Balandrán-Quintana, R.; López-Franco, Y.; Rascón-Chu, A. Pectin and Pectin-Based Composite Materials: Beyond Food Texture. *Molecules* **2018**, *23*, 942. [CrossRef] [PubMed]
8. Odun-Ayo, F.; Reddy, L. Potential Biomedical Applications of Modified Pectin as a delivery system for bioactive substances. *Polysaccharides* **2023**, *4*, 1–32. [CrossRef]
9. Belkheiri, A.; Forouhar, A.; Ursu, A.V.; Dubessay, P.; Pierre, G.; Delattre, C.; Djelveh, G.; Abdelkafi, S.; Hamdami, N.; Michaud, P. Extraction, Characterization, and Applications of Pectins from Plant By-Products. *Appl. Sci.* **2021**, *11*, 6596. [CrossRef]
10. Ropartz, D.; Ralet, M.-C. Pectin structure. In *Pectin: Technological and Physiological Properties*; Kontogiorgos, V., Ed.; Springer International Publishing: Cham, Switzerland, 2020; pp. 17–36.
11. Cao, L.; Lu, W.; Mata, A.; Nishinari, K.; Fang, Y. Egg-box model-based gelation of alginate and pectin: A review. *Carbohydr. Polym.* **2020**, *242*, 116389. [CrossRef]
12. Zhang, B.; Hu, B.; Nakauma, M.; Funami, T.; Nishinari, K.; Draget, K.I.; Phillips, G.O.; Fang, Y. Modulation of calcium-induced gelation of pectin by oligoguluronate as compared to alginate. *Food Res. Int.* **2018**, *116*, 232–240. [CrossRef]
13. Makarova, E.N.; Shakhmatov, E.G.; Belyy, V.A. Structural characteristics of oxalate-soluble polysaccharides of Sosnowsky's hogweed (Heracleum sosnowskyi Manden). *Carbohydr. Polym.* **2016**, *153*, 66–77. [CrossRef] [PubMed]
14. Patova, O.A.; Golovchenko, V.V.; Vityazev, F.V.; Burkov, A.A.; Belyi, V.A.; Kuznetsov, S.N.; Litvinets, S.G.; Martinson, E.A. Physicochemical and rheological properties of gelling pectin from Sosnowskyi's hogweed (*Heracleum sosnowskyi*) obtained using different pretreatment conditions. *Food Hydrocoll.* **2017**, *65*, 77–86. [CrossRef]
15. Popov, S.; Paderin, N.; Khramova, D.; Kvashninova, E.; Patova, O.; Vityazev, F. Swelling, Protein Adsorption, and Biocompatibility In Vitro of Gel Beads Prepared from Pectin of Hogweed *Heracleum sosnówskyi Manden* in Comparison with Gel Beads from Apple Pectin. *Int. J. Mol. Sci.* **2022**, *23*, 3388. [CrossRef] [PubMed]
16. Chan, S.Y.; Choo, W.S.; Young, D.J.; Loh, X.J. Pectin as a rheology modifier: Origin, structure, commercial production and rheology. *Carbohydr. Polym.* **2017**, *161*, 118–139. [CrossRef] [PubMed]
17. Mala, T.; Anal, A.K. Protection and Controlled Gastrointestinal Release of Bromelain by Encapsulating in Pectin–Resistant Starch Based Hydrogel Beads. *Front. Bioeng. Biotechnol.* **2021**, *9*, 757176. [CrossRef]
18. Tarifa, M.C.; Piqueras, C.M.; Genovese, D.B.; Brugnoni, L.I. Microencapsulation of *Lactobacillus casei* and *Lactobacillus rhamnosus* in pectin and pectin-inulin microgel particles: Effect on bacterial survival under storage conditions. *Int. J. Biol. Macromol.* **2021**, *179*, 457–465. [CrossRef]
19. Lee, T.; Chang, Y.H. Structural, physicochemical, and in-vitro release properties of hydrogel beads produced by oligochitosan and de-esterified pectin from yuzu (*Citrus junos*) peel as a quercetin delivery system for colon target. *Food Hydrocoll.* **2020**, *108*, 106086. [CrossRef]
20. Alghooneh, A.; Razavi, S.M.A.; Kasapis, S. Classification of hydrocolloids based on small amplitude oscillatory shear, large amplitude oscillatory shear, and textural properties. *J. Texture Stud.* **2019**, *50*, 520–538. [CrossRef]
21. Funami, T.; Nakauma, M. Instrumental food texture evaluation in relation to human perception. *Food Hydrocoll.* **2022**, *124*, 107253. [CrossRef]

22. Brighenti, M.; Govindasamy-Lucey, S.; Jaeggi, J.J.; Johnson, M.E.; Lucey, J.A. Effects of processing conditions on the texture and rheological properties of model acid gels and cream cheese. *J. Dairy Sci.* **2018**, *101*, 6762–6775. [CrossRef]
23. Koc, H.; Vinyard, C.J.; Essick, G.K.; Foegeding, E.A. Food oral processing: Conversion of food structure to textural perception. *Annu. Rev. Food Sci. Technol.* **2013**, *4*, 237–266. [CrossRef] [PubMed]
24. De Lavergne, M.D.; Van de Velde, F.; Stieger, M. Bolus matters: The influence of food oral breakdown on dynamic texture perception. *Food Funct.* **2017**, *8*, 464–480. [CrossRef] [PubMed]
25. Abid, M.; Yaich, H.; Hidouri, H.; Attia, H.; Ayadi, M.A. Effect of substituted gelling agents from pomegranate peel on colour, textural and sensory properties of pomegranate jam. *Food Chem.* **2018**, *239*, 1047–1054. [CrossRef]
26. Basu, S.; Shivhare, U.S. Rheological, textural, micro-structural and sensory properties of mango jam. *J. Food Eng.* **2010**, *100*, 357–365. [CrossRef]
27. Puleo, S.; Valentino, M.; Masi, P.; Di Monaco, R. Hardness sensitivity: Are old, young, female and male subjects all equally sensitive? *Food Qual. Prefer.* **2021**, *90*, 104118. [CrossRef]
28. Matsuyama, S.; Nakauma, M.; Funami, T.; Hori, K.; Ono, T. Human physiological responses during swallowing of gel-type foods and its correlation with textural perception. *Food Hydrocoll.* **2021**, *111*, 106353. [CrossRef]
29. Sorapukdee, S.; Jansa, S.; Tangwatcharin, P. Partial replacement of pork backfat with konjac gel in Northeastern Thai fermented sausage (Sai Krok E-san) to produce the healthier product. *Asian-Australas J. Anim. Sci.* **2019**, *32*, 1763–1775. [CrossRef]
30. Cao, C.; Feng, Y.; Kong, B.; Xia, X.; Liu, M.; Chen, J.; Zhang, F.; Liu, Q. Textural and gel properties of frankfurters as influenced by various κ-carrageenan incorporation methods. *Meat Sci.* **2021**, *176*, 108483. [CrossRef]
31. Petcharat, T.; Benjakul, S. Effect of gellan incorporation on gel properties of bigeye snapper surimi. *Food Hydrocoll.* **2018**, *77*, 746–753. [CrossRef]
32. Tan, H.-L.; Tan, T.C.; Easa, A.M. The use of selected hydrocolloids and salt substitutes on structural integrity, texture, sensory properties, and shelf life of fresh no salt wheat noodles. *Food Hydrocoll.* **2020**, *108*, 105996. [CrossRef]
33. Nepovinnykh, N.V.; Kliukina, O.N.; Ptichkina, N.; Bostan, A. Hydrogel based dessert of low calorie content. *Food Hydrocoll.* **2019**, *86*, 184–192. [CrossRef]
34. Ford, C.G.; Bolhuis, D. Interrelations Between Food Form, Texture, and Matrix Influence Energy Intake and Metabolic Responses. *Curr. Nutr. Rep.* **2022**, *11*, 124–132. [CrossRef] [PubMed]
35. Ngouémazong, D.E.; Kabuye, G.; Fraeye, I.; Cardinaels, R.; Van Loey, A.; Moldenaers, P.; Hendrickx, M. Effect of debranching on the rheological properties of Ca2+ pectin gels. *Food Hydrocoll.* **2012**, *26*, 44–53. [CrossRef]
36. Zheng, J.; Chen, J.; Zhang, H.; Wu, D.; Ye, X.; Linardt, R.J.; Chen, S. Gelling mechanism of RG-I enriched citrus pectin: Role of arabinose side-chains in cation- and acid-induced gelation. *Food Hydrocoll.* **2020**, *101*, 105536. [CrossRef]
37. Khramova, D.S.; Popov, S.V. A secret of salivary secretions: Multimodal effect of saliva in sensory perception of food. *Eur. J. Oral Sci.* **2022**, *130*, e12846. [CrossRef]
38. Devezeaux de Lavergne, M.; Strijbosch, V.M.G.; Van den Broek, A.W.M.; Van de Velde, F.; Stieger, M. Uncoupling the Impact of fracture properties and composition on sensory perception of emulsion-filled gels. *J. Texture Stud.* **2016**, *47*, 92–111. [CrossRef]
39. Devezeaux de Lavergne, M.; Van de Velde, F.; Van Boekel, M.A.J.S.; Stieger, M. Dynamic texture perception and oral processing of semi-solid food gels: Part 2: Impact of breakdown behaviour on bolus properties and dynamic texture perception. *Food Hydrocoll.* **2015**, *49*, 61–72. [CrossRef]
40. Bongaerts, J.H.H.; Rossetti, D.; Stokes, J.R. The Lubricating Properties of Human Whole Saliva. *Tribol Lett.* **2007**, *27*, 277–287. [CrossRef]
41. Funami, T.; Nakao, S.; Isono, M.; Ishihara, S.; Nakauma, M. Effects of food consistency on perceived intensity and eating behavior using soft gels with varying aroma inhomogeneity. *Food Hydrocoll.* **2016**, *52*, 896–905. [CrossRef]
42. Krop, E.M.; Hetherington, M.M.; Holmes, M.; Miquel, S.; Sarkar, A. On relating rheology and oral tribology to sensory properties in hydrogels. *Food Hydrocoll.* **2019**, *88*, 101–113. [CrossRef]
43. Methacanon, P.; Gamonpilas, C.; Kongjaroen, A.; Buathongjan, C. Food polysaccharides and roles of rheology and tribology in rational design of thickened liquids for oropharyngeal dysphagia: A review. *Compr. Rev. Food Sci. Food Saf.* **2021**, *20*, 4101–4119. [CrossRef] [PubMed]
44. Le Bleis, F.; Chaunier, L.; Della Valle, G.; Panouillé, M.; Réguerre, A.-L. Physical assessment of bread destructuration during chewing. *Food Res. Int.* **2012**, *50*, 308–317. [CrossRef]
45. Panouillé, M.; Saint-Eve, A.; Déléris, I.; Le Bleis, F.; Souchon, I. Oral processing and bolus properties drive the dynamics of salty and texture perceptions of bread. *Food Res. Int.* **2014**, *62*, 238–246. [CrossRef]
46. Kurotobi, T.; Hoshino, T.; Kazami, Y.; Hayakawa, F.; Hagura, Y. Relationship between sensory analysis for texture and instrument measurements in model strawberry jam. *J. Texture Stud.* **2018**, *49*, 359–369. [CrossRef]
47. Stribitcaia, E.; Krop, E.M.; Lewin, R.; Holmes, M.; Sarkar, A. Tribology and rheology of bead-layered hydrogels: Influence of bead size on sensory perception. *Food Hydrocoll.* **2020**, *104*, 105692. [CrossRef]
48. Yang, X.; Yuan, K.; Descallar, F.B.A.; Li, A.; Yang, X.; Yang, H. Gelation behaviors of some special plant-sourced pectins: A review inspired by examples from traditional gel foods in China. *Trends Food Sci. Technol.* **2022**, *126*, 26–40. [CrossRef]
49. Huang, T.; Tu, Z.C.; Shangguan, X.; Wang, H.; Sha, X.; Bansal, N. Rheological behavior, emulsifying properties and structural characterization of phosphorylated fish gelatin. *Food Chem.* **2018**, *246*, 428–436. [CrossRef]

50. Holdsworth, S.D. Applicability of rheological models to the interpretation of flow and processing behavior of fluid food products. *J. Texture Stud.* **1971**, *2*, 393–418. [CrossRef]
51. Ramkumar, D.H.S.; Bhattacharya, M. Relaxation behavior and the application of integral constitutive equations to wheat dough. *J. Texture Stud.* **1996**, *27*, 517–544. [CrossRef]
52. Gabriele, D.G.; Migliori, M.; Sanzo, R.D.; Rossi, C.O.; Ruffolo, S.A.; Cindio, B. Characterization of dairy emulsions by NMR and rheological techniques. *Food Hydrocoll.* **2009**, *23*, 619–628. [CrossRef]
53. Popov, S.; Smirnov, V.; Paderin, N.; Khramova, D.; Chistiakova, E.; Vityazev, F.; Golovchenko, V. Enrichment of Agar Gel with Antioxidant Pectin from Fireweed: Mechanical and Rheological Properties, Simulated Digestibility, and Oral Processing. *Gels* **2022**, *8*, 708. [CrossRef] [PubMed]
54. Lawless, H.; Heymann, H. Introduction. In *Sensory Evaluation of Food*; Food Science Text Series; Springer: New York, NY, USA, 2010.
55. Choy, J.Y.M.; Goh, A.T.; Chatonidi, G.; Ponnalagu, S.; Wee, S.M.M.; Stieger, M.; Forde, C.G. Impact of food texture modifications on oral processing behaviour, bolus properties and postprandial glucose responses. *Curr. Res. Food Sci.* **2021**, *4*, 891–899. [CrossRef] [PubMed]
56. Khin, M.N.; Goff, H.D.; Nsor-Atindana, J.; Ahammed, S.; Liu, F.; Zhong, F. Effect of texture and structure of polysaccharide hydrogels containing maltose on release and hydrolysis of maltose during digestion: In vitro study. *Food Hydrocoll.* **2021**, *112*, 106326. [CrossRef]
57. Usov, A.I.; Bilan, M.I.; Klochkova, N.G. Polysaccharides of algae. 48. Polysaccharide composition of several Calcareous red algae: Isolation of alginate from Corallina Pilulifera P. et R. (Rhodophyta, Corallinaceae). *Bot. Mar.* **1995**, *38*, 43–51. [CrossRef]

gels

Article

Wet Grinder-Treated Okara Improved Both Mechanical Properties and Intermolecular Forces of Soybean Protein Isolate Gels

Yuya Arai [1], Katsuyoshi Nishinari [2,3] and Takao Nagano [1,*]

1 Department of Food Science, Faculty of Bioresources and Environmental Sciences, Ishikawa Prefectural University, 1-308, Nonoichi 921-8836, Japan
2 Glyn O. Phillips Hydrocolloids Research Centre, School of Food and Biological Engineering, Hubei University of Technology, Wuhan 430068, China
3 Department of Food and Human Health Sciences, Graduate School of Human Life Science, Osaka City University, 3-3-138 Sugimoto Sumiyoshi, Osaka 558-8585, Japan
* Correspondence: naganot@ishikawa-pu.ac.jp; Tel.: +81-76-227-7455

Abstract: The application of okara treated by a wet-type grinder (WG) is discussed in this paper. We examined the effect of WG-treated okara on the mechanical properties and intermolecular forces in soybean protein isolate (SPI) gels. SPI gels were prepared with varying amounts of WG-treated okara, and compression tests were performed. Protein solubility was also examined by homogenizing the gel in four different solutions (S1, 0.6 M sodium chloride (NaCl); S2, 0.6 M NaCl and 1.5 M urea; S3, 0.6 M NaCl and 8.0 M urea; and S4, 1.0 M sodium hydroxide). The gel with WG-treated okara had higher breaking stress but not breaking strain. In contrast, the protein solubility in S3 was lower than those of the gel without okara or with WG-untreated okara. A negative correlation ($R^2 = 0.86$) was observed between breaking stress and protein solubility in S3. These results suggest that WG-treated okara enhanced the hydrophobic interactions of SPI gels because protein solubilization by S3 is caused by the differences in hydrophobic interactions.

Keywords: wet-type grinder; okara; soybean protein isolate; gel; intermolecular forces

Citation: Arai, Y.; Nishinari, K.; Nagano, T. Wet Grinder-Treated Okara Improved Both Mechanical Properties and Intermolecular Forces of Soybean Protein Isolate Gels. *Gels* **2022**, *8*, 616. https://doi.org/10.3390/gels8100616

Academic Editors: Baskaran Stephen Inbaraj, Kandi Sridhar, Minaxi Sharma and Miguel A. Cerqueira

Received: 31 August 2022
Accepted: 26 September 2022
Published: 27 September 2022

1. Introduction

Okara is a soybean residue and rich in dietary fiber, which contains mainly insoluble forms such as cellulose and hemicellulose [1]. In contrast, soybean protein isolate (SPI) is often used to replace animal proteins in food production because of sustainability assurance [2].

Many studies have been conducted to improve the physicochemical properties of cellulose and okara [3–5]. These studies suggested the usefulness of decreasing the particle size of okara and cellulose to improve the properties of protein-based gel food [6–12]. Ultra-high-pressure homogenization (UHPH) was applied to soybean flour for tofu production. UHPH decreased the particle size of soybean flour, and the hardness of tofu prepared from UHPH-treated soybean flour was similar to that of the control tofu [6]. The effect of nano-sized and micro-sized okara on the gel properties of tofu and silver carp surimi has been investigated [7,8]. Nano-sized okara was well distributed in the gel matrices and provided a less gritty mouthfeel than that of micro-sized okara [7]. The breaking force and penetration distance of the surimi gels increased with the increase in concentration of nano-sized okara. Light microscopy images showed that nano-sized okara was well distributed in surimi gels, whereas micro-sized okara was not [8]. Additionally, the effects of regenerated cellulose, bacterial cellulose (BC) microfibrils, and cellulose nano-crystals (CNC) on the gel properties of whey protein gels were studied. The addition of regenerated cellulose to acid-induced whey protein gels increased their storage modulus G' [9]. BC microfibrils also increased the storage modulus G' of whey protein gels [10]. The water-holding capacity and gel strength of whey protein gels increased with the increase in CNC concentration from 0% to 1.0% [11].

In our previous study, we used a wet-type grinder (WG) to improve the physicochemical properties of okara because WG is used to produce nano-cellulose [4,13,14]. We examined its effect on the mechanical properties and water-holding capacity of magnesium chloride-induced SPI gels. When the suspensions of okara were treated with a WG, the viscosity increased, and the dispersion performance improved with increasing WG passage. The breaking stress, strain, and water-holding capacity of SPI gels also increased with the addition of WG-treated okara [12]. These results demonstrate that WG-treated okara improves the gel properties of SPI. However, the mechanism of action remains unclear.

In this study, the effects of WG-treated okara on the mechanical properties and intermolecular forces of WG-treated okara–SPI gels were investigated to explore the mechanism by which WG-treated okara improves the gel properties of SPI. SPI gels were prepared with varying amounts of WG-treated okara, and compression measurements were performed. Protein solubility was examined by homogenizing the gels in four different solutions to understand the intermolecular forces required to form WG-treated okara–SPI gels.

2. Results and Discussion

2.1. Effect of WG-Treated Okara on the Mechanical Properties of SPI Gels

Heat-induced SPI gels were prepared by adding 3%, 5%, 10%, and 15% WG-untreated or treated low-protein okara, and their mechanical properties were measured using compression measurements (Figures 1 and 2). The breaking stress of the SPI gels after adding WG-untreated and treated okara increased, whereas the breaking strain of these gels did not increase. The breaking stress of the SPI gels containing WG-treated okara was higher than that of the SPI gels containing WG-untreated okara. These results indicate that WG-treated okara increases the breaking stress of the SPI gels but not the breaking strain. In our previous study, we demonstrated that the addition of WG-treated okara increases the breaking stress and strain of magnesium chloride-induced SPI gels. This effect increased with the increase in number of WG treatments. The breaking stress and strain of WG-treated okara–SPI gels also increased with the increase in concentration of WG-treated okara [12]. In addition, the effect of CNC and microcrystalline cellulose (MCC) on the gel properties of glucono-δ-lactone-induced SPI gels was investigated. The gel strength and storage modulus G' of CNC-SPI gels increased with the increase in CNC concentration from 0% to 0.75%, whereas the gel strength and storage modulus G' of MCC-SPI gels decreased [15]. The results in the present study are in line with these studies.

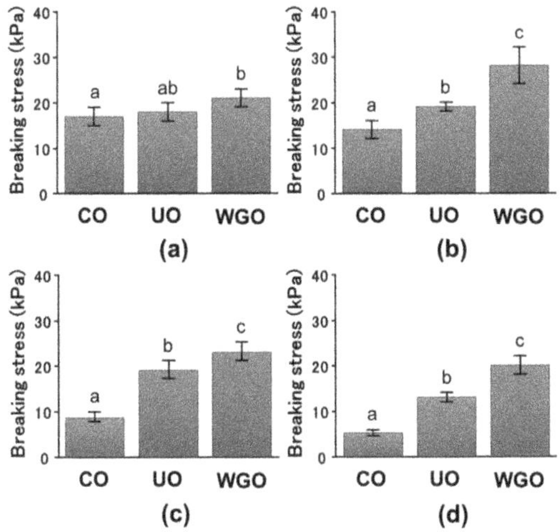

Figure 1. Breaking stress in okara–soybean protein isolate (SPI) gels. (**a**), 3% okara–SPI gels; CO, control SPI gel (15.52% SPI); UO, wet-type grinder (WG)-untreated okara–SPI gel (15.52% SPI and 0.48% WG-untreated okara); WGO, WG-treated okara–SPI gel (15.52% SPI and 0.48% WG-treated okara). (**b**), 5% okara–SPI gels; CO, control SPI gel (15.20% SPI); UO, WG-untreated okara–SPI gel (15.20% SPI and 0.80% WG-untreated okara); WGO, WG-treated okara–SPI gel (15.20% SPI and 0.80% WG-treated okara). (**c**), 10% okara–SPI gels; CO, control SPI gel (14.40% SPI); UO, WG-untreated okara–SPI gel (14.40% SPI and 1.60% WG-untreated okara); WGO, WG-treated okara–SPI gel (14.40% SPI and 1.60% WG-treated okara). (**d**), 15% okara–SPI gels; CO, control SPI gel (13.60% SPI); UO, WG-untreated okara–SPI gel (13.60% SPI and 2.40% WG-untreated okara); WGO, WG-treated okara–SPI gel (13.60% SPI and 2.40% WG-treated okara). Different letters indicate significant differences ($p < 0.05$).

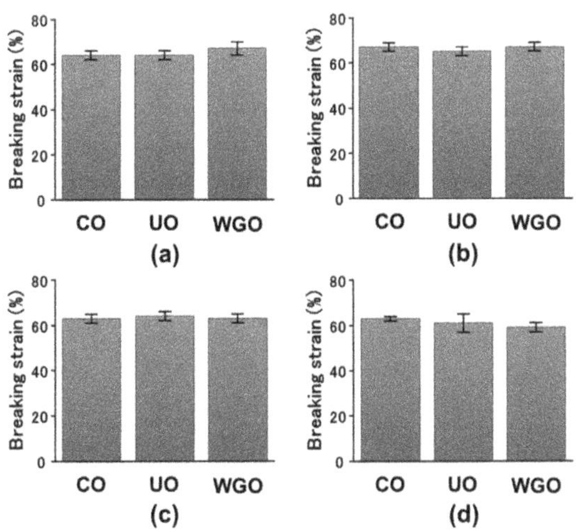

Figure 2. Breaking strain in okara–soybean protein isolate (SPI) gels. (**a**), 3% okara–SPI gels; (**b**), 5% okara–SPI gels; (**c**), 10% okara–SPI gels; (**d**), 15% okara–SPI gels. CO, control SPI gel; UO, wet-type grinder (WG)-untreated okara–SPI gel; WGO, WG-treated okara–SPI gel. SPI concentrations and WG-untreated and treated okara in 3%, 5%, 10%, and 15% okara–SPI gels are indicated in the Figure 1 legend.

2.2. *Effect of WG-Treated Okara on the Intermolecular Forces of SPI Gels*

Protein solubility in salts, denaturants, and reducing agents was investigated to understand molecular forces involved in the formation of protein gels [16]. Deng et al. [17] assessed protein solubility by homogenizing egg white gels in four different solutions (S1, 0.6 M NaCl; S2, 0.6 M NaCl and 1.5 M urea; S3, 0.6 M NaCl and 8.0 M urea; and S4, 0.6 M NaCl, 8.0 M urea, and 0.5 M 2-mercaptoethanol (2-ME)) to evaluate electrostatic interactions, hydrogen bonds, hydrophobic interactions, and disulfide bonds, respectively [17]. In this study, we used 1.0 M NaOH as S4 and examined protein solubility by homogenizing SPI gels in four different solutions. There were no differences in protein solubility in S1, S2, and S4 among the untreated, WG-untreated okara, and WG-treated okara–SPI gels. However, the protein solubility of WG-treated okara–SPI gels in S3 was lower than that of the without okara and with WG-untreated okara added SPI gels (Figure 3).

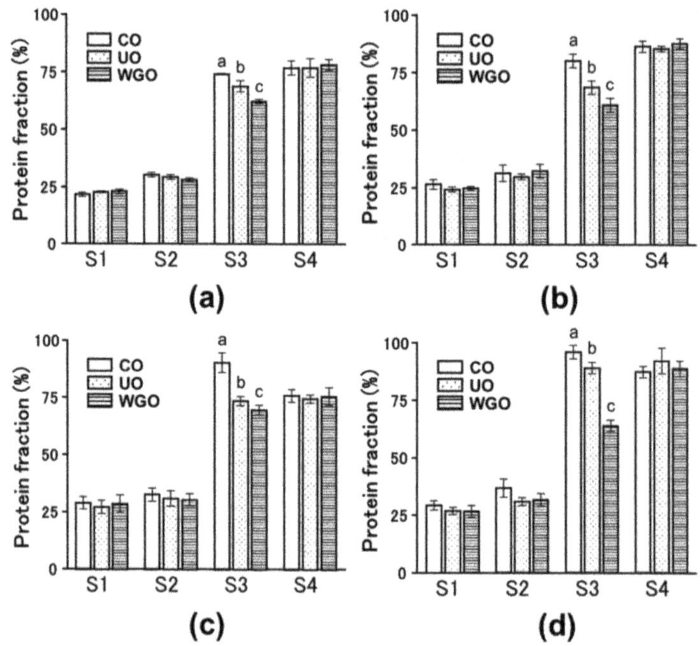

Figure 3. The protein solubility by homogenizing the okara–soybean protein isolate (SPI) gels in the four solutions (S1–S4). (**a**), 3% okara–SPI gels; (**b**), 5% okara–SPI gels; (**c**), 10% okara–SPI gels; (**d**), 15% okara–SPI gels. S1, 0.6 M sodium chloride (NaCl); S2, 0.6 M NaCl and 1.5 M urea; S3, 0.6 M NaCl and 8.0 M urea; S4, 1.0 M sodium hydroxide. CO, control SPI gel; UO, wet grinder (WG)-untreated okara–SPI gel; WGO, WG-treated okara–SPI gel. SPI concentrations and WG-untreated and treated okara in 3%, 5%, 10%, and 15% okara–SPI gels are indicated in the Figure 1 legend. Different letters indicate significant differences ($p < 0.05$).

The two major components of SPI are 7S globulin (7S) and 11S globulin (11S). The effects of various reagents on the formation of the heat-induced gels of 7S, 11S, and SPI have been previously studied [18–20]. Electrostatic interactions and disulfide bonds are involved in the formation of 11S gels, whereas hydrogen bonding and hydrophobic interactions primarily contribute to the formation of SPI gels [18]. In the 7S gelation process, gel formation was inhibited during the heating process with increasing concentrations of sodium thiocyanate, which hindered hydrophobic interactions. Furthermore, the gel was not formed in the presence of more than 4 M guanidine hydrochloride, which prevented hydrogen bonding and hydrophobic interactions. These results suggest that hydrogen bonding and hydrophobic interactions are involved in the formation of 7S gels [19]. The role of disulfide bonding in the mechanical properties and gel structures of 11S and SPI was also studied using 2-ME, which inhibited disulfide bond formation. The 11S gels showed a higher storage modulus G' and denser gel networks than those of the SPI gels. With the increase in concentration of 2-ME, the storage modulus G' and density of the gel structures of 11S gels decreased, but those of the SPI gels did not change significantly [20]. These studies indicated that hydrogen bonding and hydrophobic interactions are primarily involved in forming SPI gels.

In this study, the WG-treated okara strengthened the intermolecular forces involved in the formation SPI gels that were solubilized in S3 (0.6 M NaCl and 8 M urea). In contrast, there are no differences in protein solubility by S2 (0.6 M NaCl and 1.5 M urea) (Figure 3). According to Pérez-Mateos et al. [16], 1.5 M urea reduced hydrogen bonding, whereas 8.0 M urea decreased hydrogen bonding and hydrophobic interactions. It is suggested that during the formation of SPI gels, hydrophobic interactions are enhanced by the addition of WG-treated okara [16,17].

2.3. Relationship between Mechanical Properties and Intermolecular Forces in Okara–SPI Gels

Because upon addition of WG-treated okara the breaking stress of the SPI gel increased, and the protein solubility of the SPI gel in S3 decreased compared with those of the SPI gels without okara or with WG-untreated okara, we examined the relationship between breaking stress or strain and protein solubility in S3 (Figure 4). A negative correlation (R^2 = 0.86) was observed between breaking stress and protein solubility in S3, whereas no correlation was observed between breaking strain and protein solubility in S3.

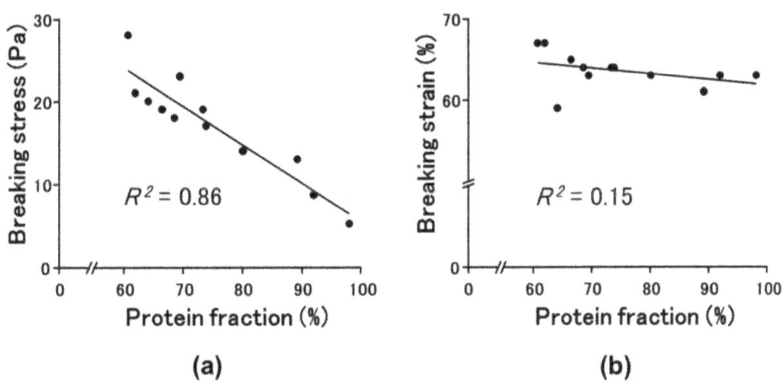

Figure 4. The relationship between breaking stress (**a**) or strain (**b**) and protein solubility in S3 (0.6 M NaCl and 8.0 M urea).

In previous studies, Fourier transform infrared analysis revealed that the band at 1618 cm^{-1} (associated with the intermolecular β-sheet structure) increased with the formation of 7S and 11S heat-induced gels. In 7S and 11S, the value of the storage modulus G' correlated well with the increase in the band at 1618 cm^{-1} (R = 0.93) [21]. Additionally, the effect of sugarcane fiber, wheat bran cellulose, or CNC on the rheological properties and structure of SPI or myofibrillar protein gels was investigated. Adding these fibers to protein gels increased their water-holding capacity and gel strength. Raman spectra showed that the formation of the β-sheet structure increased with increasing concentrations of these fibers [22–24]. These studies suggest that WG-treated okara enhances the intermolecular β-sheet structure formed in the WG-treated okara–SPI gels.

3. Conclusions

SPI gels were prepared with varying amounts of WG-treated okara, and compression measurements were performed. WG-treated okara increased the breaking stress but not the breaking strain of SPI gels. Protein solubility was examined by homogenizing the gels in four different solutions. WG-treated okara strengthened the intermolecular forces that were homogenized in S3 (0.6 M NaCl and 8.0 M urea) to form SPI gels. A significant negative correlation was observed between breaking stress and protein solubility in S3 (R^2 = 0.86). These results suggest that the WG-treatd okara increases breaking stress and enhances the hydrophobic interactions of SPI gels. This study will pave the way for future studies on developing protein-gel-based food using nano-fiber technologies.

4. Materials and Methods

4.1. Materials

Defatted okara (Newproplus 1000) and SPI (Fujipro F) were obtained from Fuji Oil (Izumisano, Japan). According to the manufacturer, Newproplus 1000 contains 6.0% water, 20.7% protein, 0.2% fat, 69.1% carbohydrates (63.7% dietary fiber), and 4.0% ash; Fujipro F contains 5.0% water, 86.3% protein, 0.2% fat, 4.0% carbohydrates, and 4.5% ash.

4.2. Preparation of Low-Protein Okara and WG-Treated Okara

The schematic diagram of the preparation procedure for low-protein okara is shown in Figure 5. A 1 M sodium hydroxide (NaOH) solution was added to the okara and stirred at 60 °C for 30 min to elute the proteins. The alkali-treated okara was centrifuged (9876× *g*, 10 min) to obtain the precipitate. Next, the precipitate was centrifuged four times with 10-fold distilled water. The amount of protein was determined using the Kjeldahl method and was found to be reduced from 17.54 ± 0.72% to 2.38 ± 0.01%. For WG-treated okara preparation, the 2 wt% low-protein okara dispersion was pulverized five times using a WG (MKCA6-2, Masuko Sangyo, Kawaguchi, Japan) with a − 0.15 mm gap at 1540 rpm of the stone disk.

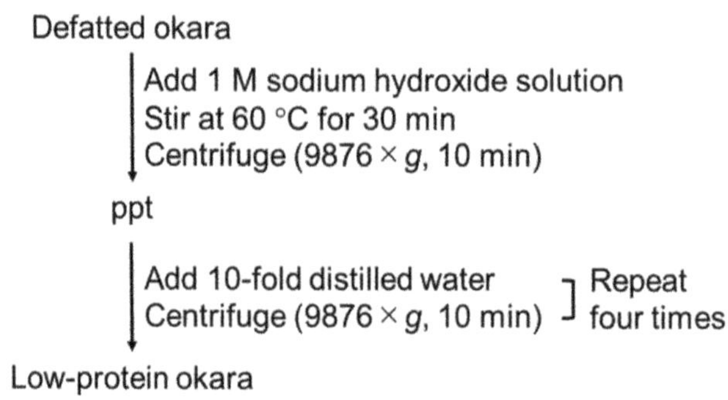

Figure 5. Schematic diagram for the preparation procedure of low-protein okara.

4.3. Preparation of Okara–SPI Gels

The preparation procedure for okara-SPI gels is shown in Figure 6. The total content of okara and SPI was adjusted to 16% and okara and concentrations of WG-treated or unteated okara and SPI for the preparation of okara–SPI gels are shown in Table 1. For okara–SPI gel preparation, SPI powder was hydrated by mixing with the WG-treated okara slurry and dispersed using a homogenizer (IKA Ultra-Turrax T8 Disperser, IKA Japan K.K., Higashiosaka, Japan). After adding 1% sodium chloride (NaCl), the sol was degassed and poured into the casing (inner diameter: 30 mm). The casing was heated at 80 °C for 30 min in a water bath (FSGPD05; Fisher Scientific International, Hampton, NH, USA).

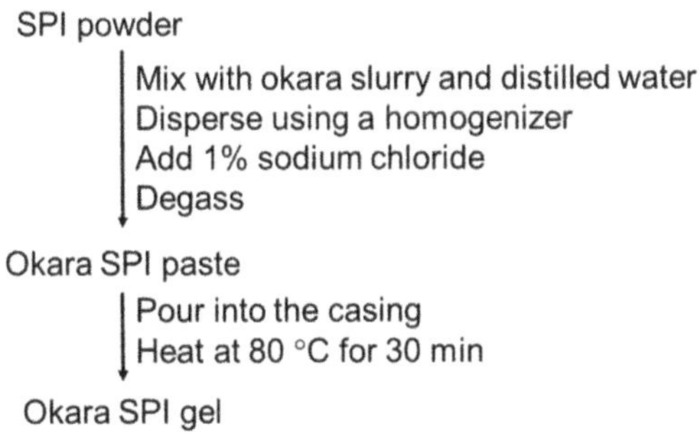

Figure 6. Schematic diagram of the preparation procedure for okara-soybean protein isolate (SPI) gels.

Table 1. Concentrations of wet-type grinder (WG)-treated or untreated okara and soybean protein isolate (SPI) for the preparation of okara–SPI gels.

	WG-Treated or Untreated Okara	SPI
3% okara–SPI	0.48%	15.52%
5% okara–SPI	0.80%	15.20%
10% okara–SPI	1.60%	14.40%
15% okara–SPI	2.40%	13.60%

The total content of okara and SPI was adjusted to 16%.

4.4. Compression Measurements

Compression measurements were performed using a texture analyzer (TA-XT2iHR, Stable Micro Systems, Godalming, Surrey, UK) attached to a 5 kg load cell at 25 °C. A cylindrical plunger with a diameter of 50 mm was used to sample the gels (30 mm diameter and 30 mm height) at a compression speed of 1 mm/s. At least six gels were examined in each experiment.

4.5. Protein Solubility

To evaluate the intermolecular forces involved in the formation of a gel, the protein solubility in four different solutions (S1, S2, S3, and S4) was examined using a protocol modified from a method described by Deng et al. [17]; S1, 0.6 M NaCl; S2, 0.6 M NaCl and 1.5 M urea; S3, 0.6 M NaCl and 8.0 M urea; and S4, 1.0 M NaOH [16].

Gels (1 g) in a 9 mL solution (S1, S2, S3, or S4) were homogenized at 12,000 rpm for 30 s and centrifuged (20,000× *g*, 10 min). The amount of protein in the supernatant was determined using the BCA protein assay kit (Thermo Fisher Scientific, Waltham, MA, USA). At least three gels were examined at each point in each experiment.

4.6. Statistical Analyses

Data represent the mean ± standard deviation. Statistical significance was calculated using one-way analysis of variance followed by Tukey's post hoc test using the Origin 2020b software (Origin Lab, Northampton, MA, USA). The data were considered statistically significant at $p < 0.05$.

Author Contributions: Conceptualization, T.N. and K.N.; methodology, T.N.; validation, Y.A., K.N., and T.N.; formal analysis, T.N.; investigation, Y.A.; data curation, T.N.; writing—original draft preparation, T.N.; writing—review and editing, T.N.; visualization, T.N.; supervision, K.N.; project administration, T.N.; funding acquisition, T.N. All authors have read and agreed to the published version of the manuscript.

Funding: This research was funded by the Fuji Foundation for Protein Research.

Institutional Review Board Statement: Not applicable.

Informed Consent Statement: Not applicable.

Data Availability Statement: Not applicable.

Conflicts of Interest: The authors declare no conflict of interest.

References

1. O'Toole, D.K. Characteristics and Use of Okara, the Soybean Residue from Soy Milk Production A Review. *J. Agric. Food Chem.* **1999**, *47*, 363–371. [CrossRef] [PubMed]
2. Nishinari, K.; Fang, Y.; Nagano, T.; Guo, S.; Wang, R. Soy as a food ingredient. In *Proteins in Food Processing*, 2nd ed.; Yada, R.Y., Ed.; Woodhead Publishing: Sawston, UK, 2018; pp. 149–186.
3. Huang, S.; He, Y.; Zou, Y.; Liu, Z. Modification of insoluble dietary fibres in soya bean okara and their physicochemical properties. *Int. J. Food Sci. Technol.* **2015**, *50*, 2606–2613. [CrossRef]
4. Nagano, T.; Arai, Y.; Yano, H.; Aoki, T.; Kurihara, S.; Hirano, R.; Nishinari, K. Improved physicochemical and functional properties of okara, a soybean residue, by nanocellulose technologies for food development—A review. *Food Hydrocoll.* **2020**, *109*, 105964. [CrossRef]
5. Pérez-López, E.; Mateos-Aparicio, I.; Rupérez, P. High hydrostatic pressure aided by food-grade enzymes as a novel approach for Okara valorization. *Innov. Food Sci. Emerg. Technol.* **2017**, *42*, 197–203. [CrossRef]
6. Liu, H.H.; Chien, J.T.; Kuo, M.I. Ultra high pressure homogenized soy flour for tofu making. *Food Hydrocoll.* **2013**, *32*, 278–285. [CrossRef]
7. Ullah, I.; Yin, T.; Xiong, S.; Zhang, J.; Din, Z.-u.; Zhang, M. Structural characteristics and physicochemical properties of okara (soybean residue) insoluble dietary fiber modified by high-energy wet media milling. *LWT* **2017**, *82*, 15–22. [CrossRef]
8. Yin, T.; Yao, R.; Ullah, I.; Xiong, S.; Huang, Q.; You, J.; Hu, Y.; Shi, L. Effects of nanosized okara dietary fiber on gelation properties of silver carp surimi. *LWT Food Sci. Technol.* **2019**, *111*, 111–116. [CrossRef]
9. Tomczyńska-Mleko, M.; Terpiłowski, K.; Mleko, S. Physicochemical properties of cellulose/whey protein fibers as a potential material for active ingredients release. *Food Hydrocoll.* **2015**, *49*, 232–239. [CrossRef]
10. Peng, J.; Calabrese, V.; Ainis, W.N.; Scager, R.; Velikov, K.P.; Venema, P.; van der Linden, E. Mixed gels from whey protein isolate and cellulose microfibrils. *Int. J. Biol. Macromol.* **2019**, *124*, 1094–1105. [CrossRef]
11. Xiao, Y.; Liu, Y.; Wang, Y.; Jin, Y.; Guo, X.; Liu, Y.; Qi, X.; Lei, H.; Xu, H. Heat-induced whey protein isolate gels improved by cellulose nanocrystals: Gelling properties and microstructure. *Carbohydr. Polym.* **2020**, *231*, 115749. [CrossRef]
12. Arai, Y.; Nishinari, K.; Nagano, T. Developing SoybeanProtein Gel-Based Foods from Okara Using the Wet-Type Grinder Method. *Foods* **2021**, *10*, 348. [CrossRef] [PubMed]
13. Iwamoto, S.; Nakagaito, A.N.; Yano, H. Nano-fibrillation of pulp fibers for the processing of transparent nanocomposites. *Appl. Phys. A* **2007**, *89*, 461–466. [CrossRef]
14. Nechyporchuk, O.; Belgacem, M.N.; Bras, J. Production of cellulose nanofibrils: A review of recent advances. *Ind. Crops Prod.* **2016**, *93*, 2–25. [CrossRef]
15. Jin, X.; Qu, R.; Wang, Y.; Li, D.; Wang, L. Effect and mechanism of acid-induced soy protein isolate gels as influenced by cellulose nanocrystals and microcrystalline cellulose. *Foods* **2022**, *11*, 461. [CrossRef]
16. Deng, C.; Shao, Y.; Xu, M.; Yao, Y.; Wu, N.; Hu, H.; Tu, Y. Effects of metal ions on the physico-chemical, microstructural and digestion characteristics of alkali-induced egg white gel. *Food Hydrocoll.* **2020**, *107*, 105956. [CrossRef]
17. Pérez-Mateos, M.; Lourenço, H.; Montero, P.; Borderías, A.J. Rheological and biochemical characteristics of high-pressure- and heat-induced gels from blue whiting (*Micromesistius poutassou*) muscle proteins. *J. Agric. Food Chem.* **1997**, *45*, 44–49. [CrossRef]
18. Utsumi, S.; Kinsella, J.E. Forces involved in soy protein gelation: Effects of various reagents on the formation, hardness and solubility of heat-induced gels made from 7S, 11S, and soy isolate. *J. Food Sci.* **1985**, *50*, 1278–1282. [CrossRef]
19. Nagano, T.; Mori, H.; Nishinari, K. Effect of heating and cooling on the gelation kinetics of 7S globulin from soybeans. *J. Agric. Food Chem.* **1994**, *42*, 1415–1419. [CrossRef]
20. Nagano, T. Contribution of disulfide bonding to viscoelastic properties and microstructures of 11S globulin gels from soybeans: Magnesium chloride-induced gels. *Food Sci. Technol. Res.* **2013**, *19*, 51–57. [CrossRef]
21. Nagano, T.; Akasaka, T.; Nishinari, K. Dynamic viscoelastic properties of glycinin and β-conglycinin gels from soybeans. *Biopolymers* **1994**, *34*, 1303–1309. [CrossRef]
22. Xiao, Y.; Li, J.; Liu, Y.; Peng, F.; Wang, X.; Wang, C.; Li, M.; Xu, H. Gel properties and formation mechanism of soy protein isolate gels improved by wheat bran cellulose. *Food Chem.* **2020**, *324*, 126876. [CrossRef] [PubMed]
23. Zhao, Y.; Zhou, G.; Zhang, W. Effects of regenerated cellulose fiber on the characteristics of myofibrillar protein gels. *Carbohydr. Polym.* **2019**, *209*, 276–281. [CrossRef] [PubMed]
24. Zhuang, X.; Zhang, W.; Liu, R.; Liu, Y.; Xing, L.; Han, M.; Kang, Z.; Xu, X.; Zhou, G. Improved gel functionality of myofibrillar proteins incorporation with sugarcane dietary fiber. *Food Res. Int.* **2017**, *100*, 586–594. [CrossRef] [PubMed]

gels

MDPI

Article

Effect of Malondialdehyde-Induced Oxidation Modification on Physicochemical Changes and Gel Characteristics of Duck Myofibrillar Proteins

Xueshen Zhu [1,*], Zhenghao Ma [1], Xinyu Zhang [1], Xuefang Huang [1], Junya Liu [1] and Xinbo Zhuang [2,*]

1 Key Lab of Biological Functional Molecules of Jiangsu Province, College of Life Science and Chemistry, Jiangsu Second Normal University, Nanjing 211200, China
2 College of Food Science and Engineering, Nanjing University of Finance & Economics, Nanjing 211171, China
* Correspondence: xueshen_zhu@163.com (X.Z.); zhuangxb@nufe.edu.cn (X.Z.)

Abstract: This paper focuses on the effect of malondialdehyde-induced oxidative modification (MiOM) on the gel properties of duck myofibrillar proteins (DMPs). DMPs were first prepared and treated with oxidative modification at different concentrations of malondialdehyde (0, 0.5, 2.5, 5.0, and 10.0 mmol/L). The physicochemical changes (carbonyl content and free thiol content) and gel properties (gel whiteness, gel strength, water holding capacity, rheological properties, and microstructural properties) were then investigated. The results showed that the content of protein carbonyl content increased with increasing MDA oxidation ($p < 0.05$), while the free thiol content decreased significantly ($p < 0.05$). Meanwhile, there was a significant decrease in gel whiteness; the gel strength and water-holding capacity of protein gels increased significantly under a low oxidation concentration of MDA (0–5 mmol/L); however, the gel strength decreased under a high oxidation concentration (10 mmol/L) compared with other groups (0.5–5 mmol/L). The storage modulus and loss modulus of oxidized DMPs also increased with increasing concentrations at a low concentration of MDA (0–5 mmol/L); moreover, microstructural analysis confirmed that the gels oxidized at low concentrations (0.5–5 mmol/L) were more compact and homogeneous in terms of pore size compared to the high concentration or blank group. In conclusion, moderate oxidation of malondialdehyde was beneficial to improve the gel properties of duck; however, excessive oxidation was detrimental to the formation of dense structured gels.

Keywords: malondialdehyde; duck meat; myofibrillar proteins; physicochemical changes; gel properties

Citation: Zhu, X.; Ma, Z.; Zhang, X.; Huang, X.; Liu, J.; Zhuang, X. Effect of Malondialdehyde-Induced Oxidation Modification on Physicochemical Changes and Gel Characteristics of Duck Myofibrillar Proteins. *Gels* **2022**, *8*, 633. https://doi.org/10.3390/gels8100633

Academic Editors: Baskaran Stephen Inbaraj, Kandi Sridhar and Minaxi Sharma

Received: 8 September 2022
Accepted: 5 October 2022
Published: 6 October 2022

1. Introduction

In general, myofibrillar proteins are important biological function proteins. The content of myosin and actin exceeds 2/3 of them. Myofibrillar proteins have better functionality and changes in their structure lead to changes that produce good textural characteristics, whose gel characteristics are the basis of processed minced meat products and directly affect the sensory properties of the final meat product [1]. In particular, oxidative modifications show the most pronounced effects. Oxidation of proteins can weaken the protein interactions and thus alter their gel formation, which has an impact on the quality of the cleavage, and intermolecular interactions lead to secondary/tertiary structural changes in the protein [2]. Side chains contain amino acids such as arginine, tyrosine, leucine, cysteine, phenylalanine, histidine, tryptophan, proline, lysine, and methionine, which are very sensitive to the action of reactive oxygen species (ROS) [3]. Among them, histidine, being highly sensitive to sulfide centers, could be oxidized at lower ROS concentrations, while leucine and proline are converted to hydroxyl derivatives. In addition, lysine is more likely to form carbonyl residues under metal catalysis [4]. Sulfur-containing amino acids such as cysteine are most susceptible to oxidation by free radicals, which is reversible, and it has some cyclic oxidation and reduction [5].

It is known that malondialdehyde (MDA) is naturally produced under meat processing conditions and is the most abundant individual aldehyde produced by lipid peroxidation. At certain ionic strengths, MDA can further promote gelation of myofibrillar proteins [6]. Zhou et al. [7] found that gels are formed by the solubilization of myofibrillar proteins at certain ionic strengths and form nondisulfide covalent bonds with MDA, a secondary fat oxidation product. Previous studies have also confirmed that the secondary and tertiary structures of myofibrillar proteins are readily altered by oxidation, leading to the unfolding of myofibrillar protein structures, which further promotes effective protein-protein interactions [8]. Thus, the oxidation of protein structures leads to changes in their physical properties, which are related to the degree of oxidation of the protein, and in general, the longer the oxidation time, the worse the functional properties of the protein [9]. It is also important to mention that, according to Wang et al. [10], mild oxidation facilitates the formation of an elastic gel lattice structure, while excessive oxidation decreases the copolymerization force and elasticity of the gel. MDA can also bind to proteins, altering protein interactions and further leading to changes in the functional properties of myofibrillar proteins in processed muscle foods. Xiong et al. [11] concluded that MDA can change the conformation of myofibrillar proteins by altering their side chains and polypeptide backbone. In addition, MDA can react with the amino group of myofibrillar proprotein to produce a strong Schiff-base-type intermolecular cross-linking, which also promotes gel formation [12]. However, few studies have been conducted on the mechanism of action of MDA on duck myofibrillar proteins (DMPs), which has been selected as an effective secondary oxidation product during lipid metabolism. Therefore, we hypothesized that direct incubation of MDA with DMPs could alter the structure of DMPs, leading to affect the formation of compact and rigid gels. The aim of this study was to investigate the effect of oxidative modification of malondialdehyde on the physicochemical changes and gel properties of DMPs, aiming to provide a basis for controlling the degree of meat oxidation and the rational use of oxidants.

2. Results and Discussion

2.1. Morphological Observation

Figure 1 shows the results of malondialdehyde-induced oxidation modification on the gel morphology of duck myofibrillar proteins (DMPs). As shown in Figure 1, the gel morphological state of the DMP samples (dissolved in 0.6 M NaCl) changed from solution to gel as the MDA concentration increased. It also clearly showed that gel samples treated at higher concentrations of MDA (0.5, 2.5, 5.0, 10 mM) became brown and gel-like and did not collapse easily, while the blank group without MDA remained sol-like and flowed easily.

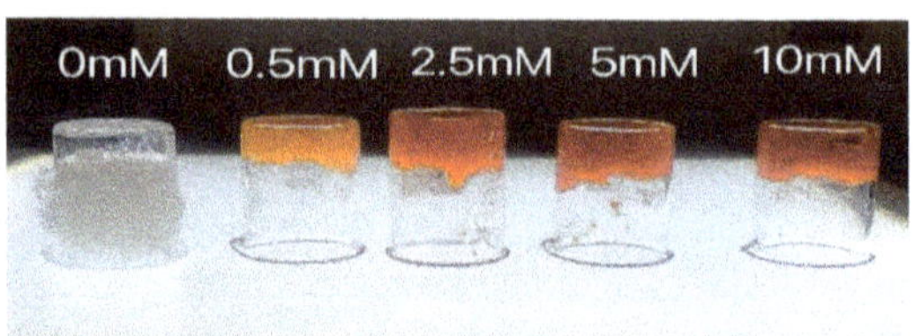

Figure 1. Morphology observation of duck myofibrillar proteins gels treated with different malondialdehyde concentrations (0, 0.5, 2.5, 5.0, and 10.0 mmol/L).

2.2. Carbonyl Content and Free Thiol Content Changes

Protein carbonylation is an important oxidation reaction of proteins that leads to the oxidation of new carbonyl groups due to their -NH- or -NH_2 groups on the side chains of amino acids [4]. Figure 2a shows the changes in the carbonyl content of the samples after oxidative treatment with different concentrations of MDA. The protein carbonyl content of the samples treated in the absence of MDA was 2.52 nmol/mg protein. The carbonyl content of DMPs increased significantly ($p < 0.05$) with increasing MDA concentration from 0 to 10 mmol/L, because MDA could be added to primary amines in protein molecules in a certain ratio to produce enamine adducts, resulting in new carbonyl groups. In addition, the molecular structure of MDA is composed of two carbonyl groups combined, where the oxidation of one MDA molecule with the protein can introduce another carbonyl group at the same time. In general, MDA could act on the nucleophilic side chain groups of cysteine, histidine, and lysine residues, which then react to form the Schiff base [13].

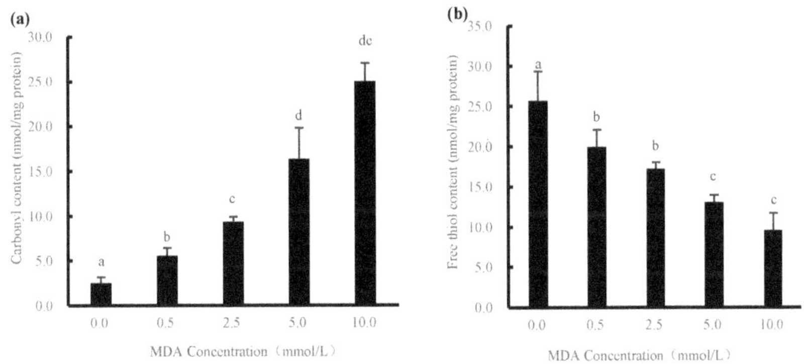

Figure 2. Changes in carbonyl content (**a**) and free thiol content (**b**) of duck myofibrillar proteins treated with different malondialdehyde (MDA) concentrations (0, 0.5, 2.5, 5.0, and 10.0 mmol/L). Different letters (a–e) indicate significant difference ($p < 0.05$).

In proteins, sulfhydryl groups are much reactive, so sulfur-containing amino acid residues can easily interact with each other to produce disulfide bonds, and the earliest oxidation of sulfhydryl groups in proteins can lead to changes in protein structure, which, in turn, can have some effects on protein function [14]. As shown in Figure 2b, the free thiol content of DMPs decreased gradually from 25.62 to 9.58 nmol/mg of protein with increasing MDA concentration, which was due to the exposure of endogenous sulfhydryl groups during the breakdown of the protein structure and exposure to pro-oxidant compounds, resulting in a constant decrease in the total amount of free thiol groups, as confirmed by the recent findings of Soglia et al. [15]. Previous reports have also indicated that the loss of sulfhydryl groups in plasma proteins occurs due to the interaction between the sulfhydryl groups of proteins and the α, β-Michael addition of unsaturated aldehydes [16]. In addition, MDA is electrophilic and can preferentially abstract hydrogen atoms from the free thiol group of cysteine, leading to a decrease in free thiol content [17].

2.3. SDS-PAGE Profile Analysis

As shown in Figure 3, the intensity of myosin heavy chain (MHC) bands was largely diminished in samples under nonreducing conditions (without Dithiothreitol, -DTT) after treatment with 2.5, 5.0, and 10.0 mmol/L MDA, respectively. Other bands of myofibrillar proteins disappeared (actin bands were slightly blurred) when the MDA concentration exceeded 2.5 mmol/L, a phenomenon indicating that with increasing oxidation concentration, protein cross-linking overloaded. Under reducing conditions (with dithiothreitol, +DDT), similar results were found as under nonreducing conditions (-DTT), where the disappeared myosin heavy chain components were not recovered and the large polymer remained in the loaded part of the gel, indicating that the cross-linking was not essentially through disulfide bonds or similar mechanisms and that the covalent bonds of this cross-linking were strong and difficult to break. It is important to mention here that fewer polymers were formed under nonreducing conditions compared to reducing conditions. We supposed that this might be due to the excessive cross-linking of proteins caused by high concentrations of MDA, forming polymers of very high molecular weight that simply cannot enter the separating gel [18]. Riley and Harding [19] reported that malondialdehyde can react with lens proteins to form covalent cross-linkages with nondisulfide bonds. Oxidative modification at different concentrations of MDA with 0.6 mmol/L NaCl resulted in the unfolding of myosin, which increased action sites, and also the formation of non-disulfide bonds was not conducive to separation between myosin and related proteins by SDS-PAGE. In the presence of MDA, the cross-linking of proteins caused by nondisulfide bonds is more likely to be associated with the formation of Schiff bases as described above. Recent findings also suggest that protein aggregation occurs in the MDA-induced oxidation system. Myosin is involved in gel formation through nondisulfide covalent bonds [20]. Disulfide bonds are responsible for most of the cross-linking and malondialdehyde also seems to contribute to cross-linking [11].

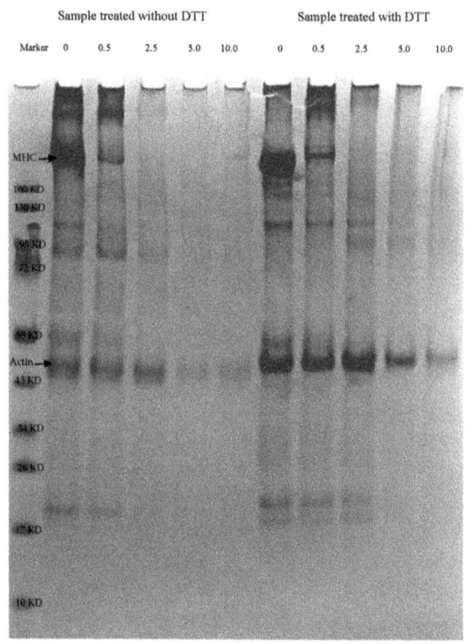

Figure 3. SDS-PAGE pattern changes of duck myofibrillar proteins treated with different malondialdehyde concentrations (0, 0.5, 2.5, 5.0, and 10.0 mmol/L).

2.4. Gel Strength and Water Holding Capacity Analysis

Gel strength is an important quality characteristic of protein gels. As shown in Figure 4a, the oxidative modification induced by MDA played an important role in affecting the gel strength. It is clearly shown that the gel strength of the sample increased significantly from 0.02 N to 0.23 N with increasing MDA concentration. This could be due to the creation of more cross-linking among proteins through nondisulfide bonds and disulfide bonds during the thermal treatment of the gel, and the formation of these covalent bonds promoted the strengthening of the gel structure; thus, the gel strength of the sample was improved [21]. However, when comparing the 5 mmol/L treatment group with the 10 mmol/L treatment group, gel strength decreased significantly with the further increase in MDA concentration ($p < 0.05$). The reason could probably be due to the oxidation at high concentrations leading to the formation of excessive covalent bonds, which, in turn, might have promoted the degradation of proteins and decrease in gel strength [22]. Previous studies have shown that disulfide and other covalent bonds are enhanced, leading to protein cross-linking and aggregation, which affects gel strength, while enhancing the degradation of actin and troponin-T [23].

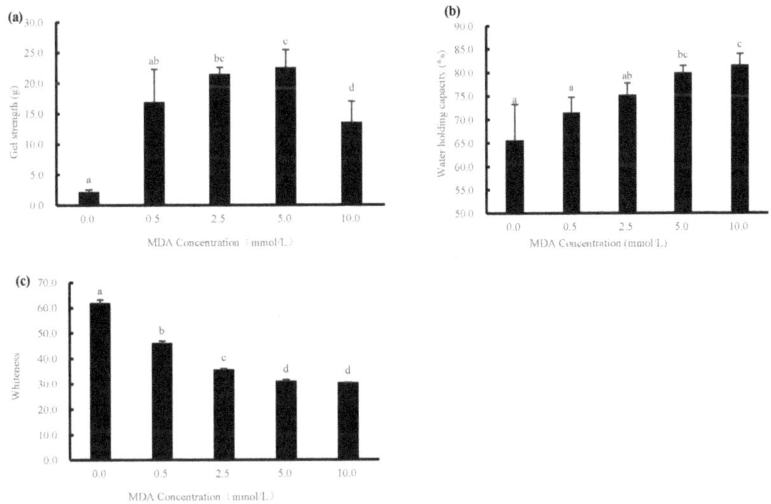

Figure 4. Changes in gel strength (**a**), water holding capacity (**b**), and whiteness (**c**) of duck myofibrillar proteins treated with different malondialdehyde (MDA) concentrations (0, 0.5, 2.5, 5.0, and 10.0 mmol/L). Different letters (a–d) indicate significant difference ($p < 0.05$).

Similar to gel strength, analysis of the water holding capacity of gels can also reflect the quality of differently treated myofibrillar protein gels. The water holding capacity of a gel is related to the spatial structure of the gel and is a response to the ability of the gel to retain water molecules [24]. As shown in Figure 4b, the water holding capacity (WHC) of the gel samples increased continuously with the increase of MDA concentration (0–10 mmol/L), from 65.6% initially to 85.55%, and the trend of its water holding capacity was basically consistent with that of the carbonyl group. It could be speculated that MDA promoted the carbonylation of proteins and was responsible for the formation of protein crosslinks throughout the incubation treatment, thus changing the reticular structure of the gels and promoting the improvement of the water holding capacity of the gels [25].

As shown in Figure 4c, whiteness decreased significantly with increasing MDA content, and the results indicate that the brightness, redness, and yellowness values of the gels were significantly affected, which corresponds to the morphological observations described previously. Xia et al. [26] also found that repeated thawing and freezing treatments of meat caused nonenzymatic oxidation reactions between fat oxidation products and amino acids in proteins, leading to a decrease in their whiteness.

2.5. Rheological Characterization

Storage modulus is an important parameter during the determination of dynamic rheological properties, which is shown in Figure 5a. First, when the MDA concentration was 0, the curve peaked from 25 °C to 46.5 °C; afterward, the curve gradually decreased until 70.1 °C, and then the curve continued to increase until the end of the test, indicating that 46.5 °C was the gel point of myofibrillar protein gel. In the temperature interval from 25 to 46.5 °C, DMPs underwent high-temperature denaturation, which exposed its amino acids and led to changes in the gel structure, resulting in the formation of a gel with high elasticity; in the temperature interval from 46.5 to 70.1 °C, the original gel matrix was destroyed, and in the interval from 70.1 to 80 °C, the ordered cross-linking of disulfide bonds continued, resulting in the formation of a stable, uniform, dense, and three-dimensional elastic network with high energy density. At low MDA concentrations (0–5 mmol/L), the storage modulus trends were similar among different MiOM groups, but the final storage modulus increased significantly with increasing MDA oxidation concentrations. It should be emphasized here that the storage modulus of the 10 mmol/L treatment group was lower when compared with the 5 mmol/L treatment group. This confirmed that when treated at relatively lower oxidation concentrations of MDA, protein–protein interactions were enhanced, producing better binding ability among the proteins, whereas excessive oxidation concentrations of MDA oxidation could lead to ultra-aggregation and shrinkage of the protein gel, which was detrimental to the gel-forming ability of DMPs, as described by the previous results on gel strength of the DMPs.

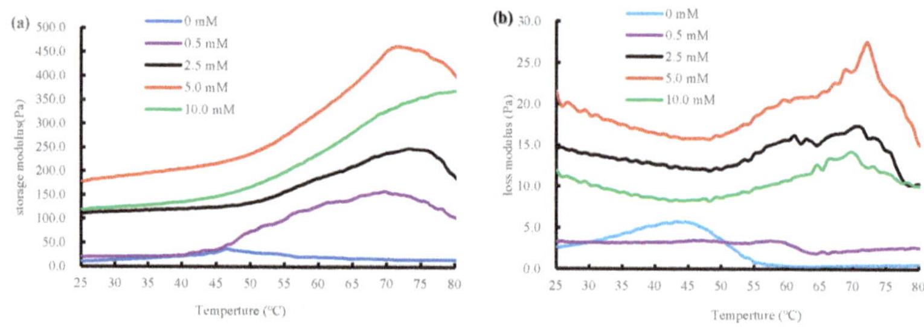

Figure 5. Storage modulus (**a**) and loss modulus (**b**) changes during heating of duck myofibrillar proteins treated with different malondialdehyde concentrations (0, 0.5, 2.5, 5.0, and 10.0 mmol/L).

The trend of the loss modulus (Figure 5b) was roughly similar with changes in the storage modulus mentioned above. The loss modulus curve (0 mmol/L) first peaked at 43.1 °C and then gradually decreased and leveled off at 56 °C. Figure 5b also clearly shows a gradual increase in the loss modulus of samples treated at different MDA concentrations (0.5–5 mmol/L) until about 70 °C, decreasing afterward, which might be due to the destruction of the previously formed gel matrix with increasing temperature [27]. In addition, the curves of the loss modulus under 10 mmol/L MDA oxidation were generally lower when compared with those of 2.5 mmol/L and 5 mmol/L treatment groups. These results coincided with the previous results of gel strength.

2.6. Nuclear Magnetic Characterization

Figure 6 shows the effect of malondialdehyde concentration on the changes in relaxation time (T_2) of the gels of DMPs. As shown in the figure, three typical peaks were found, with one peak appearing at 20–96 ms, a larger peak at 131–446 ms, and another peak after 800 ms. Our previous paper found that DMP gels generally show three peaks in the fitted NMR relaxation curves, corresponding to moderately immobilized water, immobilized water, and free water, respectively [28]. That is, T_{21} represents moderate immobilized water with peaks between 20 and 96 ms, T_{22} represents immobilized water with peaks between 131 and 446 ms, and T_{23} represents free water with peaks after 800 ms. In general, DMPs cross-link with each other to form a three-dimensional structure, locking up a large number of water molecules and forming immobilized water, which is understandable. Our results also showed that the area of T_{22} (immobilized water) increased with increasing MDA concentration (from 0 to 5.0 mmol/L). In addition, a decreasing trend of peak area could be found for the 10 mmol/L treatment group when compared with 2.5 and 5 mmol/L treatment groups. In addition, there was a decreasing trend of T_{23} (free water) with increasing MDA concentration, suggesting that especially low concentrations of MDA help to convert free water to immobilized water and reduce the free water content. These results fit well with the previous discussed results in water holding capacity, and correspond to the finding of Wang et al. [29] that the fraction of free water decreased from 7.66% to 0.15% with increasing MDA addition from 0 to 50 mM. Furthermore, the relaxation component (bound water) disappeared with the addition of MDA, mainly due to the increase in flexibility and surface hydrophobicity of the protein. Xia et al. [30] also reported that WHC had a significant positive correlation with the percentage of immobile water and a negative correlation with carbonyl content and T_{23}.

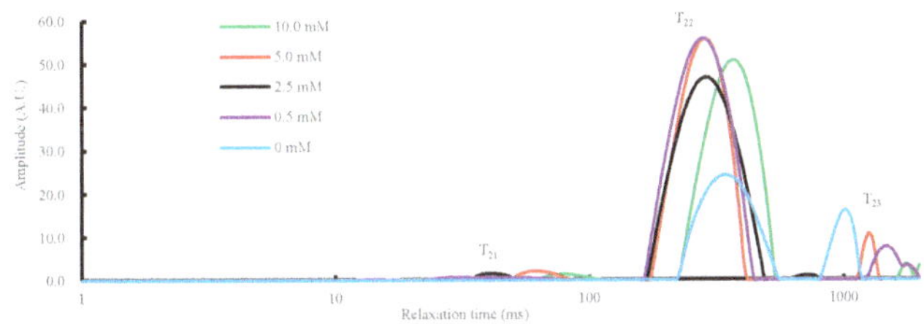

Figure 6. Effect of different malondialdehyde concentrations (0, 0.5, 2.5, 5.0, and 10.0 mmol/L) on the distribution of the T_2 of duck myofibrillar proteins gel. T_2 = spin–spin relaxation times (ms) for different types of water.

2.7. Gel Microstructure Analysis

The results of the gel microstructure analysis are shown in Figure 7. The samples treated with relatively low MDA concentrations (2.5 and 5 mmol/L) showed a better gel mesh structure, which were gradually tightened to form an ordered structure with uniform pore size. This result was consistent with the enhancement of gel strength as described previously. In contrast, the gel structure remained firm after 10 mmol/L MDA treatment, and the depression of the gel surface structure could be easily found in the figure, which could be explained by the weakening of the gel strength in our previous results. These results clearly indicated that the mesh structure of the gels could be improved at relatively mild MDA oxidation [31]. Similar results were reported by Zhou et al. [32]. Higher MDA could cause the collapse of the gel due to the presence of excessive covalent bonds.

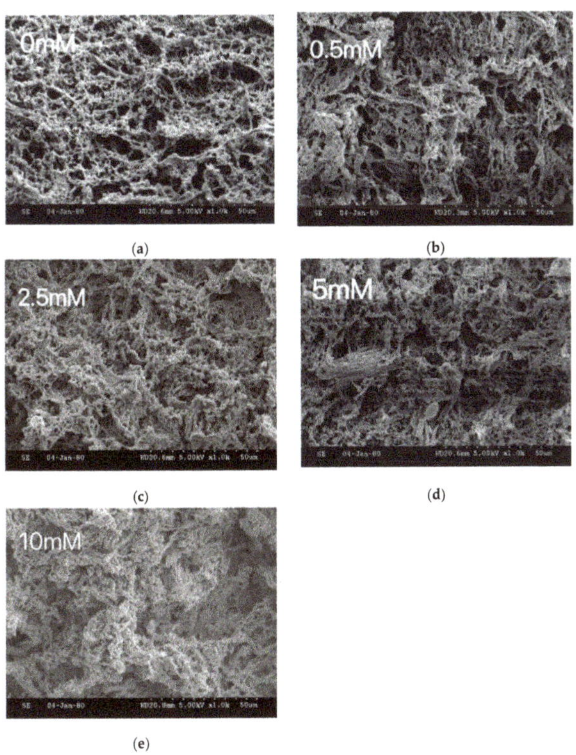

Figure 7. Gel microstructure analysis of duck myofibrillar proteins treated with malondialdehyde (MDA) concentrations of 0 mmol/L (**a**), 0.5 mmol/L (**b**), 2.5 mmol/L (**c**), 5.0 mmol/L (**d**), and 10.0 mmol/L (**e**).

3. Conclusions

MDA-induced oxidation can change the physicochemical structure of DMPs, such as a significant increase in carbonyl content and a significant decrease in free thiol content; in addition, gel whiteness and WHC showed a decreasing trend. It should be mentioned that under proper oxidation conditions, the protein gel hardness increased significantly at low concentrations of MDA (from 0 to 5 mmol/L); however, it decreased at high concentrations (10 mmol/L). The enhancement of covalent bonds promoted the consolidation of the gel structure. These results suggest that covalent bonds induced during heating at mild-low concentrations of MDA oxidation might improve the gel structure and, thus, improve the gel quality.

4. Materials and Methods

4.1. Samples and DMPs Preparation

The duck breast meat used in this experiment was obtained from the local market in Lishui, Nanjing, China. All fat was stripped off and muscles were then placed in a self-sealing bag in a −80 °C refrigerator.

To extract DMPs, 30 g of duck breast meat was added to the extract (5 times volume), homogenized at high speed for 20 s, and then placed in a refrigerated centrifuge tube and centrifuged under the following conditions: 4000× *g* at 4 °C for 10 min, repeated three times. It was then filtered through gauze, and 1% TritonX-100 (5 times volume) was used afterward to wash the above precipitate. This was repeated three times together with centrifugation methods, and it was finally washed with 0.1 mmol/L NaCl (5 times volume) to dissolve, homogenize, and centrifuge again. The supernatant was discarded and the precipitated DMPs were finally collected by filtration through gauze.

4.2. MDA Oxidation-Modified Myofibrillar Protein Treatment

The MDA stock solution was mainly prepared through 1,1,3,3-tetramethoxypropane obtained from Sigma-Aldrich Chemical Co. (St Louis, MO, USA) as described by Wu et al. [33] with minor modifications.

The MDA-induced oxidation system of DMPs was referred to Zhou et al. [7] with appropriate modifications. At first, the concentration of DMPs was adjusted to 40 mg/mL with MDA stock solution at different ratios (MDA concentration: 0, 0.5, 2.5, 5.0, and 10.0 mmol/L). The mixture was then placed in a 10 mL centrifuge tube and incubated at 25 °C for 24 h. The protein gels were then prepared as follows: protein solutions were heated directly in a water bath after the above treatment, subject to linear heating at a starting temperature of 25 °C to 85 °C at a speed of 1 °C/min, maintained at 85 °C for 5 min, and then cooled with an ice bath and stored at 4 °C before use.

4.3. Carbonyl Content Determination

The protein carbonyl content was modified according to Soglia et al. [34]. The obtained samples of DMPs were adjusted to a concentration of 20 mg/mL. During the assay, 5% SDS was used to resolve the precipitates and dinitrophenylhydrazine (DNPH) was used to label the carbonyl groups. Finally, the absorbance at 280 nm and 370 nm was measured in solution. The carbonyl content was calculated using 22,000 L/(mol·cm) as the molar extinction coefficient for conversion and expressed in units of nmol/mg protein, as shown in the equation:

$$Carbonyl\ content\ (nmol/mg\ protein) = \frac{\left[A_{370} - A_{370\,(blank)}\right] \times 10^6}{22,000 \times \left[A_{280} - \left(A_{370} - A_{370(blank)}\right) \times 0.43\right]} \tag{1}$$

4.4. Free Thiol Content Determination

The protein free thiol content was appropriately determined according to the method of Bao et al. [35]. The concentration of the oxidized DMPs obtained was adjusted to

2 mg/mL, and 300 μL of this protein concentration solution was used. Briefly, 0.5 mL of 10 mM 5,5′-dithiobis-(2-nitrobenzoic acid) was added to the solution, mixed well, and protected from light for 30 min. The molar extinction coefficient was 14,150 L/(mol·cm), and the free thiol content was calculated as follows:

$$Free\ thiols\ content\ (nmol/mg\ protein) = \frac{\left[A_{412(after)} - A_{412\ (before)}\right]}{A_{280} \times 14,150} \quad (2)$$

4.5. Sodium Dodecyl Sulfate-Polyacrylamide Gel Electrophoresis Analysis (SDS-PAGE)

DMPs with different MDA-induced oxidation modification (MiOM) were adjusted to 2 mg/mL. The DMPs were pretreated using sample buffer (Invitrogen, Thermal Fisher, Waltham, MA, USA) with or without dithiothreitol (DTT, Beyotime, Shanghai, China), and then heated in a metal bath at 99 °C for 5 min. Electrophoresis was then run at 220 V for 45 min. The Coomassie Brilliant Blue R-250 (Beyotime, Shanghai, China) solution was used to stain for 30 min and then decolorized and photoed for analysis.

4.6. Gel Hardness and Water Holding Capacity (WHC)

The gel strength of DPM samples was determined according Zhu et al. [36] using a texture analyzer (TA-XT plus Plaser, Stable Micro Systems, Surrey, United Kingdom). Measurement conditions: P/0.5R probe with a pre-measurement speed of 1 mm/s, a measurement speed of 0.5 mm/s, a post-measurement speed of 10 mm/s, compression mode, and a depth distance of 5 mm. Each treated sample was repeated 3 times.

The WHC of the gel was determined using centrifugation methods. The gel was centrifuged at 6000 r/min for 15 min at 4 °C. The centrifuge tube was then placed upside-down on absorbent paper for 30 min. Before centrifugation, the mass of the tube was m, the total mass was recorded as $m1$, and the total mass after being placed for 30 min was $m2$. Water holding capacity was calculated as follows:

$$WHC\ (\%) = \frac{m2 - m}{m1 - m} \times 100 \quad (3)$$

4.7. Gel Whiteness Determination

The gel samples were determined using a CR 400 colorimeter (Minolta Camera, Osaka, Japan). Brightness (L^*), red (a^*), and yellow (b^*) were measured three times for each sample, and the gel whiteness was calculated as follows:

$$Whiteness = 100 - \sqrt{(100 - L^*)^2 + a^{*2} + b^{*2}} \quad (4)$$

4.8. Rheological Properties Test

The freshly oxidized protein solution was measured using a rheometer (MCR-301, Anton Paar, Graz, Austria) in oscillatory mode as described by Zhuang et al. [37]. The following parameters were used: A 50 mm plate material was selected with a gap of 1 mm between the upper and lower plates, a frequency of 0.1 Hz, a strain of 2%, 25 °C/min, the heating temperature was from 25 °C to 80 °C at a speed of 2 °C/min, and the cooling rate was 5 °C/min. Before the test, paraffin oil was used to drip into the edge of the plate to isolate the sample from the outside air. The storage modulus and loss modulus during heating was then recorded.

4.9. Low-Field Nuclear Magnetic Resonance Analysis

Relaxation times (T_2) were measured according Han et al. [38] with a nuclear magnetic resonance (NMR) analyzer (MesoMR23-060H-1, Niumag elctric Co., Shanghai, China). A standard oil sample was first calibrated, a centrifuge tube of about 2 g was placed in the tester, and the spin–spin relaxation time (T_2) was selected as the Carr–Purcell–Meiboom–Gill (CPMG) sequence. The proton resonance frequency was set at 22.6 MHz and the measurement was performed at a temperature of 32 °C. The relevant parameters: the number of repetition sampling (NS) was 4 times, the repetition interval (time wait, TW) was 2000 ms, the number of echoes (NECH) was 9000, each test was performed 3 times, and the obtained curve was an exponential decay sample curve. A large number of data inversions were achieved through the data query function in the software menu.

4.10. Gel Microstructure Analysis

The gel samples were cut into squares (3 mm × 3 mm × 3 mm) and fixed with 4% malondialdehyde. Gels were then analyzed with a Hitachi S-3000N scanning electron microscope (Tokyo, Japan) at an accelerating voltage of 20 kV.

4.11. Statistical Analysis

Data were processed with SPSS 20.0 software (version 20, SPSS Inc., Chicago, IL, USA) and subjected to one-way ANOVA with Duncan's multiple range test for statistical analysis.

Author Contributions: Conceptualization, project administration, writing—review and editing, X.Z. (Xueshen Zhu); investigation, data curation, writing—original draft, Z.M.; methodology, X.Z. (Xinyu Zhang); funding acquisition, visualization, X.H. and J.L.; software, supervision, X.Z. (Xinbo Zhuang). All authors have read and agreed to the published version of the manuscript.

Funding: This research was supported by the National Natural Science Foundation of China (31301507) and the Natural Science Fund for Colleges and Universities in Jiangsu Province (13KJB550006; 19KJB550005; 21KJA550002).

Institutional Review Board Statement: Not applicable.

Informed Consent Statement: Not applicable.

Data Availability Statement: The datasets generated for this study are available on request to the corresponding author.

Acknowledgments: The authors wish to thank Lei Jin for her English revisions.

Conflicts of Interest: The authors declare no conflict of interest.

References

1. Li, C.Q.; Xiong, Y.L.L.; Chen, J. Protein Oxidation at Different Salt Concentrations Affects the Cross-Linking and Gelation of Pork Myofibrillar Protein Catalyzed by Microbial Transglutaminase. *J. Food Sci.* **2013**, *78*, C823–C831. [CrossRef] [PubMed]
2. Dean, R.T.; Fu, S.L.; Stocker, R.; Davies, M.J. Biochemistry and pathology of radical-mediated protein oxidation. *Biochem. J.* **1997**, *324*, 1–18. [CrossRef] [PubMed]
3. Hellwig, M. The Chemistry of Protein Oxidation in Food. *Angew. Chem.-Int. Ed.* **2019**, *58*, 16742–16763. [CrossRef] [PubMed]
4. Stadtman, E.R.; Levine, R.L. Free radical-mediated oxidation of free amino acids and amino acid residues in proteins. *Amino Acids* **2003**, *25*, 207. [CrossRef] [PubMed]
5. Xiong, Y.L.L.; Guo, A.Q. Animal and Plant Protein Oxidation: Chemical and Functional Property Significance. *Foods* **2021**, *10*, 40. [CrossRef]
6. Adams, A.; De Kimpe, N.; van Boekel, M. Modification of casein by the lipid oxidation product malondialdehyde. *J. Agric. Food Chem.* **2008**, *56*, 1713–1719. [CrossRef]
7. Zhou, F.; Zhao, M.; Su, G.; Cui, C.; Sun, W. Gelation of salted myofibrillar protein under malondialdehyde-induced oxidative stress. *Food Hydrocoll.* **2014**, *40*, 153–162. [CrossRef]
8. Park, D.; Xiong, Y.L.L. Oxidative modification of amino acids in porcine myofibrillar protein isolates exposed to three oxidizing systems. *Food Chem.* **2007**, *103*, 607–616. [CrossRef]
9. Lund, M.N.; Heinonen, M.; Baron, C.P.; Estevez, M. Protein oxidation in muscle foods: A review. *Mol. Nutr. Food Res.* **2011**, *55*, 83–95. [CrossRef]
10. Wang, Z.M.; He, Z.F.; Gan, X.; Li, H.J. Effect of peroxyl radicals on the structure and gel properties of isolated rabbit meat myofibrillar proteins. *Int. J. Food Sci. Technol.* **2018**, *53*, 2687–2696. [CrossRef]
11. Xiong, Y.L.; Park, D.; Zu, T. Variation in the Cross-Linking Pattern of Porcine Myofibrillar Protein Exposed to Three Oxidative Environments. *J. Agric. Food Chem.* **2009**, *57*, 153–159. [CrossRef] [PubMed]
12. Buttkus, H. Reaction of myosin with malonaldehyde. *J. Food Sci.* **1967**, *32*, 432–434. [CrossRef]
13. Burcham, P.C.; Kuhan, Y.T. Introduction of carbonyl groups into proteins by the lipid peroxidation product, malondialdehyde. *Biochem. Biophys. Res. Commun.* **1996**, *220*, 996–1001. [CrossRef]

14. Jiang, W.; He, Y.; Xiong, S.; Liu, Y.; Yin, T.; Hu, Y.; You, J. Effect of Mild Ozone Oxidation on Structural Changes of Silver Carp (Hypophthalmichthys molitrix) Myosin. *Food Bioprocess Technol.* **2017**, *10*, 370–378. [CrossRef]
15. Soglia, F.; Baldi, G.; Petracci, M. Effect of the exposure to oxidation and malondialdehyde on turkey and rabbit meat protein oxidative stability. *J. Food Sci.* **2020**, *85*, 3229–3236. [CrossRef]
16. van der Vliet, A.; Cross, C.E.; Halliwell, B.; O'Neill, C.A. Plasma protein sulfhydryl oxidation: Effect of lowmolecular weight thiols. In *Methods in Enzymology*; Academic Press: Cambridge, MA, USA, 1995; Volume 251, pp. 448–455.
17. Liao, G.; Zhang, H.; Jiang, Y.; Javed, M.; Xiong, S.; Liu, Y. Effect of lipoxygenase-catalyzed linoleic acid oxidation on structural and rheological properties of silver carp (Hypophthalmichthys molitrix) myofibrillar protein. *LWT* **2022**, *161*, 113388. [CrossRef]
18. Traverso, N.; Menini, S.; Maineri, E.P.; Patriarca, S.; Odetti, P.; Cottalasso, D.; Marinari, U.M.; Pronzato, M.A. Malondialdehyde, a Lipoperoxidation-Derived Aldehyde, Can Bring About Secondary Oxidative Damage to Proteins. *J. Gerontol. Ser. A* **2004**, *59*, B890–B895. [CrossRef]
19. Riley, M.L.; Harding, J.J. The reaction of methylglyoxal with human and bovine lens proteins. *Biochim. Biophys. Acta-Mol. Basis Dis.* **1995**, *1270*, 36–43. [CrossRef]
20. Wang, L.; Zhang, M.; Fang, Z.; Bhandari, B. Gelation properties of myofibrillar protein under malondialdehyde-induced oxidative stress. *J. Sci. Food Agric.* **2017**, *97*, 50–57. [CrossRef]
21. Wang, Z.; He, Z.; Zhang, D.; Chen, X.; Li, H. The effect of linalool, limonene and sabinene on the thermal stability and structure of rabbit meat myofibrillar protein under malondialdehyde-induced oxidative stress. *LWT* **2021**, *148*, 111707. [CrossRef]
22. Wang, H.; Song, Y.; Liu, Z.; Li, M.; Zhang, L.; Yu, Q.; Guo, Z.; Wei, J. Effects of iron-catalyzed and metmyoglobin oxidizing systems on biochemical properties of yak muscle myofibrillar protein. *Meat Sci.* **2020**, *166*, 108041. [CrossRef] [PubMed]
23. Wang, N.; Hu, L.; Guo, X.; Zhao, Y.; Deng, X.; Lei, Y.; Zhang, L.; Zhang, J. Effects of malondialdehyde on the protein oxidation and protein degradation of Coregonus Peled myofibrillar protein. *J. Food Meas. Charact.* **2022**. [CrossRef]
24. Bao, Y.; Ertbjerg, P. Effects of protein oxidation on the texture and water-holding of meat: A review. *Crit. Rev. Food Sci. Nutr.* **2019**, *59*, 3564–3578. [CrossRef]
25. Wang, Z.M.; He, Z.F.; Emara, A.M.; Gan, X.; Li, H.J. Effects of malondialdehyde as a byproduct of lipid oxidation on protein oxidation in rabbit meat. *Food Chem.* **2019**, *288*, 405–412. [CrossRef]
26. Xia, X.F.; Kong, B.H.; Xiong, Y.L.; Ren, Y.M. Decreased gelling and emulsifying properties of myofibrillar protein from repeatedly frozen-thawed porcine longissimus muscle are due to protein denaturation and susceptibility to aggregation. *Meat Sci.* **2010**, *85*, 481–486. [CrossRef] [PubMed]
27. Li, B.; Xu, Y.; Li, J.; Niu, S.; Wang, C.; Zhang, N.; Yang, M.; Zhou, K.; Chen, S.; He, L.; et al. Effect of oxidized lipids stored under different temperatures on muscle protein oxidation in Sichuan-style sausages during ripening. *Meat Sci.* **2019**, *147*, 144–154. [CrossRef] [PubMed]
28. Zhu, X.; Zhang, J.; Liu, S.; Gu, Y.; Yu, X.; Gao, F.; Wang, R. Relationship between Molecular Structure and Heat-Induced Gel Properties of Duck Myofibrillar Proteins Affected by the Addition of Pea Protein Isolate. *Foods* **2022**, *11*, 1040. [CrossRef]
29. Wang, L.; Zhang, M.; Bhandari, B.; Gao, Z.X. Effects of malondialdehyde-induced protein modification on water functionality and physicochemical state of fish myofibrillar protein gel. *Food Res. Int.* **2016**, *86*, 131–139. [CrossRef]
30. Xia, T.; Zhao, X.; Yu, X.; Li, L.; Zhou, G.; Han, M.; Xu, X.-l. Negative impacts of in-vitro oxidative stress on the quality of heat-induced myofibrillar protein gelation during refrigeration. *Int. J. Food Prop.* **2018**, *21*, 2205–2217. [CrossRef]
31. Chen, B.; Zhou, K.; Wang, Y.; Xie, Y.; Wang, Z.; Li, P.; Xu, B. Insight into the mechanism of textural deterioration of myofibrillar protein gels at high temperature conditions. *Food Chem.* **2020**, *330*, 127186. [CrossRef]
32. Zhou, F.; Sun, W.; Zhao, M. Controlled Formation of Emulsion Gels Stabilized by Salted Myofibrillar Protein under Malondialdehyde (MDA)-Induced Oxidative Stress. *J. Agric. Food Chem.* **2015**, *63*, 3766–3777. [CrossRef] [PubMed]
33. Wu, W.; Zhang, C.M.; Hua, Y.F. Structural modification of soy protein by the lipid peroxidation product malondialdehyde. *J. Sci. Food Agric.* **2009**, *89*, 1416–1423. [CrossRef]
34. Soglia, F.; Petracci, M.; Ertbjerg, P. Novel DNPH-based method for determination of protein carbonylation in muscle and meat. *Food Chem.* **2016**, *197*, 670–675. [CrossRef] [PubMed]
35. Bao, Y.; Boeren, S.; Ertbjerg, P. Myofibrillar protein oxidation affects filament charges, aggregation and water-holding. *Meat Sci.* **2018**, *135*, 102–108. [CrossRef] [PubMed]
36. Zhu, X.; Tan, B.; Li, K.; Liu, S.; Gu, Y.; Xia, T.; Bai, Y.; Wang, P.; Wang, R. The Impacts of Different Pea Protein Isolate Levels on Functional, Instrumental and Textural Quality Parameters of Duck Meat Batters. *Foods* **2022**, *11*, 1620. [CrossRef]
37. Zhuang, X.; Wang, L.; Jiang, X.; Chen, Y.; Zhou, G. Insight into the mechanism of myofibrillar protein gel influenced by konjac glucomannan: Moisture stability and phase separation behavior. *Food Chem.* **2021**, *339*, 127941. [CrossRef]
38. Han, M.; Wang, P.; Xu, X.; Zhou, G. Low-field NMR study of heat-induced gelation of pork myofibrillar proteins and its relationship with microstructural characteristics. *Food Res. Int.* **2014**, *62*, 1175–1182. [CrossRef]

gels

MDPI

Article

Enrichment of 3D-Printed k-Carrageenan Food Gel with Callus Tissue of Narrow-Leaved Lupin *Lupinus angustifolius*

Kseniya Belova [1], Elena Dushina [1], Sergey Popov [1,2,*], Andrey Zlobin [1], Ekaterina Martinson [1], Fedor Vityazev [2] and Sergey Litvinets [1]

1 Institute of biology and biotechnology, Vyatka State University, 36, Moskovskaya Str., 610000 Kirov, Russia
2 Institute of Physiology of Federal Research Centre "Komi Science Centre of the Urals Branch of the Russian Academy of Sciences", 50, Pervomaiskaya Str., 167982 Syktyvkar, Russia
* Correspondence: s.v.popov@inbox.ru; Tel./Fax: +78-21-224-1001

Abstract: The aim of the study is to develop and evaluate the printability of k-carrageenan inks enriched with callus tissue of lupin (*L. angustifolius*) and to determine the effect of two lupin calluses (LA14 and LA16) on the texture and digestibility of 3D-printed gel. The results demonstrated that the enriched ink was successfully 3D printed at concentrations of 33 and 50 g/100 mL of LA14 callus and 33 g/100 mL of LA16 callus. The feasibility of 3D printing is extremely reduced at higher concentrations of callus material in the ink. The hardness, cohesiveness, and gumminess of the 3D-printed gel with LA16 callus were weakened compared to the gel with LA14 callus. The results of rheological measurements showed that an increase in the content of LA16 callus interfered with the formation of a k-carrageenan gel network, while LA14 callus strengthened the k-carrageenan gel with increasing concentration. Gel samples at different concentrations of LA14 and LA16 calluses formed a spongy network structure, but the number of pores decreased, and their size increased, when the volume fraction occupied by LA14 and LA16 calluses increased. Simple polysaccharides, galacturonic acid residues, and phenolic compounds (PCs) were released from A-FP gels after sequential in vivo oral and in vitro gastrointestinal digestion. PCs were released predominantly in the simulated intestinal and colonic fluids. Thus, incorporating lupin callus into the hydrocolloid ink for food 3D printing can be a promising approach to developing a gelling material with new mechanical, rheological, and functional properties.

Keywords: lupin; callus culture; k-carrageenan; 3D food printing; gel; texture; rheology; simulated digestion; phenolic compounds; scanning electron microscopy

Citation: Belova, K.; Dushina, E.; Popov, S.; Zlobin, A.; Martinson, E.; Vityazev, F.; Litvinets, S. Enrichment of 3D-Printed k-Carrageenan Food Gel with Callus Tissue of Narrow-Leaved Lupin *Lupinus angustifolius*. *Gels* **2023**, *9*, 45. https://doi.org/10.3390/gels9010045

Academic Editors: Baskaran Stephen Inbaraj, Kandi Sridhar and Minaxi Sharma

Received: 12 December 2022
Revised: 29 December 2022
Accepted: 30 December 2022
Published: 6 January 2023

1. Introduction

Plant-based foods are essential for a healthy diet, and they are growing in popularity as their positive effects on human health gain wider recognition [1]. However, it is becoming more difficult to provide plant food to the constantly growing population of the planet [2]. New technologies need to be developed to produce varied and healthy plant-based foods. The cultivation of plant cells as a new approach to the production of plant-based food can become such a technology. A number of food ingredients have been registered and are currently being produced using plant cell cultures [3,4]. The possibilities of obtaining a number of valuable nutrients using cell cultures have been studied in some detail, and approaches have been developed for enriching food with ingredients beneficial to health.

Three-dimensional (3D) printing opens new possibilities for creating complex geometric structures in the individual production of dishes based on the specific nutritional needs and calorie intake of the individual [5]. The new opportunity provided by food 3D printing is extremely relevant for people with special nutritional needs, such as the elderly or patients who have difficulty eating or swallowing, children, and people with various metabolic and inflammatory disorders. This technology is easy to operate, allows for mass production, and also provides economic and environmental benefits by reducing food waste and financial costs for storing and transporting food [6]. Basically, for 3D printing food products, the extrusion method is used, in which a liquid or semi-solid material is extruded through a nozzle, creating a three-dimensional structure layer by layer, according to the digital model. Mashed potatoes, chocolate, cheese, meat, surimi, vegetable, and fruit pastes can be used as food printing materials [7]. However, with extrusion printing, it is not always possible to obtain a programmed 3D design. The decisive factor determining the quality of the printed product is the physicochemical properties of the food-grade inks (rheological and thermal properties, water retention capacity, and specific mechanisms of aggregation and gelation) [8]. Thus, both plant cell cultures and 3D printing are increasingly being used in food production. However, there is little work on the inclusion of plant cell cultures in 3D printing, although the number of original studies on plant-based 3D-printable materials has increased significantly in the past few years [9]. The question of how the inclusion of plant cell cultures can change the rheological properties of food ink and determine the structural and mechanical characteristics of the resulting 3D product remains poorly understood.

Carrot callus cells have been used as an ingredient for food 3D printing [10]. In this study, callus-based edible ink was used based on 4% alginate, the gelation of which requires divalent cations [11]. Therefore, callus-alginate ink requires curing using Ca^{2+} ions to form a hard gel after 3D printing. In the present study, we hypothesized that carrageenan-based inks would be suitable for 3D food inks, as carrageenan is a food-grade hydrocolloid with good gelling properties [12]. The callus tissue of *Lupinus angustifolius*, a narrow-leaved lupin, was chosen to demonstrate the applicability of callus material in 3D printing. *Lupinus* species is a genus of a widely consumed leguminous plant of the Fabaceae family. Several health-promoting properties of Lupinus species, including *L. angustifolius*, have been reported in preclinical and clinical human and animal studies, including antioxidant, anti-inflammatory, hypolipidemic, hypoglycemic, and hypotensive properties, among others. These biological activities are attributed to their human-health-beneficial chemical components, such as polyphenols, carotenoids, and other phytochemicals [13]. These properties make the development of cell cultures of lupin for functional food products promising.

The study's purpose was to develop the proposed k-carrageenan ink for 3D printing with callus tissue from narrow-leaved lupin, *L. angustifolius*, evaluate its printability, and determine the effect of callus on the texture and digestibility of 3D-printed gel.

2. Results and Discussion

2.1. Characterization of Lupin Callus

The cell biomass of two lines of lupine callus was used to obtain food ink for 3D printing. Callus LA14 was grown in a medium containing 2,4-dichlorophenoxyacetic acid (2,4-D) and 6-benzylaminopurine, whereas callus LA16 was grown in a medium containing kinetin and naphthylacetic acid. Callus LA14 had loose, moderately wet grayish-white tissue (Figure 1A), while the LA16 callus tissue was less loose and consisted of cells of a browner color than the LA14 tissue (Figure 1B).

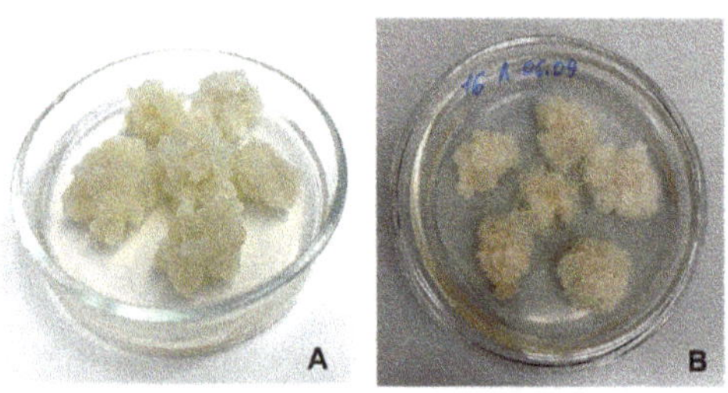

Figure 1. Appearance of the calluses LA14 (**A**) and LA16 (**B**).

Callus LA14 consisted of rounded, thin-walled parenchymal cells with an average size of 129 ± 48 μm. LA16 callus cells had an elongated and curved shape with an average size of 166 ± 116 μm (Figure 2A,B). The sphericity factors of callus cells LA14 and LA16 were 0.24 ± 0.11 and 0.61 ± 0.10 ($p < 0.05$), respectively.

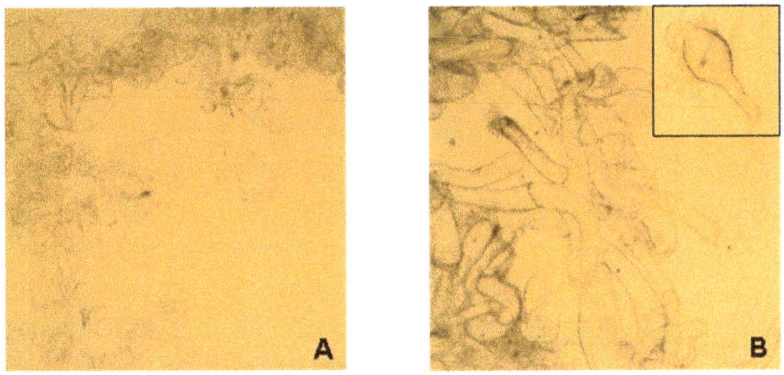

Figure 2. Micrograph of the cells of the callus LA14 (**A**) and LA16 (**B**). Magnification 10×.

The LA16 callus had 1.6 times the soluble solid content (SSC) of the LA14 callus (Table 1). LA14 and LA16 were similar in protein content but had different polysaccharide compositions. Hardly hydrolyzable polysaccharides predominated in both calluses, which included mainly cellulose and some hemicelluloses. The content of easily hydrolyzable polysaccharides (EHP), which included pectins, arabinogalactans, etc., was 1.6 times higher in LA16 than in LA14. The content of phenolic compounds (PCs) was 2.4 times lower in LA16 than in LA14 (Table 1). The residual content of 2,4-D in LA14 callus was 0.88 mg/kg, which is unacceptable under food safety requirements. Therefore, LA16 callus was considered a promising food ink ingredient for the rest of the study, while LA14 callus was used only for comparison.

Table 1. The composition of callus of narrow-leaved lupin *L. angustifolius*.

Callus	SSC (%)	Protein [a] (%)	EHP [b] (%)	HHP [b] (%)	PCs [b] (%)
LA14	2.68 ± 0.01	0.28 ± 0.01	9.2 ± 0.1	25.0 ± 2.6	1.25 ± 0.02
LA16	4.33 ± 0.11 #	0.30 ± 0.00 #	14.9 ± 0.4 #	30.2 ± 0.4 #	0.53 ± 0.01 #

[a]—per fresh weight; [b]—per dry matter contents. SSC—soluble solids content; EHP—easily hydrolyzable polysaccharides; HHP—hardly hydrolyzable polysaccharides; PCs—phenolic compounds. The data are presented as the mean ± SD ($n = 8$). #—$p < 0.05$ vs. LA14.

2.2. 3D Printing

Three ink formulations, labeled LA14-33, LA14-50, and LA14-66, were prepared by mixing 33, 50, and 66 g of LA14 callus biomass with 3 g of k-carrageenan dispersed in 100 mL of peach juice. Food inks prepared with 33, 50, and 66 g/100 mL of EA16 callus were labeled as LA16-33, LA16-50, and LA16-66, respectively. The formulations were heated to 90 °C and loaded into a food capsule for a 3D printer. After extrusion from the capsule, the ink solidified on the platform in the form of a 3D gel structure. The printability was qualitatively evaluated in terms of ease and uniformity of extrusion, precision and accuracy of printing, and stability of the geometry after printing. The use of LA14-33 and LA14-50 inks allowed for both easy extrusion from the nozzle and good shape retention even under the load of 30 successive layers (Figure 3A,B). However, the LA14-66 ink could not be 3D printed. First, the thicker ink of LA14-66 clogged the capsule nozzle, causing the ink to be squeezed out unevenly during printing; towards the end of printing, the capsule nozzle became completely clogged, and printing stopped. Second, the structure printed with LA14-66-containing ink collapsed once printed (Figure 3C).

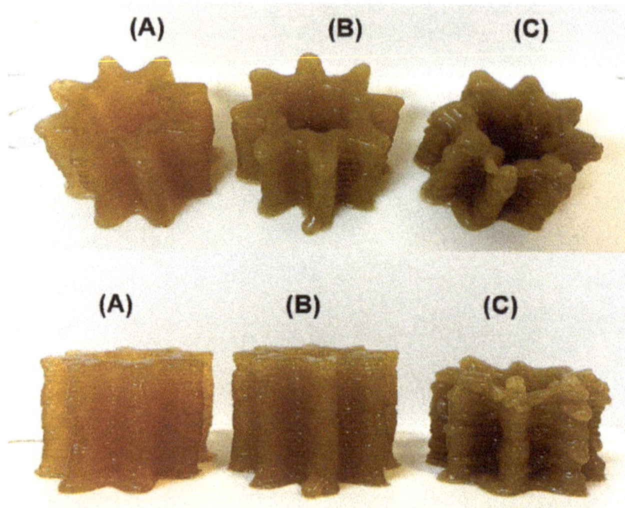

Figure 3. Representative results of the 3D printing experiment using LA14-33 (**A**), LA14-50 (**B**), and LA14-66 (**C**) inks. The top row of photographs is an angled top view; the bottom row is a side view.

Representative results of the printing experiment with food inks containing LA16 callus are shown in Figure 4. The consistency of LA16-33 ink allowed 3D printing to start, but printing stopped due to an ink clot in the capsule nozzle, so the 3D object was not printed completely (Figure 4A). 3D printing with LA16-50 and LA16-66 inks was disturbed due to the formation of thick portions that were squeezed out unevenly from the capsule nozzle. It was not possible to finish printing with LA16-50 and LA16-66 inks (Figure 4B).

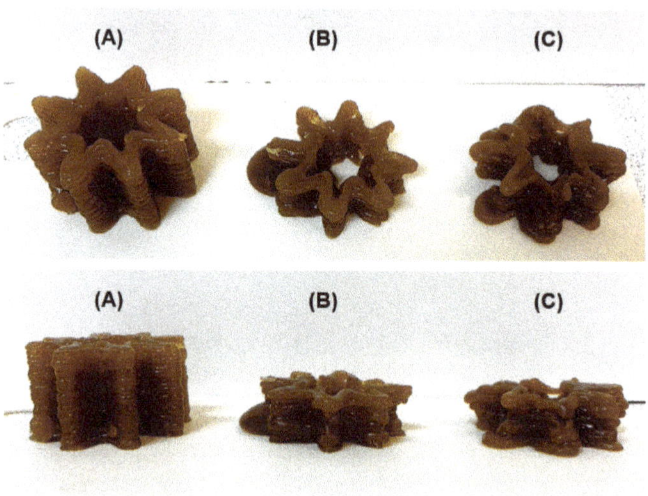

Figure 4. Representative results of the 3D printing experiments using LA16-33 (**A**), LA16-50 (**B**), and LA16-66 (**C**) inks. The top row of photographs is an angled top view; the bottom row is a side view.

2.3. *Rheological and Mechanical Properties of 3D-Printed Gels*

Angular frequency sweep measurements are conducted to predict the structural integrity and mechanical strength of a 3D-printed material. All of the 3D-printed samples, excluding LA16-66, exhibited a greater storage modulus (G′) value than loss modulus (G″) (Figure 5). This suggests the network integrity of molecules in carrageenan-based gel samples. The obtained high G′ values (40,000 > G′ > 90,000 Pa) and middle elastic character (tan δ ≤ 1.0) demonstrated that all 3D-printed gels, with the exception of the LA16-66 sample, were strong physical gels (Figure 5D).

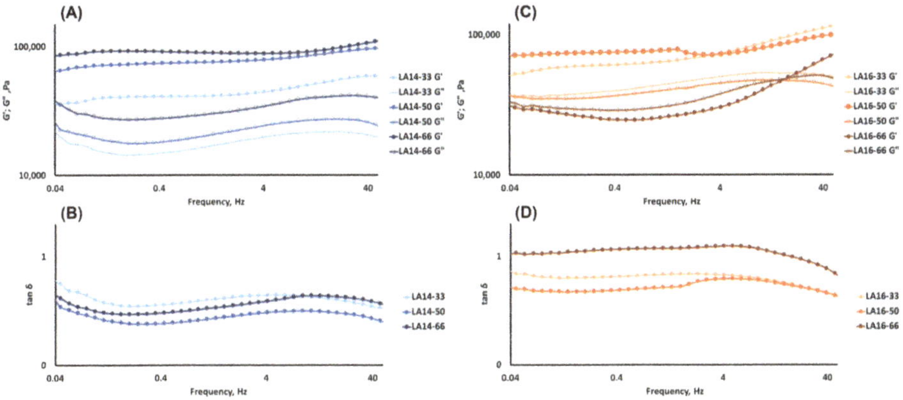

Figure 5. Rheological properties of hydrogels: storage modulus (G′) and loss modulus (G″) test results (**A**,**C**); tan δ test results (**B**,**D**) represented as a function of frequency. Void symbols represent the storage modulus G′, filled symbols represent the viscous modulus G″.

The rheological properties of the 3D-printed gel depended on the type of callus it contained. The values of G′ and G″ and complex viscosity increased with an increase in the content of LA14 in the ink (Figure 5A and Table 2). However, the values of G′ and G″, as well as the values of complex viscosity, decreased with an increase in the content of LA16 callus (Figure 5B and Table 2). For the LA16-66 sample, G′ was lower than G″ at an oscillation frequency of less than 15 Hz, indicating that the sample was a transitional sol-gel phase rather than a gel. G′ became greater than G″ at oscillation frequencies greater than 15 Hz, indicating that the LA16-66 sample behaved as a non-Newtonian fluid whose viscosity is known to change to become more solid at high oscillation frequencies. The non-Newtonian behavior of the LA16-66 sample appeared to be due to the large number of callus cells that it contained.

The mechanical behavior of the 3D-printed gels in a double compression test at room temperature is presented in Figure 6. The parameters evaluated in this study were hardness, springiness, cohesiveness, and gumminess. The hardness is the peak force during the first compression cycle and is related to the strength of the gel's structure when it is under compression. The hardness of the LA-14 callus-containing gels increased with the increase in callus tissue content (Figure 6A). The hardness of the LA14-66 gel was 1.5 times higher than that of the LA14-30 and LA14-50 gels. After the deformation of the sample, the springiness (extent to which the sample springs back after it has been deformed during the first compression) was measured. The springiness of the LA14 gel decreased slightly as the amount of callus in the ink increased (Figure 6B). The next parameter to be measured was cohesiveness, which is related to how well the gel withstands a second deformation relative to its resistance under the first deformation. LA14-50 gel exhibited the highest levels of cohesiveness. The cohesiveness of the LA14-33 and LA14-66 gels was 5 and 46% lower than that of the LA14-50 gel (Figure 6C). Lastly, gumminess, which indicates the amount of mastication needed to make a food item ready to swallow, was the same regardless of the content of LA14 callus in the gel (Figure 6D).

Table 2. Summary of power law parameters for relationship between storage modulus or viscosity and frequency (0.01 < ω < 50.00 or at 0.1/10.5 Hz) of 3D-printed gel.

Gels	Storage Modulus			Viscosity	
	A (Pa)	B (Slope)	R^2	η_{app} 0.045, Hz	η_{app} 10.500, Hz
LA14-33	69,196	0.1047	0.926	145,473.3 ± 29,131.1 [a]	834.9 ± 134.9 [a]
LA14-50	77,682	0.0355	0.623	300,836.7 ± 18,894.7 [b]	1702.3 ± 2.9 [b]
LA14-66	31,914	0.1159	0.603	359,313.3 ± 59,577.9 [b]	1601.8 ± 324.1 [c]
LA16-33	43,765	0.0635	0.893	278,873.3 ± 36,029.6 [a]	1887.0 ± 328.1 [a]
LA16-50	77,554	0.0474	0.935	162,349.0 ± 30,465.1 [b]	962.0 ± 189.0 [b]
LA16-66	92,050	0.0181	0.406	154,626.0 ± 63,583.1 [b]	855.9 ± 372.0 [b]

The data are presented as the mean ± SD (n = 8). Different letters indicate significant differences (p < 0.05) among means for different callus concentrations.

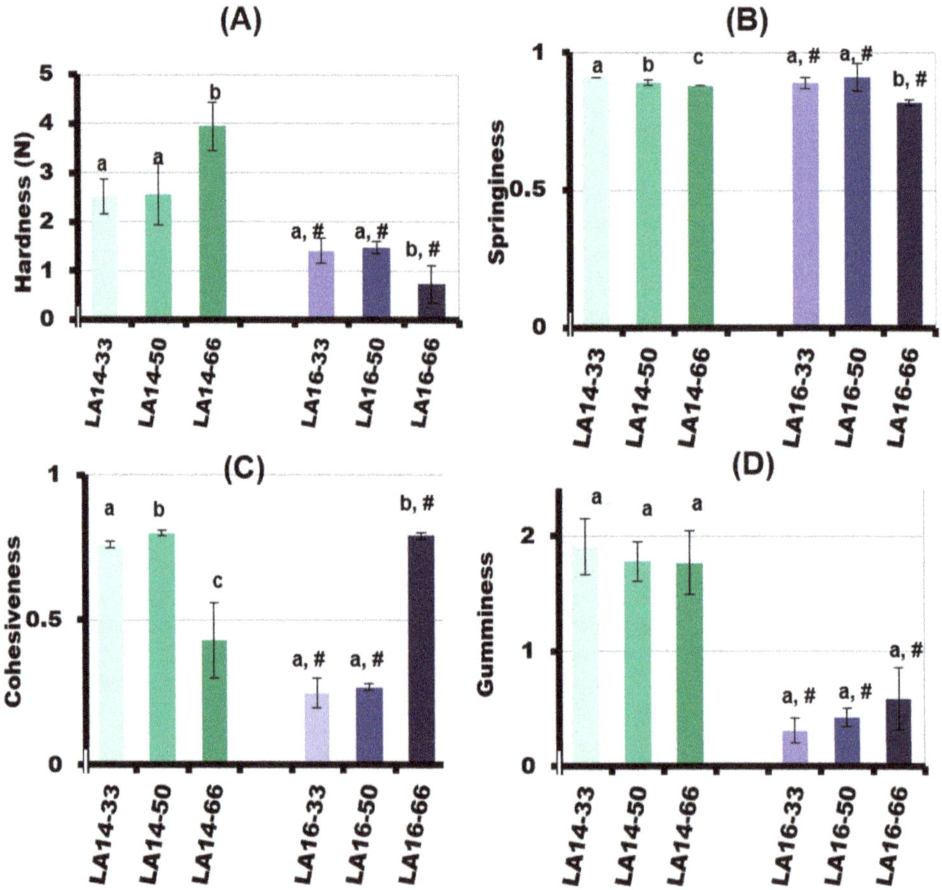

Figure 6. Mechanical properties of the 3D-printed gels. Hardness (**A**), springiness (**B**), cohesiveness (**C**), and gumminess (**D**) are given as the mean ± SD (n = 8). Different lowercase letters indicate significant differences (p < 0.05) among means for different callus concentrations; #—p < 0.05 vs. LA16 callus.

The values of the textural parameters of the LA16 gels were significantly lower than those of the LA14 gels. Increasing the callus content resulted in a decrease in the hardness of the LA16 gel such that the hardness of the LA16-66 gel was two times lower than the hardness of the LA16-50 gel and 5.4 times lower than the hardness of the LA14-66 gel (Figure 6A). The springiness of the LA16 gel decreased slightly when the callus content was increased from 50 to 66 g/100 mL (Figure 6B). The cohesiveness of the LA16 gel decreased 2.9 times with an increase in callus content from 50 to 66 g/100 mL (Figure 6C). The gumminess of the LA16-33, LA16-50, and LA16-66 gels was the same and was lower by 6, 4, and 3 times than the gumminess of LA14-33, LA14-50, and LA14-66 gels, respectively (Figure 6D).

2.4. Microstructure of 3D-Printed Gels

Figure 7 shows the morphology of 3D-printed gels. Gel samples at different concentrations of LA14 and LA16 callus formed a spongy network structure, but the callus content notably affected sample morphology. The number of pores decreased and their size increased when the volume fraction occupied by calluses LA14 and LA16 increased. Callus inclusion in the gel seemed to hinder the ice crystals' growth during the freezing process since the porosity formed when the water content was pulled away during the freeze-drying process. 3D-printed gels containing LA16 callus had smaller pores than gels containing LA14 callus. In the structure of the gel containing LA16-66 callus, dense, folded formations were found (Figure 7F). These formations can be clearly seen in the micrographs obtained using magnifications of 500× and 1000× (Figure 8). It can be assumed that these are dense conglomerates of callus cells.

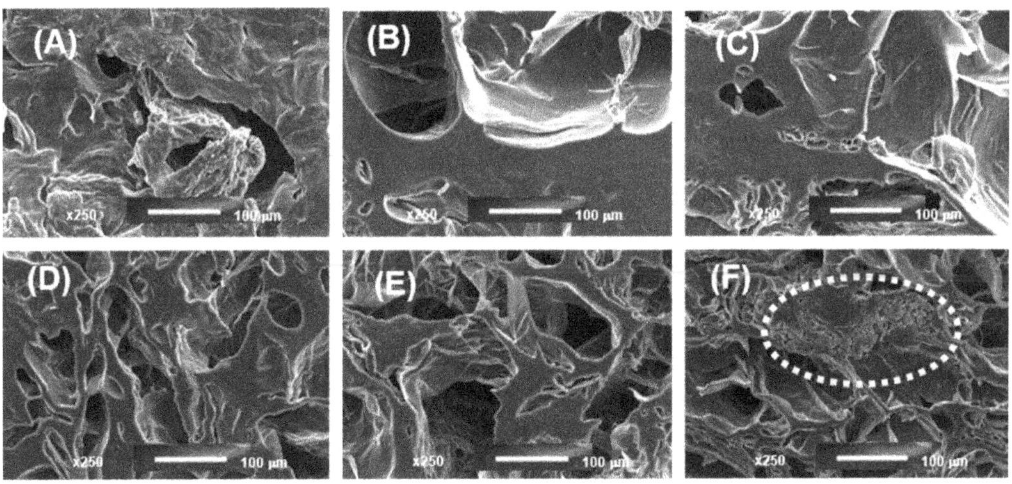

Figure 7. Scanning electron micrographs of the 3D-printed gel containing LA14 (**A–C**) and LA16 (**D–F**) callus. The content of callus in the ink was 33 (**A**,**D**), 50 (**B**,**E**), and 66 g/100 mL (**C**,**F**). Magnification 250×; scale bar 100 μm. The area of dense folded formation is outlined by a dotted line.

EDS analysis revealed similar contents of oxygen (87–92 wt%) in the 3D-printed gels containing LA14 and LA16 calluses. It has been established that the content of the K^+ cation in the gel increased with an increase in the concentration of callus in the ink. The K^+ cation content was 3.9 ± 0.0, 5.0 ± 0.2, and 6.5 ± 0.0 wt% in LA14-33, LA14-50, and LA14-66 gel samples, respectively. The K^+ cation content was 4.7 ± 0.5, 5.1 ± 0.4, and 7.5 ± 0.3 wt% in LA16-33, LA16-50, and LA16-66 gel samples, respectively.

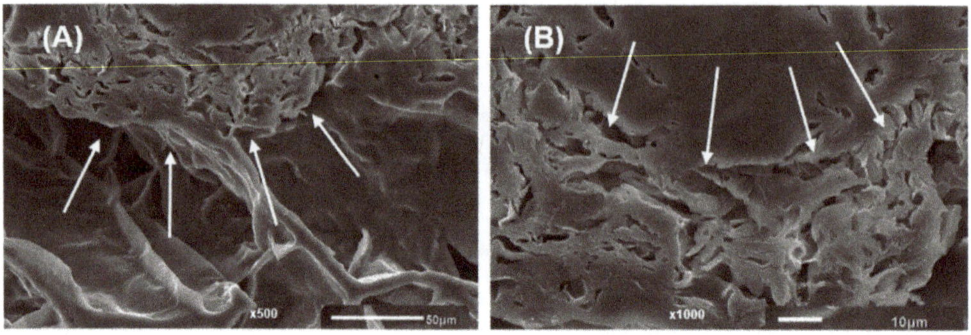

Figure 8. Scanning electron micrographs of the 3D-printed gel containing LA16-66 callus at (**A**) magnification 500×; scale bar 50 μm and (**B**) magnification 1000×; scale bar 10 μm. The arrows indicate the area of dense folded formation.

2.5. Simulated Digestibility of 3D-Printed Gel Containing LA16-33 Callus

The content of SSC in the incubation medium after each phase of simulated digestion gradually decreased, indicating good digestibility of the 3D-printed gel (Figure 9A). The incubation medium after each phase of simulated digestion of LA16-33 gel contained PCs. The most PCs were released in the SIF and SCF media. Two times fewer PCs were released during the destruction of the gel during in vivo chewing and during the subsequent in vitro SGF phase than during the SIF and SCF phases (Figure 9B). The content of PCs in the undigested residue was 21 ± 7 μg/mL.

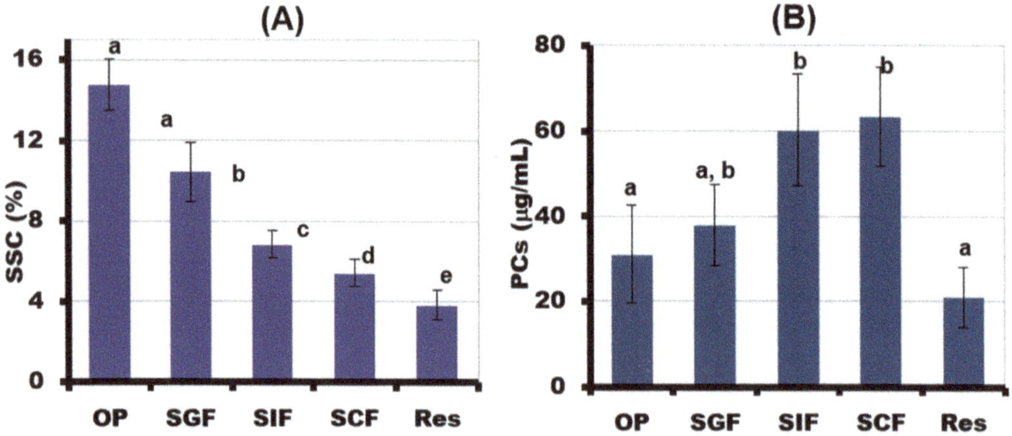

Figure 9. The amount of soluble solids content (SSC) (**A**) and total phenolic compounds (PCs) (**B**) released from LA16-33 gel during successive oral in vivo (OP) and gastrointestinal in vitro phases of digestion. SGF, SIF, and SCF—simulated gastric, intestinal, and colonic fluids, respectively. The data are presented as the mean ± SD. Different lowercase letters indicate significant differences among means for the different phases of digestion ($n = 6$, $p < 0.05$).

During in vivo chewing (OP), the gel containing LA16-33 released the most sugars. The content of neutral sugars in the incubation medium was 146 ± 20, 81 ± 7, 45 ± 4, and 22 ± 5 mg/mL after the OP, SGF, SIF, and SCF phases, respectively. The main neutral sugars that were released during digestion were fructose and glucose (Figure 10). Sucrose was not detected in phases OP, SGF, and SIF, but it was contained in the SCF incubation medium at a concentration of 6.6 ± 1.0 mg/mL. In addition to neutral monosaccharides, residues of galacturonic acid were found in the digestion medium.

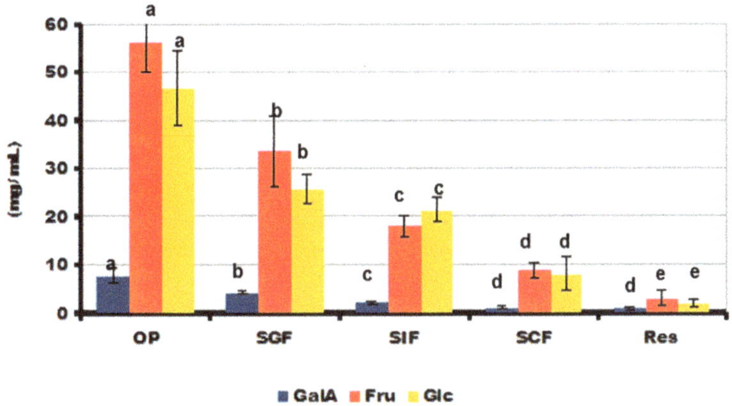

Figure 10. The amount of galacturonic acid (GalA), fructose (Fru), and glucose (Glc) released from LA16-33 gel during successive oral in vivo (OP) and gastrointestinal in vitro phases of digestion. SGF, SIF, and SCF—simulated gastric, intestinal, and colonic fluids, respectively. The data are presented as the mean ± SD. Different lowercase letters indicate significant differences among means for the different phases of digestion of each sugar (n = 6, $p < 0.05$).

The present study was devoted to the enrichment of hydrocolloid ink with callus material for 3D-printed food gel. Inks for 3D printing were prepared from a mixture of 3% k-carrageenan and biomass of lupine calluses LA14 and LA16. The results demonstrated that the enriched ink was successfully 3D printed at concentrations of 33 and 50 g/100 mL of LA14 callus and 33 g/100 mL of LA16 callus.

k-carrageenan is a linear sulfated polysaccharide derived from red algae, which is widely used in the food industry due to its ability to form thermoreversible gels [12]. In 3D food printing, k-carrageenan has been used as part of multicomponent food inks along with gelatin [14], starch and xanthan [15], beeswax and xanthan [16], freeze-dried vegetable powders [17], surimi [18], and others. However, unless the 3D printing parameters were changed, one-component k-carrageenan ink was only marginally printable [19,20]. Based on existing data, a 3% solution of k-carrageenan did not allow us to obtain 3D objects (data not shown). The mechanism of k-carrageenan gelation involves a conformational transition from a random coil structure when heated to double helices, followed by the aggregation of helices during the cooling process [21]. The reason for the low printability of k-carrageenan is that the transition from random coils to double helices is extremely fast, whereas the subsequent aggregation of double helices to complete the formation of a gel network takes several hours [19]. The 3% k-carrageenan ink supplemented with lupin calluses was successfully 3D printed. Therefore, the addition of callus materials seemed to accelerate the gelation of k-carrageenan. This effect may be related to the gel-promoting action of ions and polysaccharides contained in the callus. It has been found that as the content of callus increased, printing deteriorated due to the thickening of the ink in the printing capsule and difficulty in extrusion. At the same time, there was an increase in the content of potassium ions in the ink. It is well known that the gelation of carrageenan is highly dependent on potassium ion concentration [22]. The printability of ink with LA16 callus at concentrations of 50 and 66 g/100 mL was lower than that of LA14 callus, possibly due to the higher content of polysaccharides in LA16 callus. Indeed, the content of easily hydrolyzable polysaccharides in LA16 callus was 1.6 times higher than in LA14 callus. One of the main easily hydrolyzable polysaccharides is pectin [23], which is a gel-forming polysaccharide of the plant cell wall [24]. The content of pectin in plant callus was previously shown in a number of papers [25–28]. Therefore, we hypothesize that callus polysaccharides may alter the printability of carrageenan inks. However, the composition and gelling properties of lupine callus polysaccharides were not established in the present study.

The effect on the textural, rheological, and morphological properties of a 3D-printed gel was found to depend on the type and concentration of callus in the ink composition. Rheological measurements and a double compression test showed that the structure of the 3D-printed gel changed with increasing LA14 callus content. The increase in gel strength resulting from the addition of LA14 callus may be related to the gel-promoting action of K^+ ions and polysaccharides contained in the callus. Carrageenans are formed of alternate disaccharide units of D-Gal and 3,6-anhydro-Gal joined by α-1,3- and β-1,4-glycosidic linkage; hydrogen bonding is well known to be the primary mechanism involved in carrageenan gelation (Figure 11A). Kappa-carrageenan contains one sulfate group (OSO_3^-) per disaccharide repeating unit; therefore, mono- and divalent cations cross-link double helices with outward-oriented OSO_3^- groups in addition to hydrogen bonding [29]. K^+ ions from callus material are supposed to provide ionic cross-links between adjacent κ-carrageenan chains formed by sulfated galactose residues (Figure 11B, K^+). Gelling polysaccharides of the callus cell wall, such as pectins, could also enhance the 3D-printed gel (Figure 11B, CPs). The size and shape of LA14 callus cells evidently did not interfere with the formation of cross-links (Figure 11B, LA14 callus cells). In the case of the LA16-containing ink, callus cell conglomerates may have prevented the convergence of the κ-carrageenan chains and the formation of a three-dimensional network (Figure 11B, LA16 callus cells). As a result, a less strong gel was formed compared to the ink that contained LA14 callus. Further, the 3D object that was printed with the highest concentration of LA16 (66 g/100 mL) was not a "gel" and had the lowest hardness. We hypothesize that the elongated shape of LA16 callus cells may have contributed to the formation of cell conglomerates, the presence of which can be inferred from SEM images of gels printed with LA16 callus inks (Figure 6).

The effect of callus cells on the mechanical properties of the gel matrix was previously shown in the study [10], where the multiplication of carrot callus cells in agar gel reduced the gel strength of the fabricated structure. The authors considered that the decrease in the homogeneity of the gel matrix affected its mechanical properties. The encapsulation of living lettuce leaf cells in pectin-based bio-ink had a negative effect on mechanical properties [30]. A decrease in product hardness with the addition of cellular material has previously been shown when incorporating microalgae powder into cookies by conventional methods [31]. Conversely, Vieira et al. [32] incorporated different powder amounts of microalgae into shortbread biscuits, which promoted an increase in cookie hardness directly proportional to the microalgae powder content. A decrease in the hardness of the gel containing LA16 callus was accompanied by an increase in cohesiveness and gumminess. Similar mechanical behavior was observed when button mushroom powder was added to wheat dough [33].

The mechanical behavior of food gels during oral processing, such as chewing, bolus formation, and swallowing, is determined by their textural and rheological properties [34]. Furthermore, hardness, cohesiveness, gumminess, and springiness are critical parameters for predicting the sensory perception of food gels. Consumers consider hardness to be an important indicator because it can simulate the force required for food to compress between teeth or between the tongue and palate [35]. Softening the k-carrageenan gel induced by LA16 incorporation would presumably make it easier to chew. Therefore, LA16-containing gel may be required by patients with dysphagia [36]. Semi-solid foods with low hardness and high cohesiveness values are assumed to be ideal for dysphagic diets [37]. It has been shown that gel containing LA16 callus at a concentration of 66 g/100 mL had a low hardness and a high cohesiveness. However, the feasibility of this ink composition for 3D printing was low. Therefore, further modification of the lupin callus ink composition and printing parameters is required to improve fluidity during 3D printing.

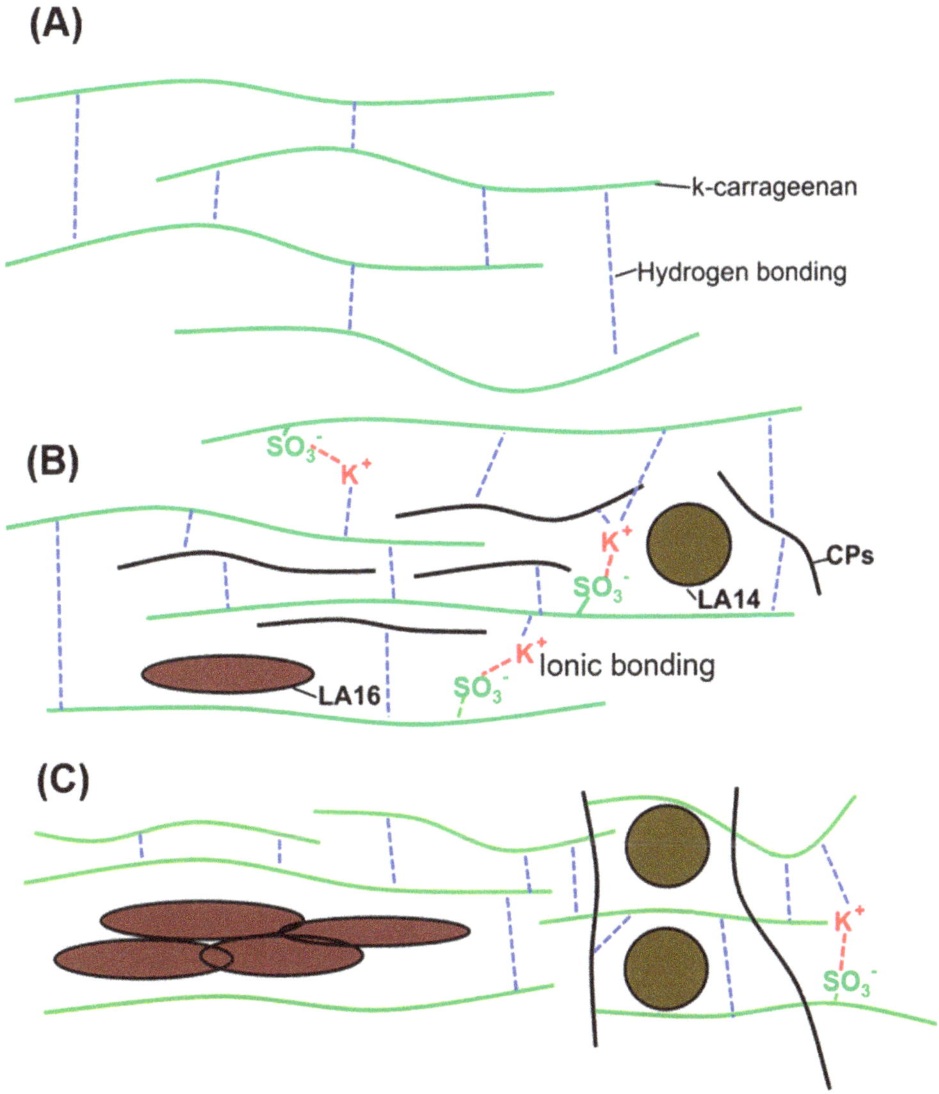

Figure 11. Schematic illustration of a proposed gel network in 3% k-carrageenan (**A**), 3% k-carrageenan enriched with LA callus at 33 g/100 mL (**B**), and 3% k-carrageenan enriched with LA callus at 66 g/100 mL (**C**) inks. CPs–callus polysaccharides. LA14 and LA16–LA14 and LA16 callus cells, respectively. Ionic and hydrogen bonds are shown as red and blue dotted lines, respectively. The polysaccharide chain of k-carrageenan is shown as a solid green line.

In terms of nutritional quality, the lupin calluses contained a high concentration of polysaccharides and PCs. In order to preliminarily assess the nutritional value of the 3D gel containing LA16-33 callus, it was subjected to simulated digestion. Three-dimensional gels containing LA14 callus were excluded from this experiment due to food safety reasons—callus LA14 was grown in a medium containing 2,4-D, and the residual content of 2,4-D in LA14 callus was detected, which is not allowed in food products.

Successive oral in vivo and gastrointestinal in vitro digestion of LA16-33-containing gel in SGF, SIF, and SCF fluids demonstrated a decrease in the SSC content during digestion and the release of simple sugars. The data indicate good digestibility of the printed gel; however, it should be noted that some of the simple sugars come from peach juice, which was part of the ink. The detection of GalA in the incubation medium indicates partial hydrolysis of the polysaccharides of the callus cell wall. GalA is the main monosaccharide residue that forms the pectin macromolecule. Pectin is a complex plant polysaccharide that is resistant in the human stomach and small intestine and is fermented by colonic bacteria in the large bowel. So, pectin is a dietary fiber with good prebiotic properties [38,39]. Furthermore, pectin is highly valued as a functional food ingredient due to its hypolipidemic, hypoglycemic, satiating, antibacterial, and antitumor biological effects [40]. Therefore, it can be assumed that LA16-containing food gel will be a good source of pectin with health-promoting properties.

PCs represent secondary metabolites of plants and are highly valuable food components [41]. Many studies have shown that callus cultures are a good source of PCs [42–46]. PCs possess diverse bioactivities, including antioxidant, anti-inflammatory, and anticancer effects [47–49]. To predict the part of the gastrointestinal tract where LA16 callus would bring biological activity, PCs released from LA16-33 gels were evaluated during successive OP and SGF, SIF, and SCF digestion. The preferential release of PCs in the SIF and SCF indicates that digesting LA16-33 gel will preserve its biological potential until the lower intestines. Therefore, lupine callus may be of interest for the development of food gels for the prevention and treatment of inflammatory bowel diseases (IBD), which has emerged as a public health challenge worldwide. IBD is characterized by segmental inflammation anywhere in the intestine (Crohn's disease) or superficial inflammation of the mucosal layer of the colon (ulcerative colitis) [50]. PCs have been shown to confer symptomatic and health-related quality of life improvements in IBD patients [51].

3. Conclusions

In this study, lupin calluses LA14 and LA16 were efficiently incorporated into k-carrageenan-based food ink for 3D printing. The printability of k-carrageenan ink with LA14 callus was higher than that with LA16 callus. 3D-printed food gel can contain up to 50 g per 100 mL of LA14 callus and 33 g per 100 mL of LA16 callus. The feasibility of 3D printing is extremely reduced at higher concentrations of callus material in the ink. The hardness, cohesiveness, and gumminess of 3D-printed gels with LA16 callus were weakened compared to gels with LA14 callus. The results of rheological measurements showed that an increase in the content of LA16 callus interfered with the formation of a k-carrageenan gel network, while LA14 callus strengthened the k-carrageenan gel with increasing concentration. The digestibility of 3D-printed gel with LA16 callus in a simulated digestion model showed its nutritional value as a source of simple sugars, cell wall polysaccharides, and PCs. Wherein PCs were released mainly into the SIF and SCF, demonstrating the biological potential for the lower intestines. Based on its combination of textural properties and digestive profile, lupine callus LA16 may be considered an ingredient in functional food gels for patients with dysphagia and intestinal inflammation. Incorporating lupin callus into the hydrocolloid ink for food 3D printing can be a promising approach to developing gelling material with new mechanical and rheological properties.

4. Materials and Methods

4.1. Materials

Refined KA120R k-carrageenan was purchased from Greenfresh Food Co., Ltd. (Longhai City, Zhangzhou, Fujian, China). The (-SO_3-) content in k-carrageenan was 11.7 ± 0.7%. Peach juice concentrate was purchased from a local supermarket. The reagents, including the Folin and Ciocalteu's phenol reagent, pectinase from *Aspergillus niger* (>1 U/mg), D-(+)-galacturonic acid monohydrate (99.1%), and ferulic acid (99.8%) were purchased from Sigma-Aldrich (St. Louis, MO, USA). To build calibration curves, we used: 2,4-dichlorophenoxyacetic acid

(GSO 9105-2008, Ecolan, Russia), D-(+)-glucose monohydrate (99.9%; Panreac Química SLU, Barcelona, Spain); and 3,5-dimethylphenol (99.8%, Acros Organics, Waltham, MA, USA).

4.2. Obtaining and Characterizing Calluses

The callus culture of narrow-leaved lupine (*Lupinus angustifolius* L.) of the legume family (Fabaceae) was obtained in the laboratories of Vyatka State University's Department of Biotechnology. Callus line No. 14 (LA14) was bred from stem segments of a native plant. Callus was induced on Murashige-Skoog medium with the addition of phytohormones: 2,4-dichlorophenoxyacetic acid (2,4-D)–2 mg/L and 6-benzylaminopurine (6-BAP)–0.2 mg/L. The optimal combination of phytohormones to support the growth of angustifolia callus during long-term cultivation is a combination of 2,4 D, 2.0 mg/L + 6-BAP, 0.1 mg/L. During several passages, lupine calli were grown on media with various combinations of phytohormones, with a visual assessment of cell growth and viability. The best transition from medium No. 14 to medium No. 16 (LA16) was observed with the following phytohormone contents: 1.0 mg/L naphthylacetic acid (NAA) and 0.1 mg/L kinetin.

The specific growth rates of calluses LA14 and LA16 were 0.46 ± 0.04 and 0.46 ± 0.04 g/d, respectively. The callus cultures were subcultured for 21 days at 26 °C in the darkness. A freshly harvested callus was microscopically examined and chemically analyzed. A portion of the harvested biomass was frozen for the subsequent preparation of ink for 3D printing.

The size and shape of the cells were studied using a Motic BA300 optical microscope (China) with a built-in video eyepiece. Each analysis was performed 10 times under the same optical conditions. Cell shape was determined quantitatively using dimensionless shape indicators. The sphericity factor (SF) was calculated using the following formula:

$$SF = (dmax - dmin)/(dmax + dmin), \quad (1)$$

where dmax is the maximum diameter (μm), and dmin is the minimum diameter perpendicular to dmax (μm).

The content of dry matter in the plant raw materials was determined by the thermogravimetric method. Protein content in callus tissues was determined according to Barnstein's method.

The content of easily and hardly hydrolyzable polysaccharides was determined using the Kizel and Semiganovsky method. A sample was first prepared for the analysis of easily hydrolyzable polysaccharides: 100 mL of 2% HCl was added to 2.5 g of dry callus tissue, the resulting mixture was boiled on a stove in a flask under reflux for 3 h, the contents of the reaction flask were filtered through a paper filter, the filtrate and the washings were combined in a 250 mL volumetric flask, and the residue on the filter was used. After cooling, both samples were thoroughly mixed, and the concentration of reducing substances in them was determined by the ebulliostatic method of Nizovkin and Emelyanova by titrating copper-alkaline solutions with the obtained samples. The titer of copper-alkaline solutions was determined by glucose.

The concentration of reducing substances in the analyzed samples was calculated as a percentage of glucose according to the formula:

$$X = T \times n \times 100/V \times 100, \quad (2)$$

where X is the concentration of sugar in the analyzed solution, %; T is the titer of the copper-alkaline solution for glucose, mg; V is the volume of the analyzed solution used for titration, mL. The content of EHP was calculated as follows:

$$XL = CL \times VL \times k \times 100/m \times 100, \quad (3)$$

where m is the mass of absolutely dry plant tissue taken for analysis, g; CL is the concentration of reduced substances in the hydrolyzate of EHP, %; VL is the volume of the hydrolyzate, mL; and k is the coefficient of conversion of monosaccharides into polysaccharides, which is 0.89.

The content of HHP was calculated by the formula:

$$XT = CT \times VT \times n \times K \times 100/m \times 100, \quad (4)$$

where CT is the concentration of reducing substances in the diluted neutralized hydrolysate; VT is the volume of acid hydrolysate, in mL; n is the dilution of the hydrolysate during neutralization; K is the conversion factor for monosaccharides in HHP, which is 0.9.

For PC content in callus tissue, 200.0 ± 0.1 mg of freeze-dried callus tissue was placed in a 25-mL test tube, to which 5.0 mL of a 70% methanol solution was added, closed with a cork, and mixed on a vortex-type shaker. The test tube was placed in a water bath at a temperature of 70 °C for 10 min, stirring the contents on a shaker after 5 and 10 min. The tube was then cooled to room temperature and centrifuged for 10 min at 9000 rpm. The supernatant was decanted into a 25 mL graduated tube. The extraction of polyphenols with a 70% methanol solution was carried out twice more (in 5 mL portions). The extracts were combined, the extraction volume in a graduated test tube was adjusted to 15.0 mL with a 70% methanol solution, and it was kept in a refrigerator at 4 °C for 24 h [52]. A sample of 0.4 mL of the extract was transferred into a test tube, and 4.0 mL of a 0.2 M NaHCO3 solution in 0.1 M NaOH was added. Folin–Ciocalteu reagent, mixed for 10 min at room temperature. A sample of 0.4 mL of the Folin–Ciocalteu reagent was added to the solution, mixed, and incubated for 30 min in the dark at room temperature. The absorbance of solutions was measured in a cuvette with an optical path length of 10 mm at a wavelength of 765 nm. The concentration of polyphenols in the solution was determined from a calibration curve constructed using ferulic acid (10–100 μg/mL)

Quantitative determination of 2,4-D in callus extracts was determined in the form of methyl ester by chromato-mass spectrometry on a G2589A gas chromatograph (Agilent Tech., Santa Clara, CA, USA) with an HP-1MS capillary column (0.25 mm × 30 m) (Hewlett-Packard, Palo Alto, CA, USA). The carrier gas was helium (1 mL/min). The volume of the injected sample was 1 μL (flow split 10:1). The evaporator temperature was 250 °C. Temperature conditions of the column thermostat: 50 °C (0.5 min) → 100 °C (gradient–25 °C/min) → 220 °C (gradient–12 °C/min). Mass spectrometer 5973 INERT (Agilent Tech., Santa Clara, CA, USA): ionization chamber temperature, 230 °C; the energy of ionizing electrons was 70 mV. Quantitative determination of 2,4-D was carried out by extracted ions (*m*/*z*: 175.00; 199.00; 234.00) using the method of absolute calibration using standard solutions of 2,4-D methyl esters (0.025–0.50 mg/kg).

4.3. Inks Preparation and 3D Printing

To obtain the ink ingredient, the frozen callus biomass was thawed and passed through a metal sieve with a mesh size of 1 mm to eliminate cell aggregates. Callus biomass (33, 50, or 66 g) and 3 g of k-carrageenan were dispersed in 100 mL of peach juice and left for one hour with magnetic stirring. The dispersion was heated to 90 °C in a slow cooker for 20–25 min. The hot solution was vigorously shaken and transferred into a printing capsule for a 3D printer. A FOODINI (Natural machines, Barcelona, Spain), an extrusion-based commercial 3D food printer was used to print 3D samples of different callus-containing inks. For the experiments in this study, a nozzle size of 1.5 mm was used. To assess the printability of the food inks, a "flower tower" of 30 layers was printed on a silicone mat. When the printed samples were able to maintain the structure for at least 15 min, they were considered printable. For rheological and textural measurements, samples in the form of a plate 1 cm × 1 cm × 0.3 cm (width, length, and height), and cube-shaped samples with a side of 5 mm were printed separately.

4.4. Measurement of Mechanical Properties

For two-cycle compression tests, gel samples (5 mm height, 5 mm length, and 5 mm width) were placed on the platform of the texture analyzer (Texture Technologies Corp.,

Stable Micro Systems, Godalming, UK). The tests were performed using a cylindrical aluminum probe P/25 (25-mm diameter). The gels were compressed twice at room temperature. The pre- and post-test speed was 5.0 mm/s and the test speed was 1 mm/s until a 100% strain. Destructive 100% strain was used to represent gel behavior during the chewing process. Eight replicates were made for each type of gel. For two cycles, compression-decompression provided a force-time graph and led to the extraction of eight parameters: hardness, cohesiveness, springiness, gumminess, chewiness, and resilience. All calculations were performed using Texture Exponent 6.1.4.0 software (Stable Micro Systems, Godalming, UK) according to the manufacturer's instructions. One-way ANOVA with Tukey's honest significance test was applied to determine statistically significant differences. Values of $p \leq 0.05$ were considered statistically significant.

4.5. Measurement of Rheological Properties

The rheological property of the samples was determined in a rotational-type rheometer (Anton Paar, Physica MCR 302, Graz, Austria) equipped with a parallel plate geometry (diameter 25 mm) and a gap of 4 mm between the two plates. Four repetitions were performed for each sample. The sample loading area was preheated to 20 °C before gel loading. After loading, the samples were equilibrated at 20 °C for 5 min before the measurement. The obtained mechanical spectra were characterized by the values of G′ and G″ (Pa) as a function of frequency in the range of 0.05–50.00 Hz at 20 °C and a constant stress of 9.0 Pa, which was within the linear viscoelastic region. The loss factor tan δ was calculated as the ratio of G″ and G′. The degree of frequency dependence for G′ was determined by the power-law parameters, and is expressed as follows:

$$G' = A \times \omega \times B, \quad (5)$$

where G′ is the storage modulus, ω is the oscillation frequency (Hz), and A is a constant.

4.6. In Vivo Oral Phase (OP) and Static In Vitro Gastrointestinal Digestion

OP digestion was carried out according to the method proposed earlier [53]. Each type of gel (~4 g) was chewed 20 times by six healthy people (three men and three women) to simulate the maximum destruction of the sample in the oral cavity according to preliminary testing in which the largest number of chews was 20. Thereafter, the bolus was spat three times into a beaker to reduce sample loss. Immediately after this, 4.0 mL of water was added to the beaker, mixed by shaking, and all the liquid part was separated for analysis. The gel pieces were transferred to a 20 mL jacketed glass vessel (reactor) for further in vitro digestion.

After OP, gels were sequentially incubated in 4.0 mL of the pre-heated SGF (pH 1.5, 0.08 M HCl, and 0.03 M NaCl), SIF (pH 6.8, 0.05 M KH_2PO_4, and 0.05 M NaOH), and SCF (0.07 M KH_2PO_4, 0.07 M $NaHPO_4$, and pectinase: 1.7 mg/mL) at 37 °C and under continuous shaking (250 rpm) for 2, 4, and 18 h, respectively. Before replacing it with another, the medium was completely separated from the gel by a grid (mesh size 350 µm) and used for analysis. After incubation in simulated colonic fluid, the gel residue was destroyed by heating (95 °C, 10 min) in water (16 mL). The data from SCF and gel residue were summed up.

The contents of SSC, GalA, neutral monosaccharides, and PCs were determined in the fluid after each (OP, SGF, SIF, and SCF) phase of digestion. For this, aliquots (1–2 mL) of incubation medium were taken and centrifuged, and the resulting supernatant was precipitated with a fourfold volume of 96% ethanol. The precipitate was washed twice with 96% ethanol before being dissolved in 3 mL of H_2O; the resulting solution was used to calculate the molecular weight of soluble polysaccharides and the content of GalA by reacting the sample with 3,5-dimethylphenol in the presence of concentrated H_2SO_4 [54]. The alcohol supernatant was used to determine the total amount of sugars using the phenol-sulfur method. The PC content was determined in the incubation medium as described above.

4.7. Statistical Analysis

The significance of the differences among the means in the PC content, mechanical parameters, and digestion study was estimated with one-way ANOVA and Fisher's least significant difference (LSD) post hoc test. Statistical differences with p-values lower than 0.05 were considered significant. All calculations were performed using the statistical package Statistica 10.0 (StatSoft, Inc., Tulsa, OK, USA). The data presented were expressed as the means $\pm$ SD.

Author Contributions: Conceptualization, S.L.; methodology, S.P.; software, F.V.; validation, S.P. and S.L.; formal analysis, S.P.; investigation, K.B., E.D., A.Z., E.M. and F.V.; resources, S.L.; data curation, S.P.; writing—original draft preparation, S.P. and K.B.; writing—review and editing, S.P.; visualization, F.V.; supervision, S.L.; project administration, S.L.; funding acquisition, S.L. All authors have read and agreed to the published version of the manuscript.

Funding: This study was supported by the Russian Science Foundation (grant number 22-24-01060).

Institutional Review Board Statement: Part of the research (scanning electron microscopy) was carried out within the framework of the strategic project "Health Technologies" of Vyatka State University (Program Priority 2030, Agreement #075-15-2021-1200). A further part of the research (rheological analysis) was carried out within the framework of the research topic of the Institute of Physiology of Komi Science Centre of The Urals Branch of the Russian Academy of Sciences FUUU-2022-0066 (1021051201895-9).

Informed Consent Statement: Informed consent was obtained from all subjects involved in the study.

Data Availability Statement: The data that support the findings of this study are available from the corresponding author upon reasonable request.

Acknowledgments: Ekaterina Martinson performed the scanning electron microscopy and Fedor Vityazev performed the rheological tests. The authors are grateful to Nikita Paderin and Vasily Smirnov from the Institute of Physiology of the FRC Komi Science Centre of the Urals Branch of the Russian Academy of Sciences for technical assistance.

Conflicts of Interest: The authors declare no conflict of interest.

References

1. Key, T.; Papier, K.; Tong, T. Plant-based diets and long-term health: Findings from the EPIC-Oxford study. *Proc. Nutr. Soc.* **2022**, *81*, 190–198. [CrossRef] [PubMed]
2. Nordlund, E.; Lille, M.; Silventoinen, P.; Nygren, H.; Seppänen-Laakso, T.; Mikkelson, A.; Aura, A.-M.; Heiniö, R.-L.; Nohynek, L.; Puupponen-Pimiä, R.; et al. Plant cells as food—A concept taking shape. *Food Res. Int.* **2018**, *107*, 297–305. [CrossRef]
3. Krasteva, G.; Georgiev, V.; Pavlov, A. Recent applications of plant cell culture technology in cosmetics and foods. *Eng. Life Sci.* **2021**, *21*, 68–76. [CrossRef] [PubMed]
4. Gubser, G.; Vollenweider, S.; Eibl, D.; Eibl, R. Food ingredients and food made with plant cell and tissue cultures: State-of-the art and future trends. *Eng. Life Sci.* **2021**, *21*, 87–98. [CrossRef]
5. Escalante-Aburto, A.; Trujillo-de Santiago, G.; Álvarez, M.M.; Chuck-Hernández, C. Advances and prospective applications of 3D food printing for health improvement and personalized nutrition. *Compr. Rev. Food Sci. Food Saf.* **2021**, *20*, 5722–5741. [CrossRef] [PubMed]
6. Jayaprakash, S.; Paasi, J.; Pennanen, K.; Flores Ituarte, I.; Lille, M.; Partanen, J.; Sozer, N. Techno-Economic Prospects and Desirability of 3D Food Printing: Perspectives of Industrial Experts, Researchers and Consumers. *Foods* **2020**, *9*, 1725. [CrossRef]
7. Tejada-Ortigoza, V.; Cuan-Urquizo, E. Towards the Development of 3D-Printed Food: A Rheological and Mechanical Approach. *Foods* **2022**, *11*, 1191. [CrossRef]
8. Outrequin, T.C.R.; Gamonpilas, C.; Siriwatwechakul, W.; Sreearunothai, P. Extrusion-based 3D printing of food biopolymers: A highlight on the important rheological parameters to reach printability. *J. Food Eng.* **2023**, *342*, 111371. [CrossRef]
9. Wang, M.; Li, D.; Zang, Z.; Sun, X.; Tan, H.; Si, X.; Tian, J.; Teng, W.; Wang, J.; Liang, Q. 3D food printing: Applications of f-based materials in extrusion-based food printing. *Crit. Rev. Food Sci. Nutr.* **2021**, *62*, 7184–7198. [CrossRef]
10. Park, S.M.; Kim, H.W.; Park, H.J. Callus-based 3D printing for food exemplified with carrot tissues and its potential for innovative food production. *J. Food Eng.* **2020**, *271*, 109781. [CrossRef]
11. Cao, L.; Lu, W.; Mata, A.; Nishinari, K.; Fang, Y. Egg-box model-based gelation of alginate and pectin: A review. *Carbohydr. Polym.* **2020**, *242*, 116389. [CrossRef] [PubMed]
12. Seisun, D.; Zalesny, N. Strides in food texture and hydrocolloids. *Food Hydrocoll.* **2021**, *117*, 106575. [CrossRef]

13. Okagu, I.U.; Ndefo, J.C.; Aham, E.C.; Obeme-Nmom, J.I.; Agboinghale, P.E.; Aguchem, R.N.; Nechi, R.N.; Lammi, C. Lupin-Derived Bioactive Peptides: Intestinal Transport, Bioavailability and Health Benefits. *Nutrients* **2021**, *13*, 3266. [CrossRef] [PubMed]
14. Warner, E.L.; Norton, I.T.; Mills, T.B. Comparing the viscoelastic properties of gelatin and different concentrations of kappa-carrageenan mixtures for additive manufacturing applications. *J. Food Eng.* **2019**, *246*, 58–66. [CrossRef]
15. Liu, Z.; Bhandari, B.; Prakash, S.; Mantihal, S.; Zhang, M. Linking rheology and printability of a multicomponent gel system of carrageenan-xanthan-starch in extrusion based additive manufacturing. *Food Hydrocoll.* **2019**, *87*, 413–424. [CrossRef]
16. Tian, H.; Wang, K.; Lan, H.; Wang, Y.; Hu, Z.; Zhao, L. Effect of hybrid gelator systems of beeswax-carrageenan-xanthan on rheological properties and printability of litchi inks for 3D food printing. *Food Hydrocoll.* **2021**, *113*, 106482. [CrossRef]
17. Pant, A.; Lee, A.Y.; Karyappa, R.; Lee, C.P.; An, J.; Hashimoto, M.; Tan, U.-X.; Wong, G.; Chua, C.K.; Zhang, Y. 3D food printing of fresh vegetables using food hydrocolloids for dysphagic patients. *Food Hydrocoll.* **2021**, *114*, 106546. [CrossRef]
18. Kim, S.M.; Wen, Y.; Kim, H.W.; Park, H.J. Textural and sensory qualities of low-calorie surimi with carrageenan inserted as a protein substitute using coaxial extrusion 3D food printing. *J. Food Eng.* **2022**, *333*, 111141. [CrossRef]
19. Diañez, I.; Gallegos, C.; Fuente, E.B.; Martínez, I.; Valencia, C.; Sánchez, M.C.; Diaz, M.J.; Franco, J.M. 3D printing in situ gelification of κ-carrageenan solutions: Effect of printing variables on the rheological response. *Food Hydrocoll.* **2019**, *87*, 321–330. [CrossRef]
20. Gholamipour-Shirazi, A.; Norton, I.T.; Mills, T. Designing hydrocolloid based food-ink formulations for extrusion 3D printing. *Food Hydrocoll.* **2019**, *95*, 161–167. [CrossRef]
21. Beaumont, M.; Tran, R.; Vera, G.; Niedrist, D.; Rousset, A.; Pierre, R.; Shastri, V.P.; Forget, A. Hydrogel-Forming Algae Polysaccharides: From Seaweed to Biomedical Applications. *Biomacromolecules* **2021**, *22*, 1027–1052. [CrossRef] [PubMed]
22. Zhong, H.; Gao, X.; Cheng, C.; Liu, C.; Wang, Q.; Han, X. The Structural Characteristics of Seaweed Polysaccharides and Their Application in Gel Drug Delivery Systems. *Mar. Drugs* **2020**, *18*, 658. [CrossRef] [PubMed]
23. Kharina, M.; Emelyanov, V.; Mokshina, N.; Ibragimova, N.; Gorshkova, T. Pretreatment of Sugar Beet Pulp with Dilute Sulfurous Acid is Effective for Multipurpose Usage of Carbohydrates. *Appl. Biochem. Biotechnol.* **2016**, *179*, 307–320. [CrossRef] [PubMed]
24. Holland, C.; Ryden, P.; Edwards, C.H.; Grundy, M.M.-L. Plant Cell Walls: Impact on Nutrient Bioaccessibility and Digestibility. *Foods* **2020**, *9*, 201. [CrossRef] [PubMed]
25. Ric-Varas, P.; Barceló, M.; Rivera, J.A.; Cerezo, S.; Matas, A.J.; Schückel, J.; Knox, J.P.; Posé, S.; Pliego-Alfaro, F.; Mercado, J.A. Exploring the Use of Fruit Callus Culture as a Model System to Study Color Development and Cell Wall Remodeling during Strawberry Fruit Ripening. *Plants* **2020**, *9*, 805. [CrossRef]
26. Camacho-Fernández, C.; Seguí-Simarro, J.M.; Mir, R.; Boutilier, K.; Corral-Martínez, P. Cell Wall Composition and Structure Define the Developmental Fate of Embryogenic Microspores in *Brassica napus*. *Front. Plant Sci.* **2021**, *2*, 737139. [CrossRef]
27. Betekhtin, A.; Rojek, M.; Milewska-Hendel, A.; Gawecki, R.; Karcz, J.; Kurczynska, E.; Hasterok, R. Spatial Distribution of Selected Chemical Cell Wall Components in the Embryogenic Callus of *Brachypodium distachyon*. *PLoS ONE* **2016**, *11*, e0167426. [CrossRef]
28. Farghaly, F.A.; Salam, H.K.; Hamada, A.M.; Radi, A.A. Alleviating excess boron stress in tomato calli by applying benzoic acid to various biochemical strategies. *Plant Physiol. Biochem.* **2022**, *182*, 216–226. [CrossRef]
29. Campo, V.L.; Kawano, D.F.; Braz da Silva, D., Jr.; Carvalho, I. Carrageenans: Biological properties, chemical modifications and structural analysis—A review. *Carbohydr. Polym.* **2009**, *77*, 167–180. [CrossRef]
30. Vancauwenberghe, V.; Mfortaw, M.W.B.M.; Vanstreels, E.; Verboven, P.; Lammertyn, J.; Nicolai, B. 3D printing of plant tissue for innovative food manufacturing: Encapsulation of alive plant cells into pectin based bio-ink. *J. Food Eng.* **2019**, *263*, 454–464. [CrossRef]
31. Onacik-Gür, S.; Zbikowska, A.; Majewska, B. Effect of Spirulina (*Spirulina platensis*) addition on textural and quality properties of cookies. *Ital. J. Food Sci.* **2018**, *30*, 1–12. [CrossRef]
32. Vieira, M.V.; Oliveira, S.M.; Amado, I.R.; Fasolin, L.H.; Vicente, A.A.; Pastrana, L.M.; Fucinos, P. 3D printed functional cookies fortified with *Arthrospira platensis*: Evaluation of its antioxidant potential and physical-chemical characterization. *Food Hydrocoll.* **2020**, *107*, 105893. [CrossRef]
33. Keerthana, K.; Anukiruthika, T.; Moses, J.A.; Anandharamakrishnan, C. Development of fiber-enriched 3D printed snacks from alternative foods: A study on button mushroom. *J. Food Eng.* **2020**, *287*, 110116. [CrossRef]
34. De Lavergne, M.D.; Van de Velde, F.; Stieger, M. Bolus matters: The influence of food oral breakdown on dynamic texture perception. *Food Funct.* **2017**, *8*, 464–480. [CrossRef] [PubMed]
35. Pascua, Y.; Koc, H.; Foegeding, E.A. Food structure: Roles of mechanical properties and oral processing in determining sensory texture of soft materials. *Curr. Opin. Colloid Interface Sci.* **2013**, *18*, 324–333. [CrossRef]
36. Raheem, D.; Carrascosa, C.; Ramos, F.; Saraiva, A.; Raposo, A. Texture-Modified Food for Dysphagic Patients: A Comprehensive Review. *Int. J. Environ. Res. Public Health* **2021**, *18*, 5125. [CrossRef]
37. Sungsinchai, S.; Niamnuy, C.; Wattanapan, P.; Charoenchaitrakool, M.; Devahastin, S. Texture modification technologies and their opportunities for the production of dysphagia foods: A review. *Compr. Rev. Food Sci. Food Saf.* **2019**, *18*, 1898–1912. [CrossRef]
38. Elshahed, M.S.; Miron, A.; Aprotosoaie, A.C.; Farag, M.A. Pectin in diet: Interaction with the human microbiome, role in gut homeostasis, and nutrient-drug interactions. *Carbohydr. Polym.* **2021**, *255*, 117388. [CrossRef]
39. Naqash, F.; Masoodi, F.A.; Rather, S.A.; Wani, S.M.; Gani, A. Emerging concepts in the nutraceutical and functional properties of pectin—A review. *Carbohydr. Polym.* **2017**, *168*, 227–239. [CrossRef]

40. Minzanova, S.T.; Mironov, V.F.; Arkhipova, D.M.; Khabibullina, A.V.; Mironova, L.G.; Zakirova, Y.M.; Milyukov, V.A. Biological activity and pharmacological application of pectic polysaccharides: A review. *Polymers* **2018**, *10*, 1407. [CrossRef]
41. Lama-Muñoz, A.; Contreras, M.d.M. Extraction Systems and Analytical Techniques for Food Phenolic Compounds: A Review. *Foods* **2022**, *11*, 3671. [CrossRef] [PubMed]
42. Nieto-Trujillo, A.; Cruz-Sosa, F.; Luria-Pérez, R.; Gutiérrez-Rebolledo, G.A.; Román-Guerrero, A.; Burrola-Aguilar, C.; Zepeda-Gómez, C.; Estrada-Zúñiga, M.E. Arnica montana Cell Culture Establishment, and Assessment of Its Cytotoxic, Antibacterial, α-Amylase Inhibitor, and Antioxidant In Vitro Bioactivities. *Plants* **2021**, *10*, 2300. [CrossRef] [PubMed]
43. Khan, H.; Khan, T.; Ahmad, N.; Zaman, G.; Khan, T.; Ahmad, W.; Batool, S.; Hussain, Z.; Drouet, S.; Hano, C.; et al. Chemical Elicitors-Induced Variation in Cellular Biomass, Biosynthesis of Secondary Cell Products, and Antioxidant System in Callus Cultures of Fagonia indica. *Molecules* **2021**, *26*, 6340. [CrossRef] [PubMed]
44. Maliński, M.P.; Kikowska, M.A.; Soluch, A.; Kowalczyk, M.; Stochmal, A.; Thiem, B. Phytochemical Screening, Phenolic Compounds and Antioxidant Activity of Biomass from Lychnis flos-cuculi L. In Vitro Cultures and Intact Plants. *Plants* **2021**, *10*, 206. [CrossRef]
45. Park, J.-S.; Seong, Z.-K.; Kim, M.-S.; Ha, J.-H.; Moon, K.-B.; Lee, H.-J.; Lee, H.-K.; Jeon, J.-H.; Park, S.U.; Kim, H.-S. Production of Flavonoids in Callus Cultures of Sophora flavescens Aiton. *Plants* **2020**, *9*, 688. [CrossRef]
46. Erst, A.A.; Petruk, A.A.; Erst, A.S.; Krivenko, D.A.; Filinova, N.V.; Maltseva, S.Y.; Kulikovskiy, M.S.; Banaev, E.V. Optimization of Biomass Accumulation and Production of Phenolic Compounds in Callus Cultures of *Rhodiola rosea* L. Using Design of Experiments. *Plants* **2022**, *11*, 124. [CrossRef]
47. Sahiner, M.; Yilmaz, A.S.; Gungor, B.; Ayoubi, Y.; Sahiner, N. Therapeutic and Nutraceutical Effects of Polyphenolics from Natural Sources. *Molecules* **2022**, *27*, 6225. [CrossRef]
48. Gasmi, A.; Mujawdiya, P.K.; Noor, S.; Lysiuk, R.; Darmohray, R.; Piscopo, S.; Lenchyk, L.; Antonyak, H.; Dehtiarova, K.; Shanaida, M.; et al. Polyphenols in Metabolic Diseases. *Molecules* **2022**, *27*, 6280. [CrossRef]
49. Singh Tuli, H.; Kumar, A.; Ramniwas, S.; Coudhary, R.; Aggarwal, D.; Kumar, M.; Sharma, U.; Chaturvedi Parashar, N.; Haque, S.; Sak, K. Ferulic Acid: A Natural Phenol That Inhibits Neoplastic Events through Modulation of Oncogenic Signaling. *Molecules* **2022**, *27*, 7653. [CrossRef]
50. Kofla-Dłubacz, A.; Pytrus, T.; Akutko, K.; Sputa-Grzegrzółka, P.; Piotrowska, A.; Dzięgiel, P. Etiology of IBD—Is It Still a Mystery? *Int. J. Mol. Sci.* **2022**, *23*, 12445. [CrossRef]
51. Hagan, M.; Hayee, B.H.; Rodriguez-Mateos, A. (Poly)phenols in Inflammatory Bowel Disease and Irritable Bowel Syndrome: A Review. *Molecules* **2021**, *26*, 1843. [CrossRef] [PubMed]
52. Singleton, V.L.; Rossi, J.A. Colorimetry of total phenolics with phosphomolybdic-phosphotungstic acid reagents. *Am. J. Enol. Vitic.* **1965**, *16*, 144–158.
53. Khin, M.N.; Goff, H.D.; Nsor-Atindana, J.; Ahammed, S.; Liu, F.; Zhong, F. Effect of texture and structure of polysaccharide hydrogels containing maltose on release and hydrolysis of maltose during digestion: In vitro study. *Food Hydrocoll.* **2021**, *112*, 106326. [CrossRef]
54. Usov, A.I.; Bilan, M.I.; Klochkova, N.G. Polysaccharides of algae. 48. Polysaccharide composition of several Calcareous red algae: Isolation of alginate from Corallina Pilulifera, P. et R. (Rhodophyta, Corallinaceae). *Bot. Mar.* **1995**, *38*, 43–52. [CrossRef]

gels

Article

Three-Dimensional Printing Parameter Optimization for Salmon Gelatin Gels Using Artificial Neural Networks and Response Surface Methodology: Influence on Physicochemical and Digestibility Properties

Nailín Carvajal-Mena [1], Gipsy Tabilo-Munizaga [1,*], Marleny D. A. Saldaña [2], Mario Pérez-Won [1], Carolina Herrera-Lavados [1], Roberto Lemus-Mondaca [3] and Luis Moreno-Osorio [4]

1 Department of Food Engineering, Universidad del Bío-Bío, Avenida Andrés Bello 720, Chillán 3780000, Chile; nailincm@gmail.com (N.C.-M.); mperez@ubiobio.cl (M.P.-W.); carolina.phl@gmail.com (C.H.-L.)
2 Department of Agricultural, Food and Nutritional Science, University of Alberta, Edmonton, AB T6G 2P5, Canada; marleny.saldana@ualberta.ca
3 Department of Food Science and Chemical Technology, Universidad de Chile, Santos Dumont 964, Santiago 8330015, Chile; rlemus@uchile.cl
4 Department of Basic Sciences, Universidad del Bío-Bío, Avenida Andrés Bello 720, Chillán 3780000, Chile; lmoreno@ubiobio.cl
* Correspondence: gtabilo@ubiobio.cl

Abstract: This study aimed to optimize the 3D printing parameters of salmon gelatin gels (SGG) using artificial neural networks with the genetic algorithm (ANN-GA) and response surface methodology (RSM). In addition, the influence of the optimal parameters obtained using the two different methodologies was evaluated for the physicochemical and digestibility properties of the printed SGG (PSGG). The ANN-GA had a better fit (R^2 = 99.98%) with the experimental conditions of the 3D printing process than the RSM (R^2 = 93.99%). The extrusion speed was the most influential parameter according to both methodologies. The optimal values of the printing parameters for the SGG were 0.70 mm for the nozzle diameter, 0.5 mm for the nozzle height, and 24 mm/s for the extrusion speed. Gel thermal properties showed that the optimal 3D printing conditions affected denaturation temperature and enthalpy, improving digestibility from 46.93% (SGG) to 51.52% (PSGG). The secondary gel structures showed that the β-turn structure was the most resistant to enzymatic hydrolysis, while the intermolecular β-sheet was the most labile. This study validated two optimization methodologies to achieve optimal 3D printing parameters of salmon gelatin gels, with improved physicochemical and digestibility properties for use as transporters to incorporate high value nutrients to the body.

Keywords: salmon gelatin; 3D printing; dimensional stability; artificial neural network; secondary structure; digestibility

Citation: Carvajal-Mena, N.; Tabilo-Munizaga, G.; Saldaña, M.D.A.; Pérez-Won, M.; Herrera-Lavados, C.; Lemus-Mondaca, R.; Moreno-Osorio, L. Three-Dimensional Printing Parameter Optimization for Salmon Gelatin Gels Using Artificial Neural Networks and Response Surface Methodology: Influence on Physicochemical and Digestibility Properties. *Gels* **2023**, *9*, 766. https://doi.org/10.3390/gels9090766

Academic Editors: Baskaran Stephen Inbaraj, Kandi Sridhar and Minaxi Sharma

Received: 10 August 2023
Revised: 11 September 2023
Accepted: 14 September 2023
Published: 20 September 2023

1. Introduction

Consumers are increasingly interested in ingesting proteins from alternative sources such as plant proteins, insects, algae, and food byproducts [1,2]. A recent example used a byproduct from the fish industry to obtain proteins that have high nutritional and biological value, such as gelatin extracted from fish skin [3]. Gelatin is a coiled, partially hydrolyzed form of collagen of great value for the pharmaceutical, cosmetic, and food industries due to its versatility as an enhancer in a wide range of techno-functional properties. In addition, gelatin is a hydrocolloid that solidifies when it is cold, and it is thermo-reversible, which enables it to exist as a solution or gel, depending on the temperature it is exposed to. Earlier, Carvajal-Mena et al. [3] demonstrated the potential use of salmon skin gelatin as a raw material for 3D printing.

Three-dimensional printing is a novel technology that can develop customized food products with personalized nutrition, and it can create food products easily, economically, and quickly [4]. Furthermore, 3D printing allows the use and combination of various ingredients, including drug administration; thus, adapting food for children, the elderly, and other individuals with special dietary needs [4,5]. However, there are still many reservations about how 3D-printed foods are perceived by consumers. Some studies have shown that repeated consumption of 3D-printed foods can increase acceptability, although in recent times, it has also been claimed that consumers have developed awareness of the health and environmental benefits of these foods compared to conventional production techniques [6]. In the meantime, even though the development and knowledge of this technology are increasing, it is necessary to promote the dissemination of the nutritional advantages and health benefits that 3D-printed foods can offer.

Extrusion 3D printing is the most widely used in the food industry as it can print food inks by extrusion; thus, producing objects with high quality and precision [4–7]. The development of edible food inks used for the 3D printing process are currently limited. Therefore, optimization of the printing properties and parameters of these food materials is essential to achieve a customized, self-supporting matrix that guarantees smooth extrusion through the nozzle, followed by rapid resolidification to maintain the shape of the printed object [8]. Therefore, it is necessary to optimize the printing parameters for a successful 3D food printing process. Among the important mechanical parameters are the extrusion speed and nozzle diameter and height [9]. The extrusion speed directly affects the stability and quality of the printed geometry, and it has been reported in several studies that a successful process should use speed values between 10 and 70 mm/s [10]. Similarly, nozzle diameter and height influence the volume of the printed object, which is associated with the ability to support weight during printing and prevent deformed printed matrices caused by low printing efficiency. Various studies have suggested that successful printing requires a nozzle diameter between 0.41 and 2.00 mm and a height between 0.3 and 1.1 mm [9–11]. Based on this information, it is feasible to establish a methodology to predict matrix printability by optimizing the parameters of the 3D printing process.

Recently, artificial intelligence (AI) has been used in the food industry to optimize extraction processes [12] or to determine specific drying or dehydration conditions [1]. Among AI methodologies, response surface methodology (RSM) is a multivariate strategy that can address both experimental design and statistical modeling by investigating the associations between one or more response variables and a set of experimental factors [12,13]. Therefore, a more efficient method for solving complicated nonlinear problems with significantly high predictive ability is artificial neural network (ANN) modeling, which has been used to establish a more in-depth study of the food industry processes [14]. The ANN is a theoretical mathematical model that emulates the human brain and consists of linear or nonlinear processing elements. The ANN can map complex dynamic phenomena based on self-learning, fault tolerance, and neuron robustness. Meanwhile, genetic algorithms (GA) are a faster optimization technique than traditional algorithms that include a functional probabilistic global search system [15]. Some current research has applied AI to study extrusion processes and its use for predicting physical properties (color and texture) in an extruded rice paste [16] and optimizing 3D printing of chicken meat and the feasibility of the printed product [15]. Furthermore, modeling of the extrusion process of wheat and soybean flour pastes has revealed that the ANN methodology was better adapted to the experimental conditions of the extrusion process than the RSM [17].

Although the experimental optimization of several protein-rich 3D printed foods formulations has been performed, there is no study available that combines 3D printing and AI to design foods with specialized nutritional requirements [1]. In addition, there are no studies on applying these methodologies in 3D printing of protein foods to optimize and predict printing parameters that improve physicochemical properties and increase the digestibility of printed products [18]. Therefore, the main objective of this study was to optimize the 3D printing parameters of salmon gelatin gels (SGG) using artificial neural networks with a genetic algorithm (ANN-GA) and response surface methodology (RSM). In addition, the influence of the optimal parameters obtained using these two different methodologies was determined for the physicochemical and digestibility properties of the printed SGG (PSGG).

2. Results and Discussion

2.1. Optimization Using Response Surface Methodology

The ANOVA results enabled the evaluation of the statistical significance of the independent variables in the predicted quadratic model (Table 1). The regression coefficients (R^2) were 90.61, 87.43, and 96.76% for viscosity, hardness, and dimensional stability, respectively. These regression coefficient values showed that the regression model adequately defined the behavior of the printing variables. The most influential parameters in the printing process were extrusion speed and nozzle diameter because they significantly affected the three response variables to achieve an adequate printed gel. Figure 1 shows the influence of the printing parameters on the physicochemical properties (viscosity, hardness, and dimensional stability) of the 3D PSGG.

Table 1. Regression coefficients of the second-order polynomial models predicted for the response variables.

Regression Coefficient	Estimated Coefficient		
	Apparent Viscosity	Hardness	Dimensional Stability
β_0	−149.34	−7.59	35.01
β_1	36.17	0.70	3.48
β_2	−535.37	−16.64	11.48
β_3	758.42	44.17	49.23
β_{11}	−0.49	−0.009	−0.05
β_{12}	−5.33	−0.045	−0.66
β_{13}	−2.09	−0.05	−0.67
β_{22}	304.42	7.67	1.18
β_{23}	30.35	1.29	−0.21
β_{33}	−761.45	−44.96	−35.64
R^2 (%)	90.61	87.43	96.76

Linear (β_1, β_2, β_3), quadratic (β_{11}, β_{22}, β_{33}), and interaction coefficients (β_{12}, β_{13}, β_{23}); intercept (β_0); and R^2: coefficient of determination. The nozzle diameter used during the printing process is critical for a high precision surface and finish on the printed gels [19]. Selecting the appropriate nozzle size depends on the printing material and determines the pressure required to achieve gel creep [11]. Figure 1 shows that increasing the nozzle diameter negatively affects the physical properties of the printed gels. In addition, the material flowed faster with larger nozzle sizes and produced thicker lines with poor and rough surfaces (Figure 1). When comparing the three nozzle sizes, the samples printed with the 0.70 mm nozzle diameter exhibited a better appearance that was similar to the designed object, including high dimensional stability and high viscosity and hardness.

The height between the nozzle tip and the printing bed is a fundamental parameter for good quality of the printed object, and a critical height is established for each printing material. In this study, the nozzle height significantly influenced the viscosity of the printed gels. At a lower nozzle height, results showed that the extruded lines were thicker, and lumps formed on the surface of the printed object because the tip of the nozzle was submerged in the material, dragging it along with the movement, and producing deformed and unstable objects. However, when the height was excessive, the gel was deposited in an imprecise manner with irregular lines. Maintaining an adequate nozzle height is required to achieve a well-defined target geometry [19] because the distance from the printed layer to the nozzle tip significantly affects the resolution of printed gels.

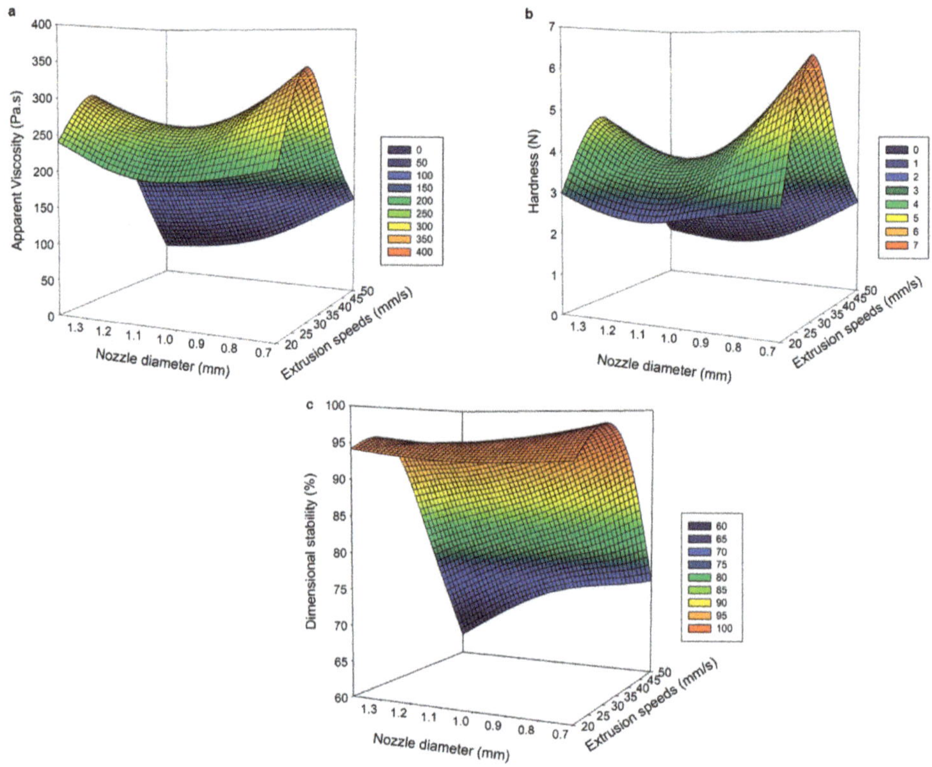

Figure 1. Response surface analysis to optimize salmon gelatin printing parameters related to extrusion speed and nozzle diameter at optimal values for nozzle height. (**a**) Apparent viscosity, (**b**) hardness, and (**c**) dimensional stability.

The extrusion speed positively affected the physical properties (viscosity, hardness, and dimensional stability) of the printed gels, as it is directly related to the quality and thickness of the extruded filament. As extrusion speed increased, more material was extruded up to a critical point at which the flow was so high that gels were deformed and unstable. Likewise, a low extrusion speed produced a discontinuous extruded filament, which resulted in insufficient structural integrity with a deposit of material that did not cover the nozzle's trajectory, showing irregular printed gels with broken internal filaments, and an erroneous arrangement of the SGG. Earlier, Southerland et al. [20] reported that an accurate chocolate print was obtained at a slow printing speed but failed to eliminate small air blockages that can occur during extrusion in the syringe. However, Derossi et al. [21] demonstrated that high flow rates were required to achieve better definition and uniformity in the printed object because fruit formulations at low speeds exhibited irregular shapes with discontinuous material lines and unwanted pores. Therefore, extruded lines were thinner and continuous at an optimal extrusion speed; thus, forming high-quality PSGG.

A multivariate optimization was performed to establish the optimal printing parameters for salmon gelatin (Figure 2a). The multiple regression model explained 86.95% of the variability experienced in the response variables. The optimal printing parameters were 31.54 mm/s for the extrusion speed, 0.82 mm for the nozzle diameter, and 0.47 mm for the nozzle height, and the response values were 323.81 Pa·s for the viscosity, 6.43 N for hardness, and 97.78% for the dimensional stability. The linear correlation between the desired optimal and experimental values in the response variables was significant, and the linear model explained 93.99% of the relationship between both variables (Figure 2b).

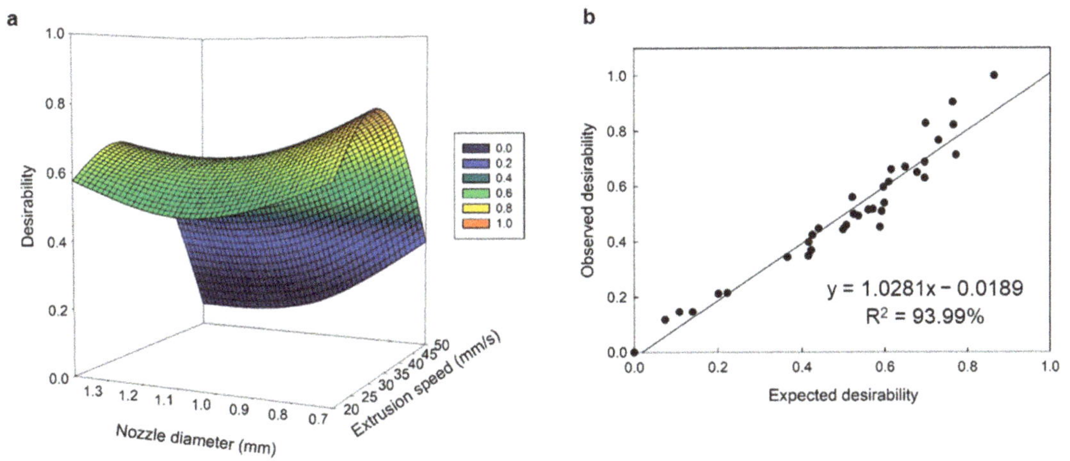

Figure 2. Multiple response optimization analysis. (**a**) Desirability response surface, and (**b**) linear regression analysis between observed and predicted desirability model.

This study shows how printing parameters significantly influenced the physical properties of the printed gels because, at high extrusion speeds and higher printing nozzle diameters and heights, gels with irregularly shaped extruded filaments and poor definition were formed; thus, leading to negative values of viscosity, hardness, and dimensional stability.

2.2. Optimization Using Artificial Neural Networks with Genetic Algorithm

The ANN-GA is a modern modeling technique for process optimization because it has high prediction and estimation reliability [21]. In this study, an ANN was trained with a multilayer feed-forward propagation learning algorithm to obtain a function that was optimized by the GA using the variables of the printing process (extrusion speed, nozzle diameter, and height) to maximize the viscosity, hardness, and dimensional stability of the 3D printed SGG (Figure 3). Errors and accuracy of the neural network prediction were performed through multiple rounds of the learning process. In addition, network performance was validated and evaluated by the RSME and R^2 values obtained at different stages of this study. The R^2 coefficients for viscosity, hardness, and dimensional stability were 99.88, 99.81, and 99.68% for training; 99.97, 99.29, and 98.94% for validation; and 99.97, 99.29, and 99.30% for testing, respectively. Similar values between training and testing corroborated the accuracy of the ANN model.

Table 2 shows the mean values for sensitivity when evaluating the predictive ability of the input neurons. Results indicated that the most significant predictor was the extrusion speed (100%), followed by the nozzle diameter (55.10%), and nozzle height (30.44%). Furthermore, extrusion speed was the most influential factor on the response variables, which closely concurred with the RSM results. Several studies have indicated that printing speed is a crucial parameter because it directly influences printing efficiency [9]. A lower resolution with unstable structures at faster printing speeds due to the formation of irregular layers was reported, while the printing process slowed down at lower speeds, which also generated structural instability [11–21]. Moreover, nozzle diameter was the second most important parameter where the printing became larger at greater diameters, which led to a poor weight distribution that destabilized the base of the printed gel and caused the figure to collapse. The main drawback of the ANN-GA methodology is that it does not consider the interactions between the different independent variables. However, the accuracy and correlation of the predicted and experimental results were higher with this methodology because it has the potential to train and learn from the experimental data of the response variables.

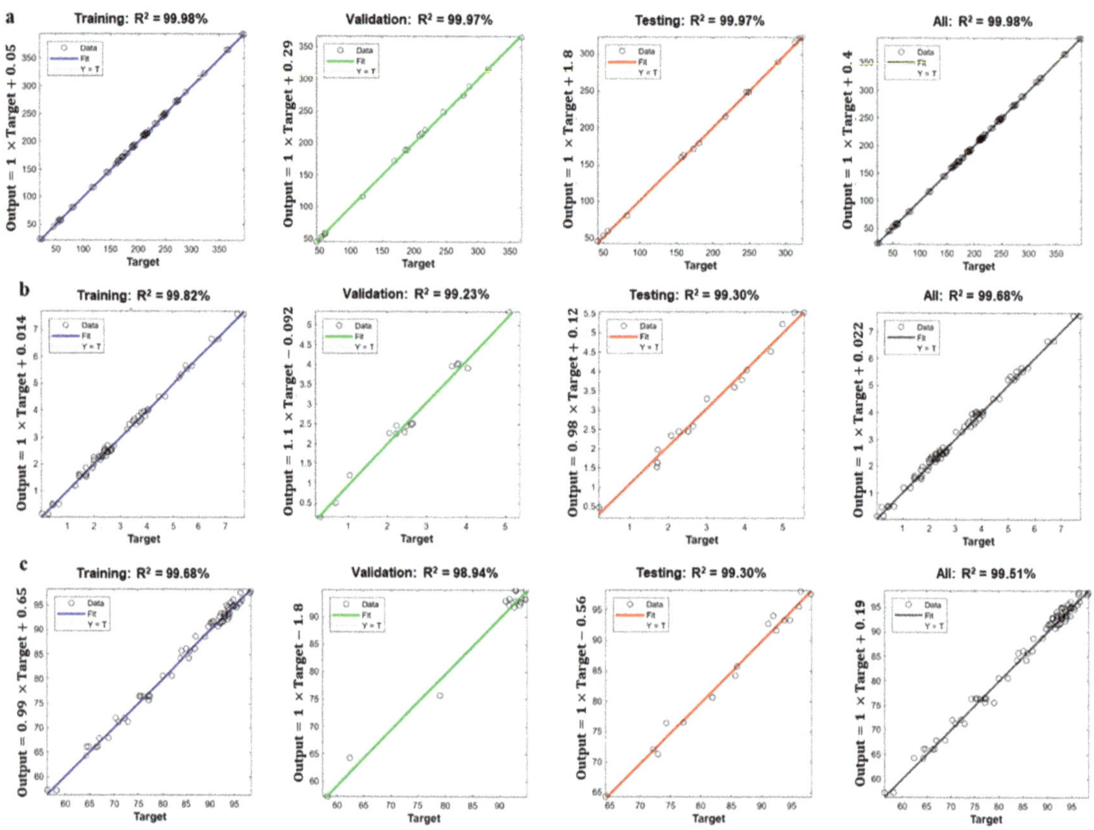

Figure 3. Regression analysis of training, validation, and testing for all the experiments of: (**a**) apparent viscosity, (**b**) hardness, and (**c**) dimensional stability. R^2 is the coefficient of determination.

Table 2. Artificial neural networks with genetic algorithm (ANN-GA) sensitivity analysis of salmon gelatin gels.

Neural Network (NN)	Extrusion Speed (mm/s)	Nozzle Diameter (mm)	Nozzle Height (mm)
NN1	0.35	0.24	0.42
NN2	0.24	0.63	0.13
NN3	0.58	0.10	0.33
NN4	0.52	0.44	0.04
NN5	0.88	0.02	0.09
NN6	0.78	0.11	0.11
NN7	0.42	0.35	0.23
NN8	0.60	0.29	0.11
NN9	0.49	0.31	0.20
NN10	0.48	0.46	0.07
NN11	0.60	0.37	0.04
NN12	0.54	0.25	0.20
Average Importance	0.54	0.30	0.16
Normalized Importance (%)	100.00	55.10	30.44

The optimal printing conditions were 24 mm/s for the extrusion speed, 0.70 mm for the nozzle diameter, and 0.50 mm for the nozzle height to achieve 380.85 Pa·s for viscosity, 7.75 N for hardness, and 98.87% for dimensional stability. These predicted values were not significantly different ($p < 0.05$) from the experimental data (R^2 = 99.98%) under the optimal printing conditions (Figure 4).

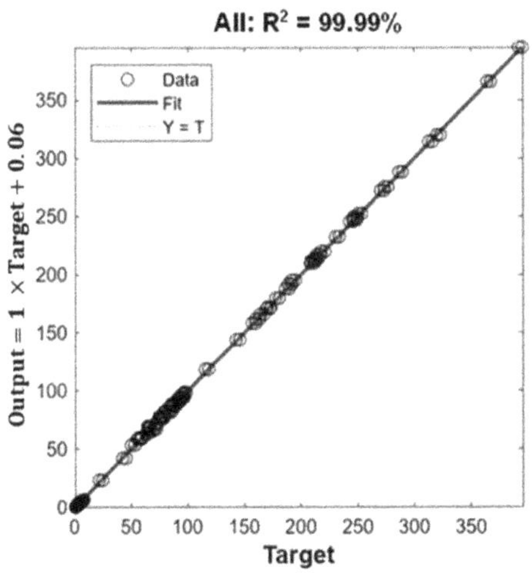

Figure 4. Regression analysis of all the experiments for the total desirability study. R^2 is the coefficient of determination.

2.3. Comparison between Response Surface Methodology and Artificial Neural Networks with Genetic Algorithm

The comparison between the RSM and ANN-GA methodologies for predictive ability and statistical accuracy was based on the statistical error parameters. The R^2 values were higher using the ANN-GA model, indicating that this methodology is more accurate, convenient, and reliable than the RSM for predicting the viscosity, hardness, and dimensional stability of the PSGG. The ANN-GA provided better optimization competence because it learned and predicted models more accurately. In addition, the ANN-GA can also perform limited experiments to process nonlinear relationships between network inputs and outputs and make high-quality predictions.

Several studies have reported the predictive efficiency of the ANN-GA model versus the RSM and have reported its high predictive ability and accuracy [22–24]. The validation and testing of the predictive ability of both models involved the predicted values of the models and the experimental data for the different response variables (Figures 1 and 3). The R^2 values for the RSM and ANN-GA were 93.99% and 99.98%, respectively, which corroborated a better performance of the ANN-GA model. Based on the results obtained, the ANN-GA model exhibited lower residuals with minor variation. In contrast, the deviations between the predicted and experimental responses were more significant in the RSM model.

Although the ANN-GA for the PSGG showed higher quality, stability, and definition with respect to the designed object, the RSM for the PSGG also reported high-correlation values with stable printed structures and no compressed definition of the layers. Therefore, we studied the physicochemical and digestibility properties of the printed gels under the optimal conditions determined using both methodologies.

2.4. Thermal Properties

Figure 5 shows the thermograms obtained using the DSC analysis of the gels before printing and after printing at optimal conditions using the RSM and ANN-GA methods.

All the gels had two peaks of thermal transitions: the first peak corresponded to the glass transition and the second peak corresponded to the protein denaturation (Figure 5).

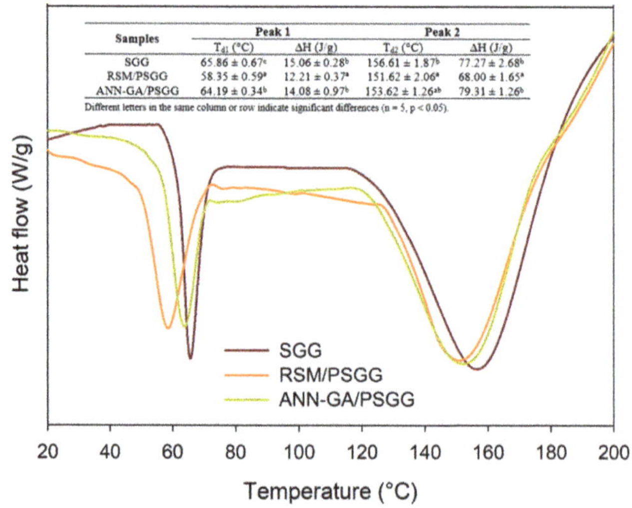

Samples	Peak 1		Peak 2	
	T_{d1} (°C)	ΔH (J/g)	T_{d2} (°C)	ΔH (J/g)
SGG	65.86 ± 0.67[c]	15.06 ± 0.28[b]	156.61 ± 1.87[b]	77.27 ± 2.68[b]
RSM/PSGG	58.35 ± 0.59[a]	12.21 ± 0.37[a]	151.62 ± 2.06[a]	68.00 ± 1.65[a]
ANN-GA/PSGG	64.19 ± 0.34[b]	14.08 ± 0.97[b]	153.62 ± 1.26[ab]	79.31 ± 1.26[b]

Different letters in the same column or row indicate significant differences ($n = 5$, $p < 0.05$).

Figure 5. Differential scanning calorimetry thermograms of salmon gelatin gels before and after printing. SGG: salmon gelatin gel; PSGG: printed salmon gelatin gel; RSM: response surface methodology; ANN-GA: artificial neural networks with genetic algorithm.

The glass transition temperatures (Tg) varied between 58.35 and 65.86 °C, and the Tg values of the gel before printing were significantly higher than the printed gels under both optimal conditions. This higher transition temperature in the SGG occurred due to the larger number of bonds and interactions between the gelatin chains that promoted thermal stability [25]. In addition, gelatin tends to absorb water from the medium, which means that evaporation of this crystallizable water is also responsible for an increase in Tg [25].

Meanwhile, the pressure exerted on the gels during the extrusion process also affects the Tg values. The ANN-GA for the PSGG gels had higher Tg values than the RSM for PSGG gels that can be associated with lower free volume and higher water loss during extrusion 3D printing. In addition, the ΔH values of SGG and ANN-GA/PSGG showed no significant differences but were significantly higher than the RSM/PSGG values. The smaller nozzle diameter used for printing the ANN-GA/PSGG gels generated higher pressure to which they were subjected; thus, causing minimal polymer expansion and creating more compact, crosslinked molecular chains, which required higher energy to achieve molecular mobility [26].

The denaturation temperature and enthalpy of the SGG were slightly higher than for the PSGG due to the more compact molecular bonds that required higher temperature and energy to achieve the helix–spiral transition of the gelatin and break the associated hydrogen bonds.

2.5. *In Vitro Digestibility*

Figure 6 shows the digestibility of the SGG and PSGG. The DH increases slightly during the gastric stage, while its increase in the intestinal stage was significant in the first 30 min for all gels studied. Then, pancreatic enzymes broke peptide bonds during the intestinal stage, which created more active sites for further hydrolysis [27]. During the gastric stage, the DH was 15.64% for the SGG and 20.06–21.49% for the PSGG, showing significant differences between the gels before and after printing under the different conditions.

The DH value of the SGG in the intestinal stage was 46.93%, with significant differences between gels printed under both conditions, indicating that the extrusion 3D printing process significantly affected the DH. The protein extrusion process caused conformational changes such as unfolding, association, aggregation, and crosslinking of molecular chains for protein degradation or oxidation [28]. In addition, the extrusion process was an efficient way to increase digestibility due to breaks in the polymeric chains that comprise the protein structure. Although there were no significant differences between both printed gels, the DH values for the RSM/PSGG were higher than for the ANN-GA/PSGG during the whole digestive process. Therefore, during gel formation under the optimal conditions provided by the RSM, gel network formation was altered, which left more empty spaces between the layers; thus, enabling the enzymes to enter and diffuse more easily.

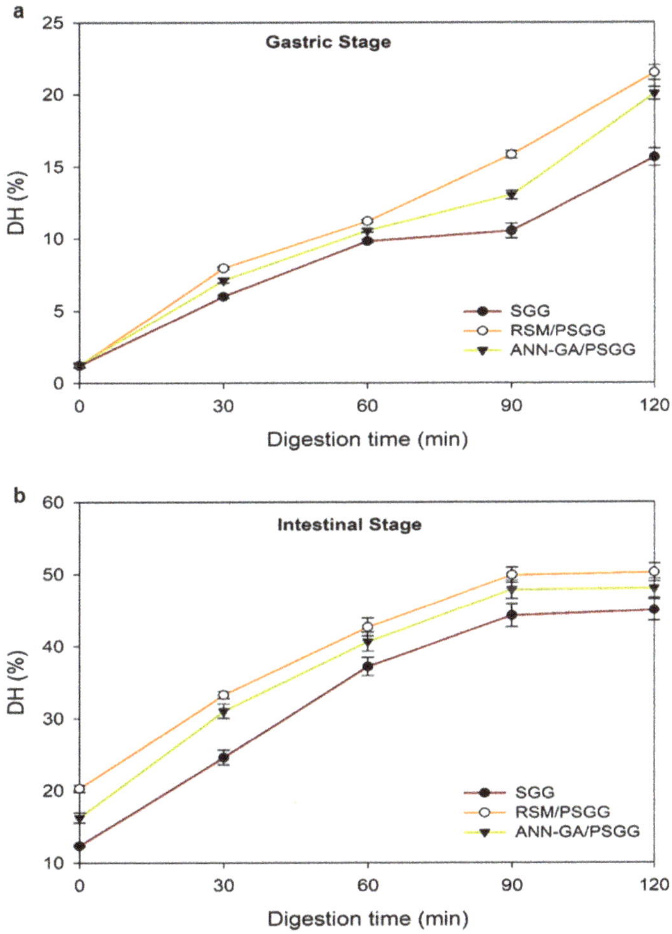

Figure 6. Degree of hydrolysis of gelatin gels at: (**a**) gastric stage, and (**b**) intestinal stage. DH: degree of hydrolysis; SGG: salmon gelatin gel; PSGG: printed salmon gelatin gel; RSM: response surface methodology; ANN-GA: artificial neural networks with genetic algorithm.

The DH for both the SGG and PSGG were higher than values reported in the literature for other gelatins of marine origin. Earlier, Wang et al. [27] reported values of 31.79% for gelatin from Nile tilapia flakes. Overall, the results of the gastrointestinal digestion analysis showed how the extrusion 3D printing process improved the digestibility of salmon gelatin by increasing the number of bioaccessible amino acids (glycine and proline) for subsequent gastrointestinal absorption.

2.6. Fourier Transform Infrared Spectroscopy with Attenuated Total Reflectance

The secondary structures of the SGG and PSGG gels printed under optimal conditions during the digestion process were analyzed by studying the area of the bands in the amide I region (1600–1700 cm^{-1}), which consisted of C=O bond stretching vibrations and, to a lesser extent, C-N stretching vibrations in the polypeptide chain (Figure 7). The quantitative analysis of the second derivative spectra showed that all the gels exhibited six structures in the amide I region: 1604–1623 cm^{-1} β-sheet intermolecular; 1641–1645 cm^{-1} random coil; 1650–1654 cm^{-1} α-helix; 1661–1668 cm^{-1} β-turn; 1681–1684 cm^{-1} intramolecular antiparallel β-sheet; and 1699–1701 cm^{-1} antiparallel β-sheet [29]. The predominant protein structure in all the gels was the antiparallel β-sheet with initial values between 35.08% and 37.49%. Meanwhile, the α-helix structure had values close to 20%, similar values to findings reported by Yang et al. [29] for other gelatins of marine origin (Salmon and Alaska pollock).

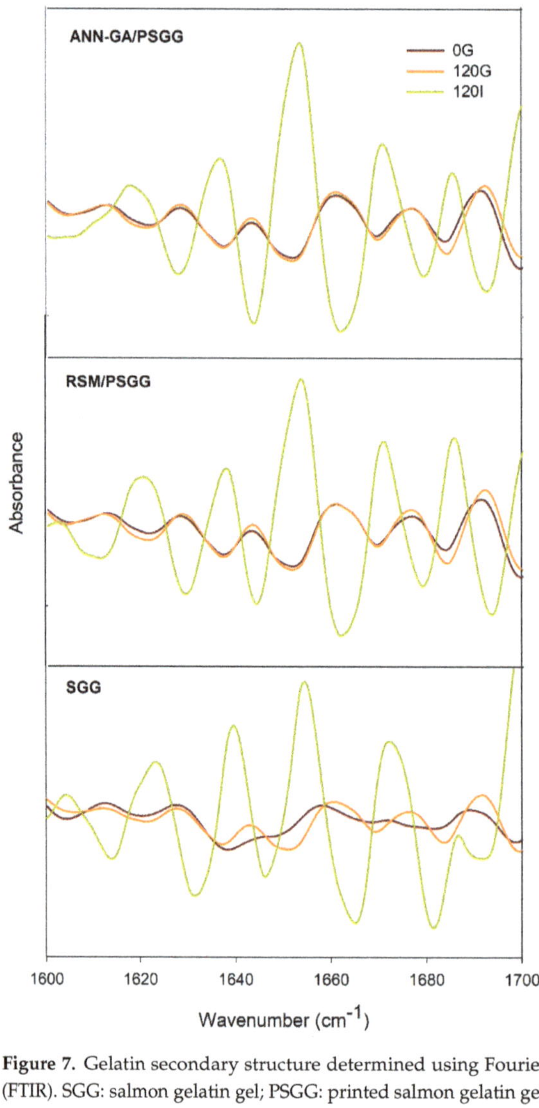

Figure 7. Gelatin secondary structure determined using Fourier transform infrared spectroscopy (FTIR). SGG: salmon gelatin gel; PSGG: printed salmon gelatin gel; RSM: response surface methodology; ANN-GA: artificial neural networks with genetic algorithm.

The variations that occurred in the secondary structures of the gelatin gels during the digestion process showed that the β-turn structure increased significantly and was the most resistant to enzymatic hydrolysis of the structures in the gels (Figure 8). The intermolecular β-sheet structures decreased significantly as the DH increased, reaching zero during the intestinal digestion stage, which demonstrated that this structure is the most labile. The behavior of the structure was directly correlated with the higher DH shown by the printed gels, which indicated that greater protein denaturation occurred due to the action of pancreatic enzymes during the intestinal stage. After completing the digestive process, all the gels still exhibited bands of various protein structures, which could be related to the remaining collagen protein that was not hydrolyzed during the digestive process.

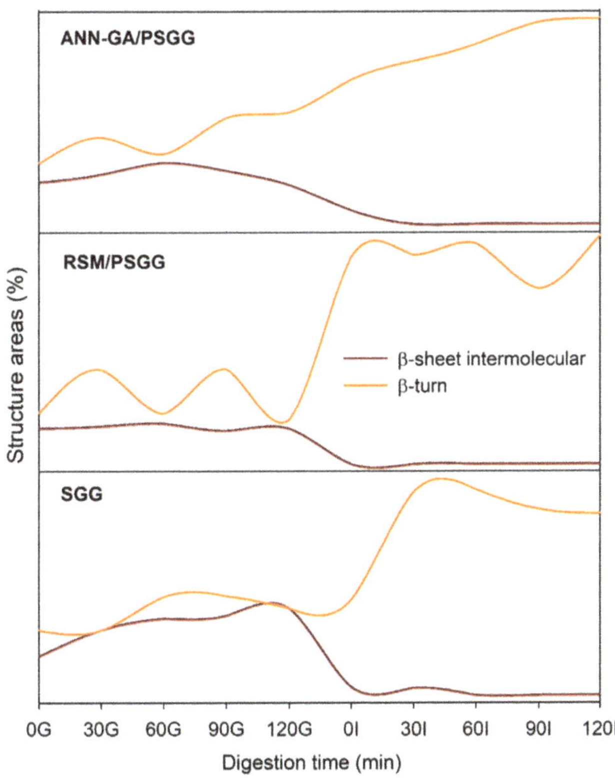

Figure 8. Percentage of the most resistant and most labile areas of the secondary structure of salmon gelatin gels during the digestive process. SGG: salmon gelatin gel; PSGG: printed salmon gelatin gel; RSM: response surface methodology; ANN-GA: artificial neural networks with genetic algorithm.

3. Conclusions

The ANN-GA methodology correlated better with the experimental conditions of the 3D printing process than with the RSM methodology. Among the three printing parameters evaluated, the extrusion speed was the most influential parameter established by both methodologies. The optimal printing parameter values established by the ANN-GA methodology were 0.70 mm for the nozzle diameter, 0.5 mm for the nozzle height, and 24 mm/s for the extrusion speed. The final printed product values were 380.85 Pa·s for the apparent viscosity, 7.75 N for hardness, and 98.87% for the dimensional stability, and there was a 99.98% correlation between predicted and experimental values. The optimization methodologies directly influenced the quality of the printed gels by affecting the thermal properties because more compact structures were formed, which required higher denaturation energy and temperatures. The extrusion 3D printing process improved the digestibility of salmon gels with values ranging from 46.93% (SGG) to 51.52% (PSGG). In addition, the gel's secondary structure was modified during the digestive process that caused higher hydrolysis of the intermolecular β-sheet structures, which were the most labile of all the gels investigated, while the β-turn structure significantly increased and was the most resistant. This study demonstrates that artificial intelligence can be used to predict 3D printing process variables of food matrices to produce high-quality printed objects with improved physicochemical and digestibility properties that can be used as a transporter for incorporating high value nutrients to the body.

4. Materials and Methods

4.1. Materials

Salmon gelatin was extracted from coho salmon (Oncorhynchus kisutch) skins obtained from the Salmones Aysén S.A. salmon processing plant (Puerto Montt, Chile). The skins were washed with water to remove muscle debris and scales, cut into squares (20 mm), and stored at −20 °C until further use. All reagents and enzymes were purchased from Sigma–Aldrich Co. (St. Louis, MO, USA).

4.2. Salmon Gelatin Extraction

Salmon gelatin was extracted following the methodology reported by Carvajal-Mena et al. [3]. First, salmon skins (100 g) were immersed in a sodium hydroxide solution (2 g/L) in a 1:3 (*w*/*v*) ratio and shaken at a constant speed of 1200 rpm for 1 h. Then, they were washed 3 times in 1 L of distilled water to remove all traces of sodium hydroxide. They were treated in an acetic acid solution (4.2 g/L) at a skin/solution ratio of 1:3 (*w*/*v*) with constant agitation at 1200 rpm for 1 h and washed 3 times with distilled water (1 L) to eliminate any acetic acid in the samples. The gelatin extraction process was conducted at 45 °C for 24 h. The supernatant liquid was filtered, and the gelatin was removed. The supernatant liquid was vacuum filtered with a paper filter (Whatman 22 mm, Merck, Darmstadt, Germany) and frozen at −80 °C for freeze-drying in a laboratory freeze dryer (model Beta 1–8 LD, Martin Christ Gefriertrocknungsanlagen, Osterode am Harz, Germany) that was maintained at −42 °C and 10.21 Pa pressure for 24 h. The freeze-dried gelatin was stored at 4 °C until further use.

4.3. Preparation of Salmon Gelatin Gels

Based on preliminary results by Carvajal-Mena et al. [3], it was established that an 8% gelatin concentration was needed to obtain a stable salmon gelatin gel (SGG) for the 3D printing process. First, salmon gelatin (8 g) was dissolved in distilled water (100 mL) at 60 °C for 30 min, and the glass container was covered with aluminum foil during dissolution to prevent water loss. The resulting solution was cooled to room temperature and stored at 6 °C until further printing.

4.4. Dimensional Printing

The SGG was printed with a 3D System 60M printer (Hyrel International Inc., Norcross, GA, USA). Stereolithography files with cubic designs (Figure 9; 20 mm width and length, and 30 mm height) were converted to G-codes with the Repretel software (Hyrel International Inc., Norcross, GA, USA). The printing process was optimized by studying the nozzle height and diameter and extrusion speed variables. The SGG were printed at four extrusion speeds (20, 30, 40 and 50 mm/s); three nozzle diameters (0.7, 1.05 and 1.4 mm); and three nozzle heights (0.3, 0.5 and 0.7 mm). The printing temperature (15 °C) and printing bed temperature (6 °C) were controlled throughout the process using Repretel software version 3.083.

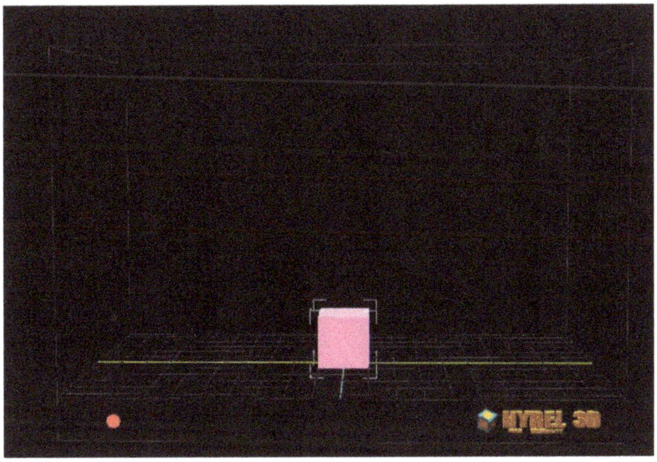

Figure 9. Image of the object to be printed on the Repetrel software platform. The pink cube shows the design of the figure to be printed, represented in a 3D coordinate system that simulates the printing platform. The blue and yellow lines mark the center of the platform on the "x" and "y" axes, respectively.

4.5. Optimization of 3D Printing Process

4.5.1. Response Surface Methodology

This study used a multifactorial design with three main variables. The second-order polynomial model represented viscosity, hardness, and dimensional deviation as a function of three independent variables (nozzle diameter, nozzle height, and extrusion speed). The experiments were randomly conducted, and data analyses were performed with the Statgraphics Centurion XVI statistical software (Statistical Graphics Corp., Herdon, VA, USA). The response surface model describing the behavior of the dependent variables (viscosity, hardness, and dimensional deviation) as a function of the three independent variables is expressed in Equation (1), as follows:

$$Y = \beta_0 + \sum_{i=1}^{3} \beta_i X_i + \sum_{i=1}^{3} \beta_{ii} X_i^2 + \sum_{i=1}^{2} \sum_{j=i+1}^{3} \beta_{ij} X_i X_j \quad (1)$$

where β_0 is the constant; β_i, β_{ii}, and β_{ij} are regression coefficients; X_i and X_j are the levels of coefficients that are independent of the variables; and Y is the dependent variable.

This experimental study design consisted of 36 experiments (Figure 10). Experiments were randomized, and averages were used for the analyses to minimize variability in the response variables. The impact of the independent variables on the dependent variables was accounted for using analysis of variance (ANOVA) of the data at $p < 0.05$.

4.5.2. Artificial Neural Networks with Genetic Algorithm

Parameter modeling and optimization of the 3D printing process of SGG were performed by artificial neural networks with the genetic algorithm (ANN-GA) using the MATLAB mathematical software (version 2023a, Natick, MA, USA). The ANN model was constructed by the backpropagation technique and feed-forward neural network. It consisted of different weights and biases between three input neurons for nozzle height, nozzle diameter, and extrusion speed and three output neurons for the viscosity, hardness, and dimensional stability response variables. The multilayer perceptron algorithm in the ANN modeling was trained with the Levenberg–Marquardt algorithm in 108 runs, including the 3 replicates of each of the 36 experiments. The ANN modeling uses replicates instead of mean values to assess model variability and variances. The runs were divided into three parts, 70% training, 15% validation, and 15% test sets. The training data were used to fit the parameters of the network model; the test data were used to calculate the estimation accuracy; and the validation data were used to ensure the robustness of the network and its parameters. The number of neurons in the hidden layers was determined by a maximum R^2 and a minimum mean square error (MSE) value that were found in 12 hidden neurons in which the transfer function was a hyperbolic sigmoid function for the hidden layer and a linear function for the output layer.

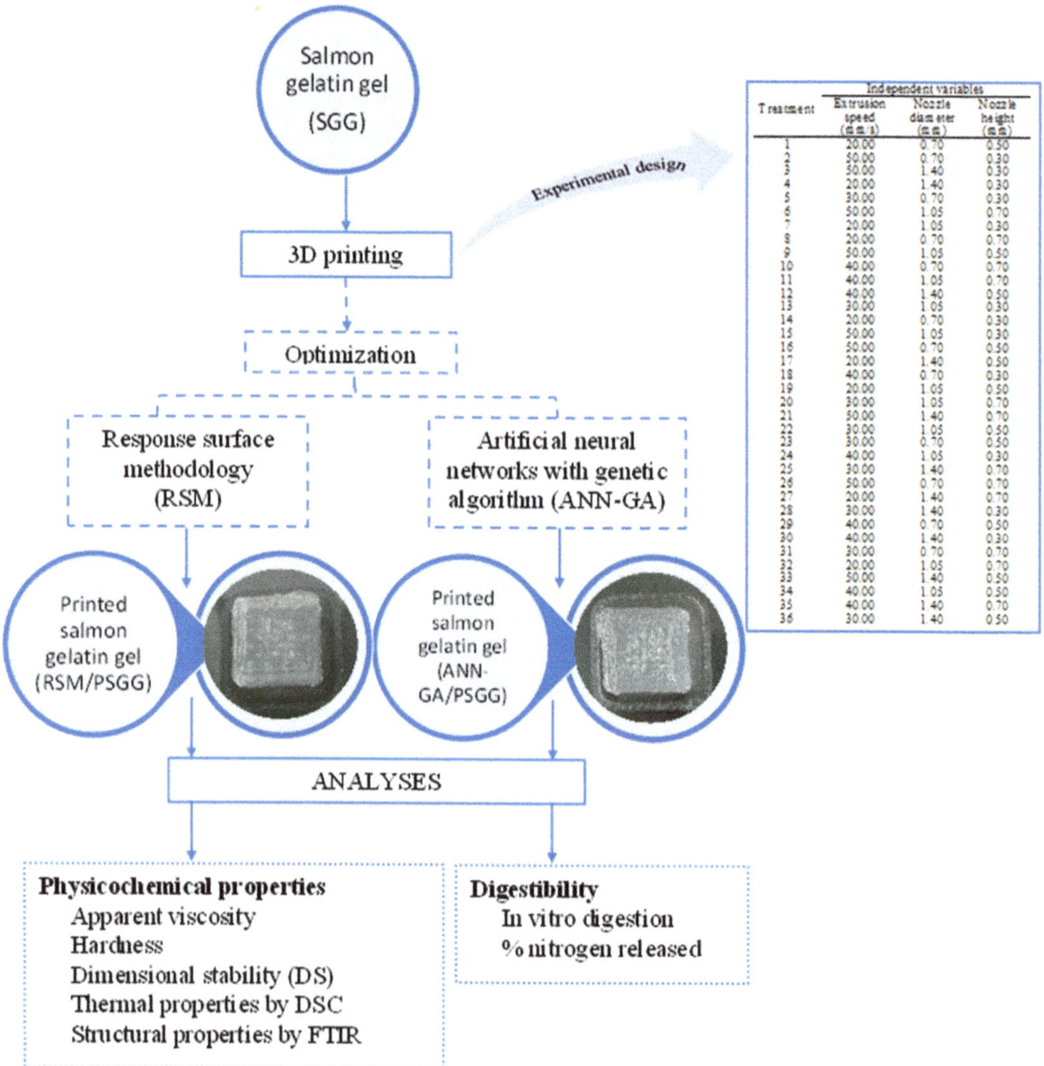

Treatment	Independent variables		
	Extrusion speed (mm/s)	Nozzle diameter (mm)	Nozzle height (mm)
1	20.00	0.70	0.50
2	50.00	0.70	0.30
3	50.00	1.40	0.30
4	20.00	1.40	0.30
5	30.00	0.70	0.30
6	50.00	1.05	0.70
7	20.00	1.05	0.30
8	20.00	0.70	0.70
9	50.00	1.05	0.50
10	40.00	0.70	0.70
11	40.00	1.05	0.70
12	40.00	1.40	0.50
13	30.00	1.05	0.30
14	20.00	0.70	0.30
15	50.00	1.05	0.30
16	50.00	0.70	0.50
17	20.00	1.40	0.50
18	40.00	0.70	0.30
19	20.00	1.05	0.50
20	30.00	1.05	0.70
21	50.00	1.40	0.70
22	30.00	1.05	0.50
23	30.00	0.70	0.50
24	40.00	1.05	0.30
25	30.00	1.40	0.70
26	50.00	0.70	0.70
27	20.00	1.40	0.70
28	30.00	1.40	0.30
29	40.00	0.70	0.50
30	40.00	1.40	0.30
31	30.00	0.70	0.70
32	20.00	1.05	0.70
33	50.00	1.40	0.50
34	40.00	1.05	0.50
35	40.00	1.40	0.70
36	30.00	1.40	0.50

Figure 10. Experimental design for salmon gelatin gel.

The genetic algorithm (GA) is a multi-objective optimization search technique based on the principles of biological evolution [15]. The ANN model was convolved with GA to optimize the variables of the 3D printing process and maximize the desired output variables. The optimization parameters had a population size of 50 and a cross probability fraction of 0.8.

4.5.3. Model Validation

The models were validated by applying the two statistical parameters, root mean square error (*RMSE*) and coefficient of determination (R^2), using Equations (2) and (3), respectively.

$$RMSE = \sqrt{\frac{\sum_{i=1}^{n}\left(Y_{ei} - Y_{pi}\right)^2}{n}} \quad (2)$$

$$R^2 = 1 - \sum_{i=1}^{n}\left(\frac{\left(Y_{ei} - Y_{pi}\right)^2}{\left(Y_m - Y_{pi}\right)^2}\right) \quad (3)$$

where n is the number of experiments; Y_{ei} is the experimental viscosity, hardness, or dimensional stability; Y_{pi} is the predicted viscosity, hardness, or dimensional stability; and Y_m is the average viscosity, hardness, or dimensional stability.

4.6. Apparent Viscosity

The apparent viscosity of the printed SGG (PSGG) under the different printing conditions was determined with a controlled strain and tension rheometer (Physica MCR 300, Anton Paar, Filderstadt, Germany). The instrument was equipped with serrated parallel plate geometry (25 mm diameter) with a 1 mm GAP; the shear rate varied between 0.1 and 100 s^{-1} at a constant temperature of 6 °C. The data of the apparent viscosity curves of the SGG were recorded with the US200 Physica software version 2.01. The viscosity value was determined in the portion of the curve where it became linear.

4.7. Hardness

The hardness of the PSGG was determined with a texture analyzer (TAXT plus100, Stable Micro Systems Ltd., Godalming, UK). The SGG underwent two compression cycles with a cylindrical probe (Code P/50, Stable Micro System Ltd., Godalming, UK). Test conditions were 2 mm s^{-1} pre-test velocity, 1 mm s^{-1} test velocity, and 5 mm s^{-1} post-test velocity, while compression was 50% of its original height at 6 ± 2 °C. The time between the first and second compression was 3 s. The hardness was determined by the force–time curve.

4.8. Dimensional Stability

Dimensional stability was determined according to the method reported by Carvajal-Mena et al. [3], using a cube (20 mm width and length, and 30 mm height) as a standard shape. The length, width, and height dimensions were measured with a Vernier caliper in five different positions for each direction. Mean values were used, and this measurement was performed after 24 h of printing at 6 ± 2 °C. The dimensional stabilities were determined according to Equations (4) and (5).

$$DS_{L,W,H} = \frac{(Measured\ value - Target\ value)}{Target\ value} \quad (4)$$

$$\%\ DS_T = \frac{DS_L + DS_W + DS_H}{3} \times 100 \quad (5)$$

where DS_T is the total dimensional stability and $DS_{L,W,H}$ is the dimensional stability for length, width, and height, respectively.

4.9. Thermal Properties

The thermal properties of the SGG were analyzed by differential scanning calorimetry (DSC) using a DSC instrument (model DSC1, Mettler Toledo AG, Analytical, Schwerzenbach, Switzerland). Approximately 10 mg of wet gels were placed in a 40 µL airtight aluminum container with distilled water and used as a reference. The gels were first subjected to an isothermal phase at 6 °C for 5 min and then scanned from 20 to 200 °C at a rate of 5 °C/min. Denaturation temperature and enthalpy were estimated by determining the area under the transition curves with the Stareware version 10.01 software (Mettler Toledo, Schwerzenbach, Switzerland).

4.10. Dynamic In Vitro Simulation of Human Gastroduodenal Digestion

The digestibility of in-mold and PSGG under the optimal conditions obtained by the RSM and ANN was determined by a dynamic model of in vitro human digestion [30]. First, the gels were subjected to the oral phase, mixed with simulated salivary fluid (1:1 v/v) to a total volume of 40 mL, and α-amylase was added. Then, the solution was vigorously shaken for 2 min and was immediately afterward placed in a gastric bioreactor (MiniBio, Applikon Biotechnology, Delft, The Netherlands). Subsequently, 60 mL of simulated gastric fluid with pepsin was added to the gastric bioreactor at 37 °C, and $CaCl_2$ was incorporated to start the "my-Control" version 1.0X control software (Applikon, Delft, The Netherlands). The software was programmed to perform a gastric pH gradient from 4.5 to 1.2 and intermittent pulses at 200 rpm. The software also controlled the passage of the digested gastric solution to the duodenal bioreactor through a peristaltic pump 10 min after the start of the digestive process. According to Levi and Lesmes [30], 10 mL of simulated duodenal fluid was added to the duodenal bioreactor (MiniBio, Applikon Biotechnology, Delft, The Netherlands), the temperature was maintained at 37 °C, pH was constant at 6.1, and a physiological rate of bile secretion was controlled by a peristaltic pump. The pancreatic enzymes trypsin and α-chymotrypsin were added in two pulses, the first at 10 min and the second at 50 min after the start of the intestinal phase. Total digestion time of the experiment was set at 2 h, and samples were collected from the gastric and duodenal bioreactors at 0, 30, 60, 90 and 120 min. Gastric solutions were neutralized by rapidly raising the pH to 7 with NaOH (1 M), while intestinal solutions were quenched with PMSF (0.5 mM phenylmethylsulfonyl fluoride). Samples were stored at −20 °C until further use.

4.11. Degree of Hydrolysis

The degree of hydrolysis (DH) of the protein was determined by the o-phthaldialdehyde (OPA) method described by Nielsen et al. [31]. The OPA reagent was prepared according to the methodology described by Opazo-Navarrete et al. [32]: 160 mg of OPA were dissolved in 4ml ethanol and then were added to a previously prepared solution containing 7.62 g borax, and 200 mg of SDS dissolved on 150 mL deionized water, mixed well and 176 mg of dithiothreitol (DTT) was added, made up to 200 mL and then stored in an amber bottle for 24 h after preparation. The L-serine standard curve was prepared at concentrations ranging from 50 to 200 mg/mL [31]. The digested samples were centrifuged for 20 min at 14,000× g, and the supernatant was used for measurements. A total of 200 μL of supernatant was mixed with 1.5 mL of OPA reagent for 3 min, after which absorbance was measured at 340 nm with a spectrophotometer (Spectroquant Pharo 300, Merck KGaA, Darmstadt, Germany). Free amino groups in the digested salmon gelatins were expressed as serine amino equivalents (serine NH2). DH was calculated using Equations (6) and (7).

$$DH = \frac{h}{h_{tot}} \times 100\% \quad (6)$$

$$h = \frac{Serine\ NH_2 - \beta}{\alpha} \quad (7)$$

where the values of the constants were α = 0.796 and β = 0.457 [28]. The h_{tot} was determined as a function of the concentration of each amino acid in the protein and had a value of 13.25 milliequivalent/g.

4.12. Fourier Transform Infrared Spectroscopy with Attenuated Total Reflectance

The Fourier transform infrared spectroscopy (FTIR) spectra of in-mold and PSGG and digests were recorded with an IRPrestige-21 spectrometer (Shimadzu 21 Corporation Pte. Ltd., Kyoto, Japan). Approximately 50 μL of the solution was placed on the attenuated total reflectance (ATR) glass and 128 scans were performed in the absorption mode with a 4.0 cm^{-1} resolution at wavelengths from 4000 to 400 cm^{-1}. The amide I region (1600–1700 cm^{-1}) was smoothed with a twelve-point Savitsky–Golay function, and the normalized second derivative spectra were obtained with the IRsolution version 1.10 software (Creon Lab Control AG, Shimadzu Corporation Pte. Ltd., Kyoto, Japan). Gaussian curve-fitting was performed on the different second derivative IR spectra with the Origin version 9.0 PRO software (OriginLab Corporation, Northampton, MA, USA). The content of each of the structures was calculated from the area under the relative peaks of the assigned bands.

4.13. Statistical Analysis

Results were analyzed with the Statgraphics Centurion XVI statistical software (Statistical Graphics Corp., Herdon, VA, USA). Each experiment was performed in triplicate with three samples in each replicate. Fisher's LSD test was used to minimize significant differences with a 95% confidence interval.

Author Contributions: N.C.-M.: formal analysis, methodology, investigation, data curation, and writing—original draft; G.T.-M.: conceptualization, resources, investigation, writing—original draft, writing—review and editing, and project administration; M.D.A.S.: conceptualization, investigation, and writing—review and editing; M.P.-W.: conceptualization and visualization; C.H.-L.: investigation and methodology; R.L.-M.: visualization; L.M.-O.: methodology and visualization. All authors have read and agreed to the published version of the manuscript.

Funding: This research was funded by the ANID, Chile, through the Advanced Human Capital Training Program and National Doctorate Scholarship 2019 (Folio 21190996) and the FONDECYT program (project No. 1201578 and 1231595).

Institutional Review Board Statement: Not applicable.

Informed Consent Statement: Not applicable.

Data Availability Statement: The data presented in this study are available upon request from the authors.

Acknowledgments: The authors gratefully acknowledge the financial support provided by the ANID, Chile, through the Advanced Human Capital Training Program and the National Doctorate Scholarship 2019 (Folio 21190996) and the FONDECYT program (project No. 1201578 and 1231595).

Conflicts of Interest: The authors declare no conflict of interest.

References

1. Bedoya, M.G.; Montoya, D.R.; Tabilo-Munizaga, G.; Pérez-Won, M.; Lemus-Mondaca, R. Promising perspectives on novel protein food sources combining artificial intelligence and 3D food printing for food industry. *Trends Food Sci. Technol.* **2022**, *128*, 38–52. [CrossRef]
2. Mekala, S.; Silva, E.K.; Saldaña, M.D.A. Stabilization mechanism, kinetics and physicochemical properties of ultrasound-produced emulsions stabilized by lentil protein: A non-dairy alternative to stabilize food systems. *Eur. Food Res. Technol.* **2022**, *248*, 185–196. [CrossRef]
3. Carvajal-Mena, N.; Tabilo-Munizaga, G.; Pérez-Won, M.; Lemus-Mondaca, R. Valorization of salmon industry by-products: Evaluation of salmon skin gelatin as a biomaterial suitable for 3D food printing. *LWT* **2022**, *155*, 112931. [CrossRef]
4. Qiu, C.; Wang, C.; Li, X.; Sang, S.; McClements, D.J.; Chen, L.; Long, J.; Jiao, A.; Wang, J.; Jin, J. Preparation of high internal phase Pickering emulsion gels stabilized by glycyrrhizic acid-zein composite nanoparticles: Gelation mechanism and 3D printing performance. *Food Hydrocoll.* **2023**, *135*, 108128. [CrossRef]
5. Wang, D.; Guo, J.; Wang, Y.; Yang, Y.; Jiang, B.; Li, D.; Feng, Z.; Liu, C. Whey protein isolate nanofibrils as emulsifying agent to improve printability of Cheddar cheese for 3D printing. *Food Hydrocoll.* **2023**, *142*, 108807. [CrossRef]
6. Tesikova, K.; Jurkova, L.; Dordevic, S.; Buchtova, H.; Tremlova, B.; Dordevic, D. Acceptability analysis of 3D-printed food in the area of the Czech Republic based on survey. *Foods* **2023**, *11*, 3154. [CrossRef]
7. Lanaro, M.; Forrestal, D.P.; Scheurer, S.; Slinger, D.J.; Liao, S.; Powell, S.K.; Woodruff, M.A. 3D printing complex chocolate objects: Platform design, optimization and evaluation. *J. Food Eng.* **2017**, *215*, 13–22. [CrossRef]
8. Feng, C.; Zhang, M.; Bhandari, B. Materials properties of printable edible inks and printing parameters optimization during 3D printing: A review. *Crit. Rev. Food Sci. Nutr.* **2018**, *59*, 3074–3081. [CrossRef]

9. Zhang, J.Y.; Pandya, J.K.; McClements, D.J.; Lu, J.; Kinchla, A.J. Advancements in 3D food printing: A comprehensive overview of properties and opportunities. *Crit. Rev. Food Sci. Nutr.* **2021**, *62*, 4752–4768. [CrossRef]
10. Li, M.; Yu, L.; Zhao, J.; Zhang, H.; Chen, W.; Zhai, Q.; Tian, F. Role of dietary edible mushrooms in the modulation of gut microbiota. *J. Funct. Foods* **2021**, *83*, 104538. [CrossRef]
11. Dankar, I.; Haddarah, A.; El Omar, F.; Sepulcre, F.; Pujolà, M. Assessing the microstructural and rheological changes induced by food additives on potato puree. *Food Chem.* **2018**, *240*, 304–313. [CrossRef]
12. Ciftci, O.N.; Cahyadi, J.; Guigard, S.E.; Saldaña, M.D.A. Optimization of artemisinin extraction from *Artemisia annua* L. with supercritical carbon dioxide + ethanol using response surface methodology. *Electrophoresis* **2018**, *39*, 1926–1933. [CrossRef]
13. Patruni, K.; Rao, P.S. Viscoelastic behaviour, sensitivity analysis and process optimization of aloe Vera/HM pectin mix gels: An investigation using RSM and ANN and its application to food gel formulation. *LWT* **2023**, *176*, 114564. [CrossRef]
14. Bhagya Raj, G.V.S.; Dash, K.K. Comprehensive study on applications of artificial neural network in food process modeling. *Crit. Rev. Food Sci. Nutr.* **2022**, *62*, 2756–2783. [CrossRef]
15. Yang, G.; Tao, Y.; Wang, P.; Xu, X.; Zhu, X. Optimizing 3D printing of chicken meat by response surface methodology and genetic algorithm: Feasibility study of 3D printed chicken product. *LWT* **2022**, *154*, 112693. [CrossRef]
16. Fan, F.H.; Ma, Q.; Ge, J.; Peng, Q.Y.; Riley, W.W.; Tang, S.Z. Prediction of texture characteristics from extrusion food surface images using a computer vision system and artificial neural networks. *J. Food Eng.* **2013**, *118*, 426–433. [CrossRef]
17. Shihani, N.; Kumbhar, B.K.; Kulshreshtha, M. Modeling of extrusion process using response surface methodology and artificial neural networks. *J. Eng. Sci. Technol.* **2006**, *1*, 31–40.
18. Goh, G.D.; Sing, S.L.; Yeong, W.Y. A review on machine learning in 3D printing: Applications, potential, and challenges. *Artif. Intell. Rev.* **2021**, *54*, 63–94. [CrossRef]
19. Anukiruthika, T.; Moses, J.A.; Anandharamakrishnan, C. 3D printing of egg yolk and white with rice flour blends. *J. Food Eng.* **2020**, *265*, 109691. [CrossRef]
20. Southerland, D.; Walters, P.; Huson, D. Edible 3D Printing. In Proceedings of the Digital Fabrication 2011 Conference, NIP 27, 27th International Conference on Digital Printing Technologies, Minneapolis, MN, USA, 2–6 October 2011.
21. Derossi, A.; Caporizzi, R.; Azzollini, D.; Severini, C. Application of 3D printing for customized food. A case on the development of a fruit-based snack for children. *J. Food Eng.* **2018**, *220*, 65–75. [CrossRef]
22. Habuš, M.; Benković, M.; Iveković, D.; Pavičić, T.V.; Mustač, N.Č.; Voučko, B.; Curic, D.; Novotni, D. Effect of oil content and enzymatic treatment on dough rheology and physical properties of 3D-printed cereal snack. *J. Cereal Sci.* **2022**, *108*, 103559. [CrossRef]
23. Monticeli, F.M.; Neves, R.M.; Ornaghi, H.L., Jr.; Almeida, J.H.S., Jr. Prediction of bending properties for 3D-printed carbon fibre/epoxy composites with several processing parameters using ANN and statistical methods. *Polymers* **2022**, *14*, 3668. [CrossRef]
24. Moradi, M.; Beygi, R.; Amiri, A.; da Silva, L.F.M.; Sharif, S. 3D printing of acrylonitrile butadiene styrene by fused deposition modeling: Artificial neural network and response surface method analyses. *J. Mater. Eng. Perform.* **2022**, *32*, 2016–2028. [CrossRef]
25. Renuka, V.; Ravishankar, C.N.R.; Zynudheen, A.A.; Bindu, J.; Joseph, T.C. Characterization of gelatin obtained from unicorn leatherjacket (Aluterus monoceros) and reef cod (Epinephelus diacanthus) skins. *LWT* **2019**, *116*, 108586. [CrossRef]
26. Krishna, M.; Nindo, C.I.; Min, S.C. Development of fish gelatin edible films using extrusion and compression molding. *J. Food Eng.* **2012**, *108*, 337–344. [CrossRef]
27. Wang, L.; Liang, Q.; Chen, Q.; Xu, J.; Shi, Z.; Wang, Z.; Ma, H. Hydrolysis kinetics and radical-scavenging activity of gelatin under simulated gastrointestinal digestion. *Food Chem.* **2014**, *163*, 1–5. [CrossRef]
28. Zhang, Y.; Zhou, R.; Liu, F.; Ng, T.B. Purification and characterization of a novel protein with activity against non-small-cell lung cancer in vitro and in vivo from the edible mushroom Boletus edulis. *Int. J. Biol. Macromol.* **2021**, *174*, 77–88. [CrossRef]
29. Yang, L.; Yang, M.; Xu, J.; Nie, Y.; Wu, W.; Zhang, T.; Wang, X.; Zhong, J. Structural and emulsion stabilization comparison of four gelatins from two freshwater and two marine fish skins. *Food Chem.* **2022**, *371*, 131129. [CrossRef]
30. Levi, C.S.; Lesmes, U. Bi-compartmental elderly or adult dynamic digestion models applied to interrogate protein digestibility. *Food Funct.* **2014**, *5*, 2402–2409. [CrossRef]
31. Nielsen, P.M.; Petersen, D.; Dambmann, C. Improved method for determining food protein degree of hydrolysis. *J. Food Sci.* **2001**, *5*, 642–646. [CrossRef]
32. Opazo-Navarrete, M.; Altenburg, M.D.; Boom, R.M.; Janssen, A.E. The effect of gel microstructure on simulated gastric digestion of protein gels. *Food Biophys.* **2018**, *13*, 124–138. [CrossRef]

gels

MDPI

Article

Enrichment of Agar Gel with Antioxidant Pectin from Fireweed: Mechanical and Rheological Properties, Simulated Digestibility, and Oral Processing

Sergey Popov *, Vasily Smirnov, Nikita Paderin, Daria Khramova, Elizaveta Chistiakova, Fedor Vityazev and Victoria Golovchenko

Institute of Physiology of Federal Research Centre "Komi Science Centre of the Urals Branch of the Russian Academy of Sciences", 50 Pervomaiskaya Str., 167982 Syktyvkar, Russia
* Correspondence: s.v.popov@inbox.ru; Tel./Fax: +78-21-224-1001

Abstract: The aims of the study were to evaluate the influence of pectin isolated from fireweed (FP) on the mechanical and rheological properties of agar (A) gel, to investigate the release of phenolic compounds (PCs) and pectin from A-FP gels at simulated digestion in vitro, and to evaluate the oral processing and sensory properties of A-FP gels. The hardness of A-FP gels decreased gradually with the increase in the concentration of FP added (0.1, 0.4, and 1.6%). The hardness of A-FP1.6 gel was 41% lower than A gel. Rheological tests found A gel was a strong physical gel (storage modulus (G′) >>loss modulus (G″)), and the addition of FP up to 1.6% did not significantly change its G′. The G″ value decreased in A-FP gels compared to A gel. The release of galacturonic acid (GalA) was 3.4 ± 0.5, 0.5 ± 0.2, 2.4 ± 1.0, and 2.2 ± 0.7 mg/mL after digestion of A-FP1.6 gel in the oral in vivo phase (OP) and subsequent incubation in simulated gastric (SGF), intestinal (SIF), and colonic (SCF) fluids in vitro. The incubation medium after OP, SGF, and SIF digestion of A-FP1.6 contained 24–64 μg GAE/mL of PCs, while SCF contained 144 μg GAE/mL, supposing a predominant release of antioxidant activity from the gel in the colon. Chewing to readiness for swallowing A-FP gel required less time and fewer chews with less activity of the masseter and temporalis muscles. A-FP1.6 gel had a lower likeness score for taste and consistency and a similar score for appearance and aroma when compared with A gel. Thus, A-FP gels were weakened compared to A gel and required less time and muscle activity for oral processing. A-FP gel had antioxidant activity due to the PCs associated with pectin, while A gel had no antioxidant activity.

Keywords: agar; pectin; hydrogel; phenolic compounds; texture; rheology; simulated digestion; electromyography; acceptability

Citation: Popov, S.; Smirnov, V.; Paderin, N.; Khramova, D.; Chistiakova, E.; Vityazev, F.; Golovchenko, V. Enrichment of Agar Gel with Antioxidant Pectin from Fireweed: Mechanical and Rheological Properties, Simulated Digestibility, and Oral Processing. *Gels* **2022**, *8*, 708. https://doi.org/10.3390/gels8110708

Academic Editors: Baskaran Stephen Inbaraj, Kandi Sridhar and Minaxi Sharma

Received: 13 October 2022
Accepted: 1 November 2022
Published: 2 November 2022

1. Introduction

A combination of hydrocolloids is often used to modify rheological characteristics and impart novel mouthfeels to food gels. Agars and pectins are important hydrocolloids obtained by extraction of marine red seaweed and higher terrestrial plants, respectively [1]. Agar is a polysaccharide in which residues of 3,6-anhydro-L-galactose and D-galactose are linked in a linear chain. Upon heating, the galactan chains take a random and rigid coil conformation, form helices, and form a gel network of combined thick bundles upon subsequent cooling [2]. Thermo-reversible agar gelation does not require additional cross-linkers, such as cations, and is convenient for use in the culinary, food, and confectionery industries [3]. The high polymerization of blocks formed by two different polysaccharides, alpha and beta bonds, makes it difficult for agar to be cleaved by carbohydrate enzymes [4]. Agar is a dietary fiber, since the human digestive system does not produce agarase, and agar can be partially fermented and metabolized when entering the large intestine, which is populated by bacteria [5,6]. Therefore, agar has a prebiotic effect, stimulating the growth of Lactobacillus spp. and Bifidobacterium spp. and increasing the production of short-chain fatty acids [7,8]. In addition, agar has a convenient effect as a bulking and laxative agent because of its gelling capacity, and it has been found to possess anti-tumor, anti-oxidative, and hypoglycemic effects [9].

Pectins are complex anionic polysaccharides that can be isolated from various plant materials. Pectins used in the food industry are mainly obtained from citrus peels and apple pomace [10]. The pectin backbone comprises 1,4-linked α-D-galacturonic acid (GalA) residues, with partly methyl- and/or acetyl-esterified carboxyl groups. Side chains composed of neutral monosaccharides (arabinose and galactose residues, predominantly) linked to rhamnose residues inserted in the galacturonic backbone, forming branched areas of the pectin macromolecule [11]. Pectin with low (<50%) and high (>50%) degrees of methyl-esterification are classified as belonging to the low- (LMP) and high-methyl-esterified (HMP) types, respectively [12]. HMP gels at low pH in the presence of sugar form a physical network. Multivalent cross-linking cations are required for gelation of LMP, which can occur over a wide pH range [13]. However, the use of sugar in large quantities may limit the use of pectin gels in a healthy diet, while the use of various cross-linkers may impair the palatability of the pectin gel. The high biological activity makes it promising to include pectin into complex food gels. In particular, pectin has been shown to possess antioxidant activity [14–18]. Which functional groups of the pectin macromolecule mediate the antioxidant activity remains unknown. However, some authors believe that the antioxidant activity is due not to polysaccharide chains but to the phenolic compounds (PCs) associated with them [19].

LMP possessing antioxidant activity higher than commercial apple pectin has previously isolated from leaves of fireweed (Epilobium angustifolium L.) [20]. Fireweed pectin (FP) fractions scavenge 1,1-diphenyl-2-picrylhydrazyl- and superoxide radicals in vitro depending on the content of associated PCs [20]. In the present study, we assumed that incorporating FP into the agar gel would make it possible to obtain a mixed food gel with antioxidant activity. It has previously been shown that incorporating other hydrocolloids into agar gel changes its physicochemical properties [21–23]. Rozek et al. [24] supplemented model agar gel with grape PCs. However, the features of agar-pectin gel and enrichment of agar gel with PCs associated with polysaccharide have not been previously studied.

The aim of the study was to evaluate the influence of FP on the mechanical and rheological properties of agar gel and to investigate the release of PCs and pectin from A-FP enriched gels at simulated digestion in vitro. Electromyography (EMG)-measured oral processing and sensory properties of A-FP gels were also evaluated with the involvement of volunteers.

2. Results

2.1. General Characterization and Mechanical Properties of Gels

Agar gels enriched with antioxidant pectin were prepared by heating/cooling a mixture of 1.5% agar and 0.1, 0.4, or 1.6% FP and named A-FP0.1, A-FP0.4, or A-FP1.6, respectively. Total PC content in A-FP0.1, A-FP0.4 and A-FP1.6 gels was 23 ± 5, 84 ± 17, and 418 ± 101 μg gallic acid equivalents (GAE)/g, respectively (Table 1).

Tg and Tm represent the gelling and melting temperatures, respectively; PCs, phenolic compounds; GAE, gallic acid equivalents; WL, the weight-loss during drying. SD, standard deviation.

The images of 1.5% agar gel (A) and A-FP mixed gels are shown in Figure 1. Evidently, A gel had a light transparent structure in comparison with A-FP gels. The color of the gel became markedly darker depending on the concentration of added FP. The dark color of pectin-containing gels is, apparently, because of the presence of PCs in FP.

Table 1. Characterization of agar gel (A) and agar gels enriched with fireweed pectin (A-FP).

	A	A-FP0.1	A-FP0.4	A-FP1.6
Total PC, μgGAE/g (n = 3)	-	23 ± 5 [a]	84 ± 17 [b]	418 ± 101 [c]
Loading efficiency, % (n = 3)	-	86 ± 18 [a]	78 ± 15 [a]	97 ± 23 [a]
Puncture test:				
Hardness, N (n = 12)	1.57 ± 0.09 [a]	1.43 ± 0.33 [a,b]	1.35 ± 0.26 [b]	0.91 ± 0.14 [c]
Consistency, mJ (n = 12)	3.18 ± 0.10 [a]	2.74 ± 0.55 [b]	2.66 ± 0.45 [b]	1.99 ± 0.29 [c]
pH (n = 3)	6.22 ± 0.17 [a]	5.45 ± 0.07 [b]	5.16 ± 0.03 [c]	4.77 ± 0.02 [d]
WL, % (n = 8)	92.2 ± 0.1 [a]	92.0 ± 0.1 [b]	91.6 ± 0.1 [c]	90.3 ± 0.2 [d]
Serum release, % (n = 12)	0.68 ± 0.08 [a]	0.67 ± 0.27 [a]	0.46 ± 0.19 [b]	0.33 ± 0.09 [c]
Tg, °C (n = 8)	39.8 ± 1.2 [a]	40.6 ± 1.6 [a]	40.1 ± 0.9 [a]	41.0 ± 0.8 [b]
Tm, °C (n = 8)	69.4 ± 4.3 [a]	69.4 ± 4.4 [a]	70.2 ± 2.6 [a]	71.0 ± 2.6 [a]

Mean ± SD. Different letters among gels—$p < 0.05$.

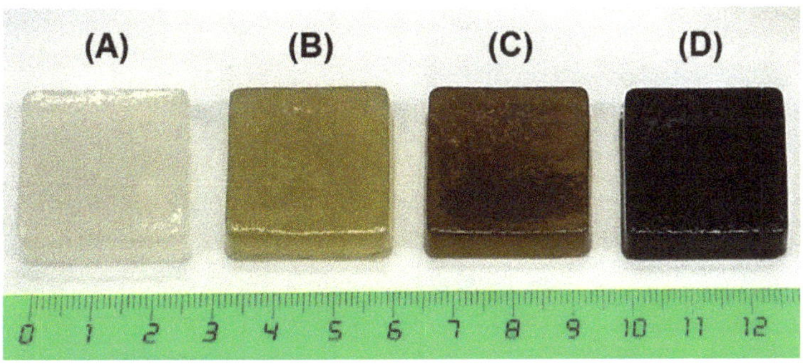

Figure 1. Appearance of A (**A**), A-FP0.1 (**B**), A-FP0.4 (**C**), and A-FP1.6 (**D**) gels.

The mechanical behavior of A and A-FP gels was first monitored in the puncture test, and all gels demonstrated force–distance curves with one peak (Figure 2A). Hardness, determined as a peak force through the puncture of A-FP0.1, A-FP0.4, and A-FP1.6 gels was 9, 14, and 44% lower than that of A gel, respectively (Table 1). The consistency of A-FP0.1, A-FP0.4, and A-FP1.6 gel was 13, 16, and 37% lower than that of A gel, respectively.

The texture profile analysis (TPA) results are presented in Figure 2B and Table 2. FP decreased the hardness of A gel. The hardness of A-FP1.6 gel was 41% lower than A gel. Similar data was obtained for cohesiveness and chewiness. The cohesiveness of A-FP0.1, A-FP0.4, and A-FP1.6 gel was 1.3, 1.6, and 2.2 times lower than that of A gel, respectively. The chewiness of A-FP0.4 and A-FP1.6 gels was 1.5 and 3 times lower, respectively, than that of A gel. However, A-FP1.6 gel exhibited the highest levels of springiness and resilience. The springiness and resilience of A-FP1.6 gel was 1.2 and 2.8-fold of A gel sample, respectively.

Table 2. The texture profile analysis (TPA) properties of A and A-FP enriched gels.

	A	A-FP0.1	A-FP0.4	A-FP1.6
Hardness, N	12.3 ± 0.3 [a]	10.7 ± 2.8 [a,b]	10.1 ± 2.0 [b]	7.3 ± 1.0 [c]
Cohesiveness	0.43 ± 0.02 [a]	0.33 ± 0.14 [b]	0.27 ± 0.12 [b]	0.20 ± 0.02 [c]
Chewiness	4.5 ± 0.4 [a]	3.5 ± 2.2 [a,b]	3.1 ± 1.8 [b]	1.5 ± 0.3 [c]
Springiness	0.83 ± 0.01 [a]	0.90 ± 0.06 [b]	0.91 ± 0.06 [b]	1.03 ± 0.5 [c]
Resilience	0.16 ± 0.02 [a]	0.27 ± 0.11 [b]	0.29 ± 0.12 [b]	0.45 ± 0.05 [c]

Mean ± SD. Different letters among gels—$p < 0.05$ (n = 12).

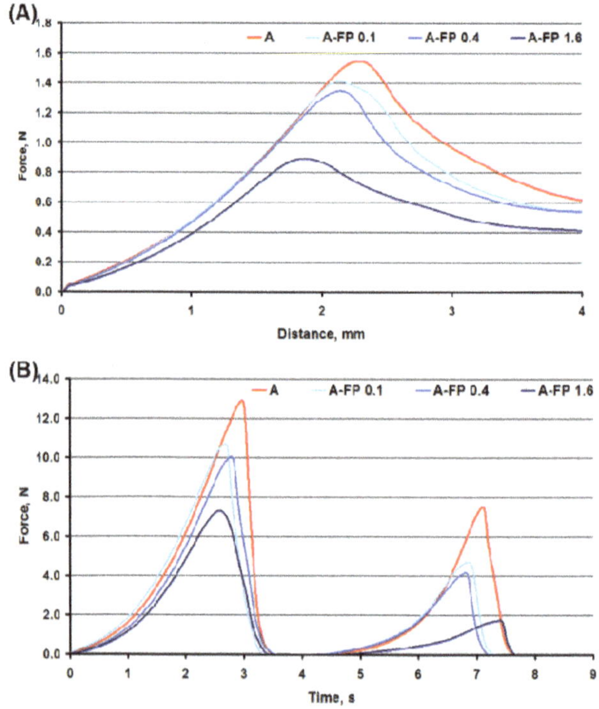

Figure 2. Mean (n = 12) force over distance (**A**) and over time (**B**) curves in puncture test and double compression tests, respectively, of A and A-FP gels.

2.2. Rheological Properties of Gels

The viscoelastic properties of A and A-FP gels were studied using oscillatory assessments with recording of the parameters of storage or elastic modulus (G′) and the loss or viscous parameters modulus (G″). Angular frequency sweep measurements are conducted to estimate the structural integrity and gel strength. The mechanical spectra are given in Figure 3A.

Both A and A-FP gels demonstrated high G′ values (10,000 > G′ > 8500 Pa) when the modulus had only a slight dependence on frequency. The loss factor tan δ << 1 confirmed the solid state of A and A-FP gels (Figure 3B). The same results were obtained for the A-FP0.1 gel (data not shown). Thus, A gel was a strong physical gel, and the addition of FP up to 1.6% did not significantly change its G′, since the G′ values for all studied gels were the same (Figure 3A). The G″ value decreased in A-FP gels compared to A gel (Figure 3A), coinciding with the viscosity values given in Table 3.

Table 3. Summary of power law parameters (0.01 < ω < 20.00 or at 0.1/10.5 Hz).

Gels	Storage Modulus			Viscosity	
	A (Pa)	B (Slope)	R^2	η_{app} 0.045, Hz	η_{app} 10.500, Hz
A	8550.0	0.054	0.951	12939.6 ± 3048.5 [a]	152.1 ± 21.9 [a]
A-FP0.1	9110.6	0.032	0.998	14143.6 ± 3235.0 [a]	149.8 ± 29.4 [a]
A-FP0.4	9478.9	0.037	0.996	14574.9 ± 3323.2 [a]	149.2 ± 29.3 [a]
A-FP1.6	7646.3	0.047	0.997	11439.0 ± 1918.9 [a]	129.3 ± 18.7 [b]

Mean ± SD. Different letters among gels—p < 0.05 (n = 8).

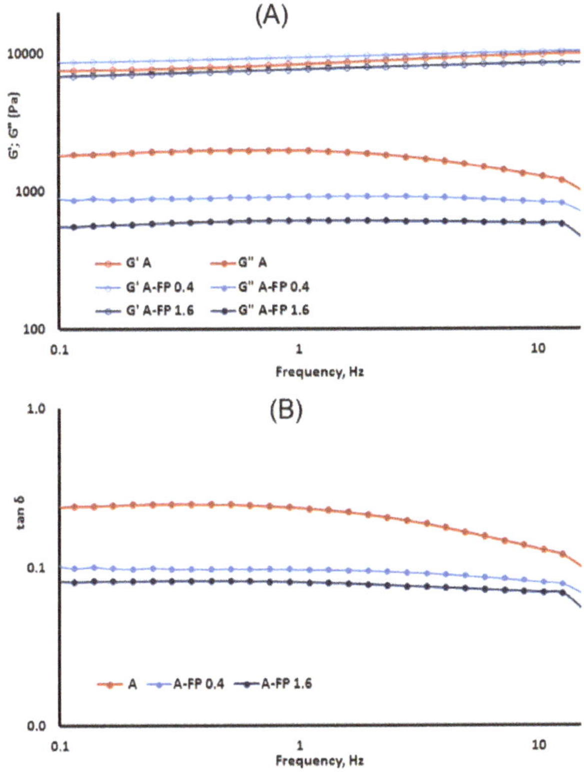

Figure 3. Storage modulus (G′, void symbols) and loss modulus (G″, filled symbols) test (**A**) and tan δ test results (**B**) represented as a function of frequency.

The temperature sweep experiments showed that the G′ values of A and A-FP gels first decreased upon heating to 69–71 °C, and then increased (Figure 4). FP failed to significantly affect the melting temperature (Tm) of agar gel (Table 1). The G′ values of A-FP gels exceeded that of A gel during the isothermal (95 °C) step. All samples demonstrated sharply increased G′ values upon cooling below 41 °C, indicating a sol–gel transition. The gelling temperature (Tg) of A-FP1.6 was slightly (0.2 °C) higher than that of A gel, as presented in Table 1. The G″ values of A gel were higher than A-FP gel at the melting step (Figure 4). However, after the phase transition, the G″ values of A sol were lower than A-FP sol during the isothermal step.

2.3. Pectin and PCs Release during Simulated Digestion of Gels

The incubation medium after each phase of simulated digestion of A-FP gels contained polysaccharide fractions. The molecular weight of polysaccharides was 225 ± 16, 154 ± 30, 216 ± 11, and 119 ± 24 kDa after successive oral in vivo (OP) and gastrointestinal in vitro digestion of A-FP1.6 in simulated gastric (SGF), intestinal (SIF), and colonic (SCF) fluids, respectively. The concentration of GalA in the incubation medium was measured to determine the release of pectin chains from the agar gel upon digestion since GalA is a major component of FP. The amount of released GalA increased in proportion to the increase in the amount of incorporated pectin at each individual phase of digestion (Figure 5A). When comparing different phases of digestion, A-FP gel released the highest level of GalA at the chewing in vivo (OP). The release of GalA from A-FP0.1, A-FP0.4 and A-FP1.6 into SGF was minimal, namely 3.0, 4.4 and 6.9 times less, respectively, than during OP. The release of GalA in SIF and SCF then increased significantly compared to SGF, but was less than during OP (Figure 5A). The incubation medium of the digestion of A gel did not contain high molecular polysaccharides or GalA (data not shown).

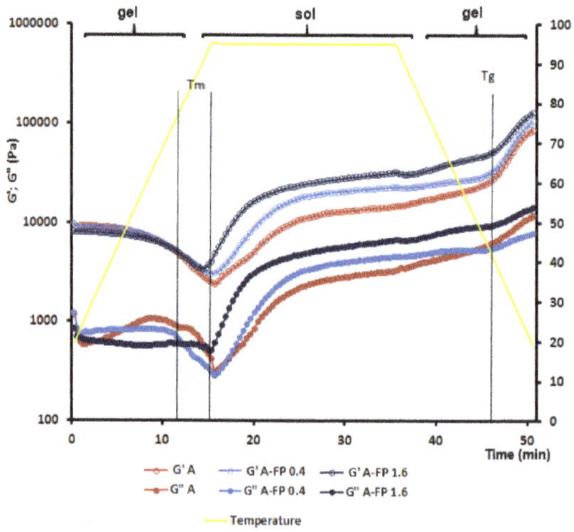

Figure 4. Temperature ramp tests performed at constant frequency (1.0 Hz) for A, A-FP0.4, and A-FP1.6 gels. An initial heating step (5 °C/min from 20 to 95 °C) was followed by an isothermal step (95 °C, 20 min) and a final cooling step (rate: 5 °C/min from 95 to 20 °C).

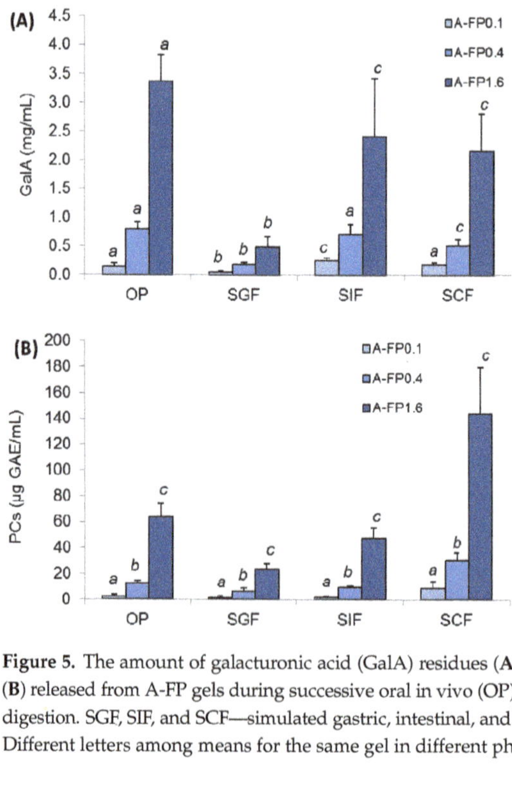

Figure 5. The amount of galacturonic acid (GalA) residues (**A**) and total phenolic compounds (PCs) (**B**) released from A-FP gels during successive oral in vivo (OP) and gastrointestinal in vitro phases of digestion. SGF, SIF, and SCF—simulated gastric, intestinal, and colonic fluid, respectively. Mean ± SD. Different letters among means for the same gel in different phases of digestion—$p < 0.05$ ($n = 6$).

PCs released from A-FP gels were evaluated during successive OP and SGF, SIF, and SCF digestion to predict the part of the gastrointestinal tract where FP would have an antioxidant effect. The incubation medium after OP, SGF, and SIF digestion of A-FP1.6 contained 24–64 μg GAE/mL of PCs, while SCF contained 144 μg GAE/mL Figure 5B.

2.4. Oral Processing of Gels

Oral processing of A and A-FP gels was investigated using EMG with a unilateral chewing style. All participants freely chewed gel samples on the preferred chewing side of the jaw (left or right) in their habitual manner and did not report any difficulty in performing the task. Typical EMG signals from rhythmic masseter and temporalis activities during chewing of A, A-FP0.4, and A-FP1.6 gels are shown in Figure 6. Oral processing of A-FP0.1 gel was not studied because of the low amount of FP incorporated and released.

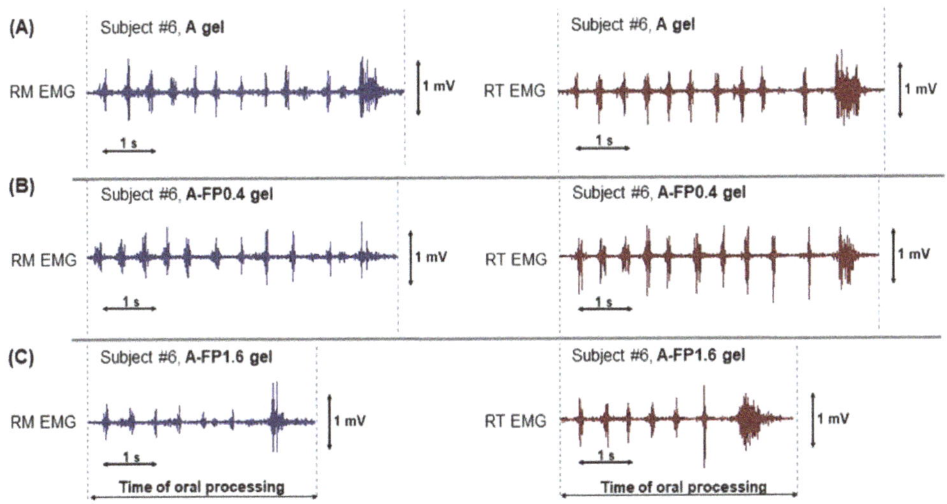

Figure 6. Example of electromyography (EMG) during chewing A (**A**), A-FP0.4 (**B**), and A-FP1.6 (**C**) gels in a representative subject. EMG signals of right masseter (RM) and right temporalis (RT) muscles are recorded. Area between the dotted lines indicates the time of oral processing.

Chewing to readiness for swallowing A-FP1.6 gel required 30% less time and 27% fewer chews (Figure 6C, Table 4). Consistent with this, masseter and temporalis activity were approximately 30% lower when chewing A-FP1.6 then when chewing A gel. A-FP0.4 failed to change the chewing pattern compared to A gel (Table 4). Despite the shortening of the chewing time over the sequence, the duration of the chewing cycle failed to change significantly. The chewing cycle lasted, on average, 700 ms for both A and A-FP gels (data not shown). The chewing frequency was similar in all treatments and was equal to 1.5 s^{-1} (data not shown).

Table 4. Oral processing parameters for chewing of A and A-FP gels.

	A	A-FP0.4	A-FP1.6
Chewing time, s	13.2 ± 8.2^{a}	12.6 ± 8.7^{a}	9.3 ± 5.1^{b}
Number of chews, times	18.3 ± 6.9^{a}	18.7 ± 12.0^{a}	13.3 ± 5.9^{b}
RM activity per sequence, mV × s	0.26 ± 0.08^{a}	0.25 ± 0.14^{a}	0.18 ± 0.93^{b}
RT activity per sequence, mV × s	0.37 ± 0.20^{a}	0.35 ± 0.20^{a}	0.19 ± 0.12^{b}
Total muscle activities per sequence, mV × s	0.63 ± 0.26^{a}	0.59 ± 0.29^{a}	0.40 ± 0.21^{b}
Total muscle activities per chew, mV × s	0.04 ± 0.01^{a}	0.03 ± 0.02^{a}	0.03 ± 0.01^{b}

Mean ± SD. Different letters among gels—$p < 0.05$ (n = 16). RM and RT—right masseter and temporalis muscles.

2.5. Acceptability of Gels

Likeness-scores of A, A-FP0.4, and A-FP1.6 gel samples are displayed in Figure 7. A-FP1.6 gel had a lower likeness-scores for taste and consistency and similar scores for appearance and aroma when compared with A gel. The acceptability of A-FP0.4 gel did not differ from that of A gel.

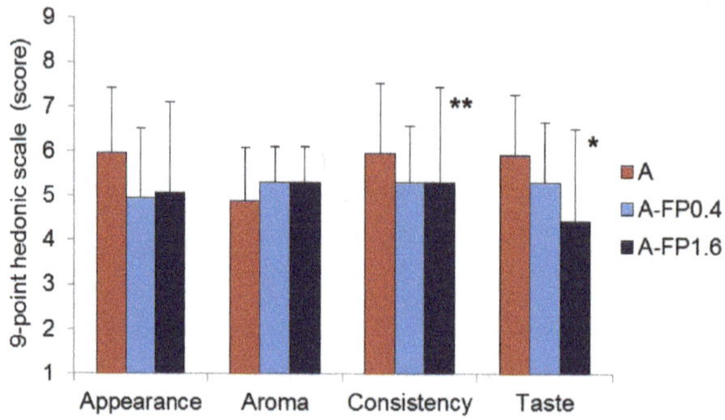

Figure 7. Sensory scores of A, A-FP0.4, and A-FP1.6 gels. Data are presented as the mean rating (n = 16). * and ** $p < 0.05$ and 0.01 vs. A gel.

3. Discussion

The results demonstrated that FP was efficiently incorporated into A gel by the method used. The efficiency of loading of pectin-associated PCs into agar gel was calculated to be equal to 78–97% as the total phenolic content in FP was earlier determined to be as much as 27 mg GAE/g [20].

The mechanical behavior observed under puncture and double compression showed that the structure of A-FP gels was weakened compared to A gel. Hydrogen bonding is well known to be involved in agarose gelation [3]. The decrease in hardness, which reflects the maximum load applied to the specimens, the consistency, which represents the required work of deformation, and the cohesiveness, which reflects the stability of the internal structure, suggests that pectin molecules interfere with the formation of hydrogen bonds between agar molecules (Figure 8). The formation of hydrogen bonds with the hydroxyl groups of pectin can decrease the density of hydrogen bonds between the adjacent D-galactose and 3,6-anhydro-L-galactose residues of the agarose chains.

The decrease in the mechanical strength of A-FP gels may also result from the decrease in pH caused by the addition of FP to agar gel. The decrease in A gel strength by FP is consistent with a previous report [25], where the authors imply carrot juice is able to weaken agar gel because of pectin content. Incorporating other polysaccharides may reduce the mechanical properties of the agar gel system. Locust bean gum (LBG), a galactomannan from the seeds of the carob tree, and salep polysaccharide, a glucomannan from the roots and tubers of orchids, have been shown to reduce strength and elongation of agar film [21]. In another study [22], the addition of a combination of LBG and xanthan gum reduced the breaking stress of the agar gel membrane. However, other studies have reported that LBG may increase the strength of agar gels [23]. According to a rheological study, FP seemed to dissolve in the aqueous phase of rigid A gel since A and A-FP gels are not significantly different in G′ value. This result was expected, as FP, being an LM pectin, needs divalent cations to form a gel network. The addition of FP reduced the G″ value of A gel, which is similar to the effect of konjac gum on agar gel [26].

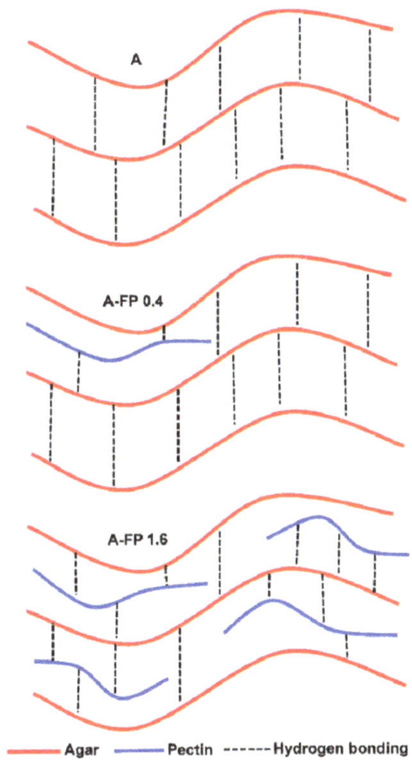

Figure 8. Schematic illustration of a proposed gel network in A and A-FP gels.

The sequential digestion model was used to predict the bioactivity of the A-FP gel in different parts of the digestive tract. It is known that the ability of pectins to bind glucose and lipids in the small intestine determines their hypoglycemic and hypolipidemic action [27,28], whereas in the large intestine, pectins stimulate beneficial microflora [29]. Chewing released a large portion of the pectin from the A-FP gel during its destruction in the oral cavity. However, the pectin is then released predominantly in SIF and SCF, thus showing a great potential for bioactivity. The minimal release of pectin in the SGF is likely because of the low pH, which promotes an excessive aggregation of polysaccharide helices. The measurement of the reactivity of incubation medium with the Folin–Ciocalteu's reagent is related to the ability to donate electrons, and therefore indicates antioxidant activity. The preferential release of PCs in the SCF indicates that digesting A-FP gel will keep its antioxidant potential until the colonic environment.

The textural and rheological properties of food gels determine their mechanical behavior during mastification, including chewing, bolus formation, and swallowing [30]. Furthermore, hardness, cohesiveness, chewiness, and springiness are critical parameters for the sensory perception of food. Hardness is considered an important indicator for the consumer, which can simulate the force required for food to compress between teeth or between tongue and palate. Cohesiveness is the extent to which a material can deform before it breaks, simulating the compression of food before teeth chew it in the mouth. Chewiness refers to the energy required for chewing solid food in a swallowable state [31]. The EMG results of the chewing of A-FP gel are consistent with the data of texture analysis. Softening A gel with FP resulted in chewing it easier has made chewing it easier. The results of EMG measuring the chewing of A-FP gel are consistent with the data of texture analysis. Softening A gel with FP made it easier to chew. Various studies have demonstrated that consumers adapt their chewing pattern to the textural properties of foods [30,32]. The observation that the harder samples required increased muscle activity for the oral processing has been reported for agar caramels [33], gellan [34] and gelatin [35] model gels.

A weaker and faster chewable A-FP gel may be required by children who prefer soft products [36], the elderly who have difficulty chewing hard foods [37], and patients of dysphagia [38]. However, the softening of A gel because of the addition of FP lowered its acceptability. In disagreement with these data, [39] and [40] reported that softer agar jelly or yogurt, respectively, with softer agar particles have higher liking scores than harder samples. In addition, overall acceptability of pectin-containing jams is higher in softer samples [41,42]. It is assumed that decreasing hardness enhances the flavor intensity, which could lead to an increase in liking of samples. In the present study, the reduced weight-loss during drying (WL) and lower serum release from A-FP1.6 gel may explain its poorer acceptability properties, because the separation of liquid from the gel network affects the perceived texture and taste [43]. It should also be taken into account that a shorter chewing time of A-FP gel can be associated with fewer signals from the oral cavity reaching the brain. In addition, shorter chewing may decrease saliva production which is important for taste perception [32]. However, saliva production was not determined in our study. In any case, the lower taste and consistency ratings of A-FP gel require further improvement in gel formulation to improve its acceptability for consumers.

4. Materials and Methods

4.1. Materials

Agar (A) powder was purchased from Zhenpai Hydrocolloids Co., Ltd. (Zhangzhou City, Fujian Province, China), with moisture 7.97 wt%, pH 6.46, and 1.5% gel strength of 1150 g/cm^2. FP was isolated from the leaves of *E. angustifolium* L. by the treatment of plant raw materials with aqueous hydrochloric acid at pH 0.8 [20]. The PC content in FP was 27 mg GAE/g. The chemical characteristics of the polysaccharides are shown in Table 5.

Table 5. Chemical characteristics of polysaccharides used.

Sample	UA [a]	Gal [a]	Xyl [a]	Glc [a]	Rha [a]	Ara [a]	M_w, kDa
A	3.9	20.0	3.0	0.9	0.2	0.7	280
FP	66.1	1.7	0.9	0.5	1.2	0.9	232

[a] Data were calculated as wt%. UA—uronic acids, Gal—galactose, Xyl—xylose, Glc—glucose, Rha—rhamnose, Ara—arabinose.

The reagents, including the Folin and Ciocalteu's phenol reagent and pectinase from *Aspergillus niger* (1.18 U/mg), were purchased from Sigma-Aldrich (St. Louis, MO, USA). Gallic acid was purchased from MP Biomedicals (Solon, OH, USA). Ethanol (95%, JSC Kirov Pharmaceutical Factory, Kirov, Russia), pyridine, boric acid, toluene and chloroform stabilized with ethyl alcohol (Vekton, Saint Petersburg, Russia), methanol (Merck, München, Germany, 99.5%), 3,5-dimethylphenol (Sigma Aldrich, USA, 99%), sodium borohydride (Sigma Aldrich, USA, 98.5%), Sulphuric acid (Sigma-tec, Moscow, Russia) were used.

4.2. Preparation and General Characterization of A and A-FP Gels

A (1.5 g), FP (0.1, 0.4, or 1.6 g), and sugar (5.0 g) were dispersed in deionized water (up to 100 g) and left for one hour under magnetic stirring. The dispersion was heated to 95 °C in a slow cooker for 45 min. The hot solution was vigorously shaken and transferred into a silicone form (10 × 26 × 26 mm), followed by cooling to room temperature for 2 h. The gels were covered with a film and kept at 4 °C before being used for further experiments.

The pH was determined using an InLab® Science Pro-ISM pH electrode and an S20 SevenEasy™ pH meter (Mettler-Toledo AG, Schwerzenbach, Switzerland) on homogenates of A and A-FP gels. Gel was weighed and mechanically broken in distilled water in a ratio of 1:10 (*w*/*v*).

WL was determined as a measure of the water-holding capacity of the gels. The weight of 12 gel samples batch before and after complete removal of moisture by drying samples was determined by AG 245 weight (Mettler Toledo International) WL was calculated as:

$$WL\ \% = (WW - WD)/WW \times 100\ \%, \quad (1)$$

where WW and WD represent the weight of the gel samples before and after drying, respectively.

Serum release was determined using the TA-XT Plus Texture Analyzer (Texture Technologies Corp., Stable Micro Systems, Godalming, UK). A pre-weighed dry filter paper was placed under the gel block to absorb the serum being exuded from the gel. The gels were compressed twice to 80% of initials their height at room temperature. The pre- and post-test speed was 5.0 mm/s and the test speed was 1 mm/s. The wet filter paper was weighed immediately after the compression. Serum release was calculated by the equation: (weight of the released serum)/(initial weight of gel). Twelve replicates were made of each type of gel.

For measurement of PC content, the gel sample (~4 g) was weighed and mechanically broken. After adding water in a ratio of 1:4 (*w*/*v*) the mix was heated in a water bath for 10 min and cooled to room temperature. The content of PCs was determined in the resulting solution with the Folin–Ciocalteu reagent using gallic acid as a standard [44]. The results were expressed as μg of gallic acid equivalents (GAE) per mL of solution or as a percentage of the total amount of released PCs and remaining in the residue.

The PC loading was calculated using the following equation:

$$\text{Loading PCs } (\mu\text{g GAE/g gel}) = (A/B) \times 100, \quad (2)$$

where A and B are the PC content (μg GAE/mL) and the concentration of gel in solution (g/mL), respectively. The encapsulate efficiency was calculated using the following equation:

$$\text{Loading efficiency } (\%) = (A/B) \times 100, \quad (3)$$

where A and B are the measured and theoretical amounts of PCs (in μg GAE/g gel), respectively. The theoretical amounts of PCs in A-FP gel were calculated according to the total PC content in FP.

4.3. Measurement of Mechanical Properties

For the puncture test, gel samples (9 mm height, 14 mm length and 14 mm width) were placed on the platform of the texture analyzer (Texture Technologies Corp., Stable Micro Systems, Godalming, UK). The test was performed using a cylindrical aluminum probe P/5 (5 mm in diameter). The method settings, including the pretest, test, and post-test speeds, were 5.0, 1.0, and 5.0 mm/s, respectively. The distance (depth of insertion) was set at 4 mm of the initial height of the samples. Several mechanical parameters were extrapolated from the force–distance curves: puncture hardness (the peak force that occurs during the puncture), elasticity (distance at the peak force), and consistency (the area of work during the puncture). The test was performed at room temperature. Each analysis was executed 12 times.

For the two-cycle compression test, gel samples (9 mm height, 14 mm length and 14 mm width) were placed on the platform of the texture analyzer (Texture Technologies Corp., Stable Micro Systems, Godalming, UK). The test was performed using a cylindrical aluminum probe P/25 (25 mm in diameter). The gels were compressed twice at room temperature. The pre- and post-test speed was 5.0 mm/s, and the test speed was 1 mm/s until a 100% strain. A destructive 100% strain was used to represent gels behavior during the chewing process. Twelve replicates were made of each type of gel. For two cycles, compression–decompression provided a force–time graph and led to the extraction of eight parameters: hardness, cohesiveness, springiness, gumminess, chewiness, and resilience.

All calculations were performed using Texture Exponent 6.1.4.0 software (Stable Micro Systems, UK) with manufacturer recommendations.

4.4. Measurement of Rheological Properties

The rheological properties of the samples were determined in a rotational-type rheometer (Anton Paar, Physica MCR 302, Graz, Austria) equipped with a parallel plate geometry (25 mm in diameter) and a gap of 1–4 mm between the two plates. Four repetitions were performed for each sample.

Temperature sweeps were carried out from 95 to 5 °C at a rate of 5.0 °C/min. The storage modulus (G′) and loss modulus (G″) values were recorded by a temperature sweep test at a constant stress of 9.0 Pa and an angular frequency of 1.0 Hz (linear viscoelastic region). Tg was determined as the temperature when G′ and G″ crossed over [45,46].

The frequency sweeps were expressed in terms of G′ and G″ through an angular frequency range of 0.01–10 Hz. The tests were performed at 20 °C and a constant stress of 9.0 Pa, which was within the linear viscoelastic region [47]. The degree of frequency dependence for the G′ was determined using the power-law parameters [48,49], expressed as:

$$G' = A \times \omega \times B \quad (4)$$

where G′ is the storage modulus, ω is the oscillation frequency (Hz), and A is a constant.

4.5. In Vivo Oral Phase (OP) and Static In Vitro Gastrointestinal Digestion

OP digestion was performed as proposed previously [43]. Each type of gel (~4 g) was chewed 20 times by six healthy people (3 men and 3 women) to simulate the maximum destruction of the sample in the oral cavity according to preliminary testing in which the largest number of chews was 20. Thereafter, the bolus was spat three times into a beaker to reduce sample loss. Immediately after this, 4.0 mL of water was added to the beaker, mixed by shaking, and all the liquid part was separated for analysis. The gel pieces were transferred to a 20 mL jacketed glass vessel (reactor) for further in vitro digestion.

In vitro gastrointestinal digestion was approached by the method [50] with minor modifications. After OP, gels were sequentially incubated in 4.0 mL of the pre-heated SGF (pH 1.5, 0.08 M HCl, and 0.03 M NaCl), SIF (pH 6.8, 0.05 M KH_2PO_4, and 0.02 M NaOH), and SCF (0.01 M KH_2PO_4, 0.05 M $NaHPO_4$, and pectinase: 1.7 mg/mL) at 37 °C and under continuous orbital shaking (250 rpm) for 2, 4, and 18 h, respectively. Before replacing it with another, the medium was completely separated from the gel by a grid (mesh size 350 µm) and used for analysis. After incubation in SCF, the gel residue was destroyed by heating (95 °C, 10 min) in water (16 mL). The data from SCF and gel residue were summed.

The contents of GalA and PCs were determined in fluid after each (OP, SGF, SIF, and SCF) phase of digestion. For this, aliquots (1–2 mL) of incubation medium were taken and centrifuged, and the resulting supernatant was precipitated with a fourfold volume of 96% ethanol. The precipitate was washed twice with 96% ethanol and dissolved with 3 mL of H_2O. The resulting solution was used to determine the molecular weight of soluble polysaccharides and the content of GalA by reaction of the sample with 3,5-dimethylphenol in the presence of concentrated H_2SO_4 [51]. The alcohol supernatant was used to determine the total amount of sugars using the phenol-sulfur method. The content of PCs was determined in the incubation medium as described above.

4.6. EMG and Acceptability Test

Sixteen volunteers (8 men and 8 women) with an average age of 36.5 participated in the study. All of the volunteers were free of masticatory or swallowing dysfunctions and did not have dentures. The experiment was conducted individually. Participants took approximately five minutes to accommodate the experimental environment. Each participant sat on a chair in a vertical but comfortable position with his head in a natural orientation. Participants were asked to rinse their mouths carefully with water between testing gel samples. The subjects were allowed to talk, rinse their mouths, and drink water freely between each trial. Two sessions were held per subject: (1) In the first session, participants scored likeness for A, A-FP0.4, and A-FP1.6 gel samples in an acceptability test (see below). (2) Deliberate unilateral chewing of the gels on the preferred chewing side with EMG recording conducted as reported previously [52,53].

Acceptability was evaluated by the same panelists using a 9-point hedonic scale, where 1 and 9 points are "extremely disliked" and "extremely liked", respectively [54]. The four samples: A gel (a training sample; it was not taken into account), A, A-FP0.4, and A-FP1.6 gel were presented subsequently. Panelists tasted samples in individual booths under standard light exposure and temperature (20 °C).

4.7. Statistics

The significance of the differences among the means in the content of PCs, mechanical parameters, and digestion studies was estimated with the one-way analysis of variance and Fisher's least significant difference post hoc test. A pared t-test for two dependent means was applied to determine statistically significant differences in EMG variables for different gels. Results from the liking test were analyzed using the Wilcoxon signed-rank test. Statistical differences with a p-value lower than 0.05 were considered significant.

5. Conclusions

In this study, FP possessing antioxidant activity was efficiently incorporated into the agar gel. The gel structure of A-FP gels was weakened compared to A gel. FP, including polysaccharide chains and their associated PCs, was released from the agar gel after sequential oral in vivo and in vitro gastrointestinal digestion. Here, PCs are released mainly into the simulated colonic fluid. Therefore, A-FP gel can be considered as a promising food gel for enhancing antioxidant protection in the colon. Chewing to readiness for swallowing A-FP gel required less time and fewer chews with less masseter and temporalis activity. Therefore, A-FP gel could be recommended for children, the elderly, patients with dysphagia, and other people who have difficulty chewing hard foods. The A-FP1.6 gel is the softest and carries the highest antioxidant potential because of PCs. However, its lower taste and consistency ratings require further improvement in gel formulation to increase its acceptability for consumers.

Thus, the advantages of agar gel enriched with FP compared to a single agar gel are demonstrated, since pectin provides its biological activity, and agar provides the necessary structural and mechanical properties. Incorporation of pectin into a gelling material, such as agar, can be a promising approach to develop a new food gel with antioxidant properties.

Author Contributions: Conceptualization and funding acquisition, S.P.; simulated digestion, sensory tests, V.S.; texture analysis, N.P.; EMG study, D.K.; phenolics measurement, E.C.; rheological tests, F.V.; pectin isolation, V.G. All authors have read and agreed to the published version of the manuscript.

Funding: This research did not receive any specific grant from funding agencies in the public, commercial, or not-for-profit sectors.

Institutional Review Board Statement: The research was carried out within the framework of the research topic of the Institute of Physiology of Komi Science Centre of The Urals Branch of the Russian Academy of Sciences FUUU-2022-0066 (1021051201895-9). The study was conducted according to the guidelines of the Declaration of Helsinki, and approved by the Ethical Committee of the Komi Science Centre of the Russian Academy of Sciences (date of approval 10 March 2022).

Informed Consent Statement: Informed consent was obtained from all subjects involved in the study.

Data Availability Statement: The data that support the findings of this study are available from the corresponding author upon reasonable request.

Conflicts of Interest: The authors declare no conflict of interest.

References

1. Manzoor, M.; Singh, J.; Bandral, J.D.; Gani, A.; Shams, R. Food hydrocolloids: Functional, nutraceutical and novel applications for delivery of bioactive compounds. *Int. J. Biol. Macromol.* **2020**, *165*, 554–567. [CrossRef] [PubMed]
2. Lomartire, S.; Gonçalves, A.M.M. Novel Technologies for Seaweed Polysaccharides Extraction and Their Use in Food with Therapeutically Applications—A Review. *Foods* **2022**, *11*, 2654. [CrossRef] [PubMed]
3. Stanley, N.F. Agars. In *Food Polysaccharides and Their Applications*; Stephen, A.M., Phillips, G.J., Williams, P.A., Eds.; Taylor & Francis Group, LLC: Boca Raton, FL, USA, 2006; pp. 217–238.
4. Hehemann, J.-H.; Smyth, L.; Yadav, A.; Vocadlo, D.J.; Boraston, A.B. Analysis of keystone enzyme in agar hydrolysis provides insight into the degradation of a polysaccharide from red seaweeds. *J. Biol. Chem.* **2012**, *287*, 13985–13995. [CrossRef]
5. Shang, Q.; Jiang, H.; Cai, C.; Hao, J.; Li, G.; Yu, G. Gut microbiota fermentation of marine polysaccharides and its effects on intestinal ecology: An overview. *Carbohydr. Polym.* **2018**, *179*, 173–185. [CrossRef] [PubMed]
6. Armisen, R.; Galatas, F. Agar. In *Handbook of Hydrocolloids*; Phillips, G.O., Williams, P.A., Eds.; Taylor & Francis Group, LLC: Boca Raton, FL, USA, 2009; pp. 82–107.
7. Hu, B.; Gong, Q.; Wang, Y.; Ma, Y.; Li, J.; Yu, W. Prebiotic effects of neoagaro-oligosaccharides prepared by enzymatic hydrolysis of agarose. *Anaerobe* **2006**, *12*, 260–266. [CrossRef] [PubMed]
8. Ramnani, P.; Chitarrari, R.; Tuohy, K.; Grant, J.; Hotchkiss, S.; Philp, K.; Campbell, R.; Gill, C.; Rowlan, I. In vitro fermentation and prebiotic potential of novel low molecular weight polysaccharides derived from agar and alginate seaweeds. *Anaerobe* **2012**, *18*, 1–6. [CrossRef] [PubMed]
9. Holdt, S.L.; Kraan, S. Bioactive Compounds in Seaweed: Functional Food Applications and Legislation. *J. Appl. Phycol.* **2011**, *23*, 543–597. [CrossRef]
10. Yang, Y.; Anderson, C.T. Biosynthesis, localization, and function of pectins in plants. In *Pectin: Technological and Physiological Properties*; Springer: Cham, Switzerland, 2020; pp. 1–15.
11. Ropartz, D.; Ralet, M.-C. Pectin structure. In *Pectin: Technological and Physiological Properties*; Kontogiorgos, V., Ed.; Springer International Publishing: Cham, Switzerland, 2020; pp. 17–36. [CrossRef]
12. Cao, L.; Lu, W.; Mata, A.; Nishinari, K.; Fang, Y. Egg-box model-based gelation of alginate and pectin: A review. *Carbohydr. Polym.* **2020**, *242*, 116389. [CrossRef]
13. Chan, S.Y.; Choo, W.S.; Young, D.J.; Loh, X.J. Pectin as a rheology modifier: Origin, structure, commercial production and rheology. *Carbohydr. Polym.* **2017**, *161*, 118–139. [CrossRef]
14. Tan, M.; Chang, S.; Liu, J.; Li, H.; Xu, P.; Wang, P.; Wang, X.; Zhao, M.; Zhao, B.; Wang, L.; et al. Physicochemical Properties, Antioxidant and Antidiabetic Activities of Polysaccharides from Quinoa (*Chenopodium quinoa* Willd.) Seeds. *Molecules* **2020**, *25*, 3840. [CrossRef]
15. Wu, D.-T.; Liu, W.; Han, Q.-H.; Wang, P.; Xiang, X.-R.; Ding, Y.; Zhao, L.; Zhang, Q.; Li, S.-Q.; Qin, W. Extraction Optimization, Structural Characterization, and Antioxidant Activities of Polysaccharides from Cassia Seed (*Cassia obtusifolia*). *Molecules* **2019**, *24*, 2817. [CrossRef] [PubMed]
16. Klosterhoff, R.R.; Bark, J.M.; Glänzel, N.M.; Iacomini, M.; Martinez, G.R.; Winnischofer, S.M.B.; Cordeiro, L.M.C. Structure and intracellular antioxidant activity of pectic polysaccharide from acerola (*Malpighia emarginata*). *Int. J. Biol. Macromol.* **2018**, *106*, 473–480. [CrossRef] [PubMed]
17. Patova, O.; Smirnov, V.; Golovchenko, V.; Vityazev, F.; Shashkov, A.; Popov, S. Structural, rheological and antioxidant properties of pectins from *Equisetum arvense* L. and *Equisetum sylvaticum* L. *Carbohydr. Polym.* **2019**, *209*, 239–249. [CrossRef] [PubMed]
18. Zeng, F.; Chen, W.; He, P.; Zhan, Q.; Wang, Q.; Wu, H.; Zhang, M. Structural characterization of polysaccharides with potential antioxidant and immunomodulatory activities from Chinese water chestnut peels. *Carbohydr. Polym.* **2020**, *246*, 116551. [CrossRef]
19. Wang, J.; Hu, S.; Nie, S.; Yu, Q.; Xie, M. Reviews on Mechanisms of In Vitro Antioxidant Activity of Polysaccharides. *Oxid. Med. Cell Longev.* **2016**, *2016*, 5692852. [CrossRef]
20. Popov, S.; Smirnov, V.; Kvashninova, E.; Khlopin, V.; Vityazev, F.; Golovchenko, V. Isolation, chemical characterization and antioxidant activity of pectic polysaccharides of fireweed (*Epilobium angustifolium* L.). *Molecules* **2021**, *26*, 7290. [CrossRef]
21. Akkaya, N.E.; Ergun, C.; Saygun, A.; Yesilcubuk, N.; Akel-Sadoglu, N.; Kavakli, I.H.; Turkmen, H.S.; Catalgil-Giz, Y. New biocompatible antibacterial wound dressing candidates; agar-locust bean gum and agar-salep films. *Intern. J. Biomacromol.* **2020**, *155*, 430–438. [CrossRef]
22. Hishikawa, Y.; Kakino, Y.; Tsukamoto, H.; Tahara, K.; Onodera, R.; Takeuchi, H. Control of drug diffusion behavior of xanthan and locust bean gum gel by agar gel. *Chem. Pharm. Bull.* **2016**, *64*, 1450–1457. [CrossRef]
23. Sousa, A.M.M.; Goncalves, M.P. Strategies to improve the mechanical strength and water resistance of agar films for food packaging application. *Carbohydr. Polym.* **2015**, *132*, 196–204. [CrossRef]
24. Rózek, A.; Achaerandio, I.; Güell, C.; López, F.; Ferrando, M. Direct formulation of a solid foodstuff with phenolic-rich multicomponent solutions from grape seed: Effects on composition and antioxidant properties. *J. Agric. Food. Chem.* **2008**, *56*, 4564–4576. [CrossRef]
25. Banerjee, S.; Ravi, R.; Bhattacharya, S. Textural characterization of gellan and agar based fabricated gels with carrot juice. *LWT-Food Sci. Technol.* **2013**, *53*, 255–261. [CrossRef]
26. Rahman, J.M.H.; Shiblee, N.I.; Ahmed, K.; Khosla, A.; Kawakami, M.; Furukawa, H. Rheological and mechanical properties of edible gel materials for 3D food printing technology. *Heliyon* **2020**, *6*, e05859. [CrossRef] [PubMed]

27. Lara-Espinoza, C.; Carvajal-Millán, E.; Balandrán-Quintana, R.; López-Franco, Y.; Rascón-Chu, A. Pectin and Pectin-Based Composite Materials: Beyond Food Texture. *Molecules* **2018**, *23*, 942. [CrossRef] [PubMed]
28. Wu, D.; Zheng, J.; Mao, G.; Hu, W.; Ye, X.; Linhardt, R.J.; Chen, S. Rethinking the impact of RG-I mainly from fruits and vegetables on dietary health. *Crit. Rev. Food Sci. Nutr.* **2020**, *60*, 2938–2960. [CrossRef] [PubMed]
29. Tan, H.; Nie, S. Deciphering diet-gut microbiota-host interplay: Investigation of pectin. *Trends Food Sci. Technol.* **2020**, *106*, 171–181. [CrossRef]
30. De Lavergne, M.D.; Van de Velde, F.; Stieger, M. Bolus matters: The influence of food oral breakdown on dynamic texture perception. *Food Funct.* **2017**, *8*, 464–480. [CrossRef] [PubMed]
31. Pascua, Y.; Koc, H.; Foegeding, E.A. Food structure: Roles of mechanical properties and oral processing in determining sensory texture of soft materials. *Curr. Opin. Colloid Interface Sci.* **2013**, *18*, 324–333. [CrossRef]
32. Koc, H.; Vinyard, C.J.; Essick, G.K.; Foegeding, E.A. Food oral processing: Conversion of food structure to textural perception. *Annu. Rev. Food Sci. Technol.* **2013**, *4*, 237–266. [CrossRef]
33. Wagoner, T.B.; Luck, P.J.; Foegeding, E.A. Caramel as a model system for evaluating the roles of mechanical properties and oral processing on sensory perception of texture. *J. Food Sci.* **2016**, *81*, S736–S744. [CrossRef]
34. Kohyama, K.; Gao, Z.; Watanabe, T.; Ichihara, S.; Nakao, S.; Funami, T. Relationships between mechanical properties obtained from compression test and electromyography variables during natural oral processing of gellan gum gels. *J. Text Stud.* **2017**, *48*, 66–75. [CrossRef]
35. Peyron, M.A.; Lassauzay, C.; Woda, A. Effects of increased hardness on jaw movement and muscle activity during chewing of visco-elastic model foods. *Exp. Brain Res.* **2002**, *142*, 41–51. [CrossRef] [PubMed]
36. Laureati, M.; Sandvik, P.; Almli, V.L.; Sandell, M.; Zeinstra, G.G.; Methven, L.; Wallner, M.; Jilani, H.; Alfaro, B.; Proserpio, C. Individual differences in texture preferences among European children: Development and validation of the Child Food Texture Preference Questionnaire (CFTPQ). *Food Qual. Prefer.* **2020**, *80*, 103828. [CrossRef]
37. Duizer, L.M.; Field, K. Changes in sensory perception during aging. In *Modifying Food Texture Volume 2: Sensory Analysis, Consumer Requirements and Preferences*; Woodhead Publishing Series in Food Science, Technology and Nutrition; Woodhead Publishing: Sawston, UK, 2015; pp. 19–44. [CrossRef]
38. Raheem, D.; Carrascosa, C.; Ramos, F.; Saraiva, A.; Raposo, A. Texture-Modified Food for Dysphagic Patients: A Comprehensive Review. *Int. J. Environ. Res. Public Health* **2021**, *18*, 5125. [CrossRef]
39. Puleo, S.; Valentino, M.; Masi, P.; DiMonaco, R. Hardness sensitivity: Are old, young, female and male subjects all equally sensitive? *Food Qual. Prefer.* **2021**, *90*, 104118. [CrossRef]
40. Mendoza, M.A.; Santagiuliana, M.; Ong, X.; Piqueras-Fiszman, B.; Scholten, E.; Stieger, M. How addition of peach gel particles to yogurt affects oral behavior, sensory perception and liking of consumers differing in age. *Food Res. Intern.* **2020**, *134*, 109213. [CrossRef] [PubMed]
41. Abid, M.; Yaich, H.; Hidouri, H.; Attia, H.; Ayadi, M.A. Effect of substituted gelling agents from pomegranate peel on colour, textural and sensory properties of pomegranate jam. *Food Chem.* **2018**, *239*, 1047–1054. [CrossRef] [PubMed]
42. Basu, S.; Shivhare, U.S. Rheological, textural, micro-structural and sensory properties of mango jam. *J. Food Eng.* **2010**, *100*, 357–365. [CrossRef]
43. Khin, M.N.; Goff, H.D.; Nsor-Atindana, J.; Ahammed, S.; Liu, F.; Zhong, F. Effect of texture and structure of polysaccharide hydrogels containing maltose on release and hydrolysis of maltose during digestion: In vitro study. *Food Hydrocoll.* **2021**, *112*, 106326. [CrossRef]
44. Ruviaro, A.R.; Barbosa, P.D.P.M.; Martins, I.M.; de Ávila, A.R.A.; Nakajima, V.M.; Dos Prazeres, A.R.; Macedo, J.A.; Macedo, G.A. Flavanones biotransformation of citrus by-products improves antioxidant and ACE inhibitory activities in vitro. *Food Biosci.* **2020**, *38*, 100787. [CrossRef]
45. Ge, S.; Liu, Q.; Li, M.; Liu, J.; Lu, H.; Li, F.; Zhang, S.; Sun, Q.; Xiong, L. Enhanced mechanical properties and gelling ability of gelatin hydrogels reinforced with chitin whiskers. *Food Hydrocoll.* **2018**, *75*, 1–12. [CrossRef]
46. Felix, M.; Romero, A.; Rustad, T.; Guerreroa, A. Rheological properties and antioxidant activity of protein gels-like systems made from crayfish concentrate and hydrolysates. *Food Bioprod. Process.* **2017**, *102*, 167–176. [CrossRef]
47. Huang, T.; Tu, Z.C.; Shangguan, X.; Wang, H.; Sha, X.; Bansal, N. Rheological behavior, emulsifying properties and structural characterization of phosphorylated fish gelatin. *Food Chem.* **2018**, *246*, 428–436. [CrossRef] [PubMed]
48. Holdsworth, S.D. Applicability of rheological models to the interpretation of flow and processing behavior of fluid food products. *J. Texture Stud.* **1971**, *2*, 393–418. [CrossRef] [PubMed]
49. Ramkumar, D.H.S.; Bhattacharya, M. Relaxation behavior and the application of integral constitutive equations to wheat dough. *J. Texture Stud.* **1996**, *27*, 517–544. [CrossRef]
50. Vityazev, F.V.; Fedyuneva, M.I.; Golovchenko, V.V.; Patova, O.A.; Ipatova, E.U.; Durnev, E.A.; Martinson, E.A.; Litvinets, S.G. Pectin-silica gels as matrices for controlled drug release in gastrointestinal tract. *Carbohydr. Polym.* **2017**, *157*, 9–20. [CrossRef] [PubMed]
51. Usov, A.I.; Bilan, M.I.; Klochkova, N.G. Polysaccharides of algae. 48. Polysaccharide composition of several Calcareous red algae: Isolation of alginate from Corallina Pilulifera P. et R. (Rhodophyta, Corallinaceae). *Bot. Mar.* **1995**, *38*, 43–51. [CrossRef]
52. Peng, J.; Lu, L.; Zhu, K.X.; Guo, X.N.; Chen, Y.; Zhou, H.M. Effect of rehydration on textural properties, oral behavior, kinetics and water state of textured wheat gluten. *Food Chem.* **2022**, *376*, 131934. [CrossRef]

53. Kohyama, K.; Gao, Z.; Ishihara, S.; Funami, T.; Nishinari, K. Electromyography analysis of naturalmastication behavior using varying mouthful quantities of two types of gels. *Physiol. Behav.* **2016**, *161*, 174–182. [CrossRef]
54. Lawless, H.; Heymann, H. Introduction. In *Sensory Evaluation of Food*; Food Science Text Series; Springer: New York, NY, USA, 2010. [CrossRef]

gels

MDPI

Article

Development of Composite Edible Coating from Gelatin-Pectin Incorporated Garlic Essential Oil on Physicochemical Characteristics of Red Chili (*Capsicum annnum* L.)

Windy Heristika [1], Andriati Ningrum [1,*], Supriyadi [1], Heli Siti Helimatul Munawaroh [2] and Pau Loke Show [3]

[1] Department of Food and Agricultural Product Technology, Faculty of Agricultural Technology, Universitas Gadjah Mada, Flora Street No. 1, Bulaksumur, Yogyakarta 55281, Indonesia
[2] Study Program of Chemistry, Department of Chemistry Education, Faculty of Mathematics and Science Education, Universitas Pendidikan Indonesia, Bandung 40154, Indonesia
[3] Department of Chemical and Environmental Engineering, University of Nottingham Malaysia, Jalan Broga, Semenyih 43500, Selangor Darul Ehsan, Malaysia
* Correspondence: andriati_ningrum@ugm.ac.id

Abstract: Red chili is a climacteric fruit that still undergoes respiration after harvest. During storage, it is susceptible to mechanical, physical, and physiological damage and decay incidence, therefore a method is needed to protect it so that the quality losses can be minimized. One way this can be achieved is by applying edible coatings that can be made from hydrocolloids, lipids, or composites of both, in addition to antimicrobial agents that can also be added to inhibit microbial growth. In this study, we detail the application of an edible coating made of gelatin composite from tilapia fish skin, which has a transparent color and good barrier properties against O_2, CO_2, and lipids. To increase its physicochemical and functional qualities, it must be modified by adding composite elements such as pectin as well as hydrophobic ingredients such as garlic essential oil. This study was conducted to determine the effect of a gelatin–pectin composite edible coating (75:25, 50:50, 25:75), which was incorporated with garlic essential oil (2% and 3%) on the physicochemical properties of red chili at room temperature (±29 °C), RH ± 69%) for 14 days. The best treatment was the 50–50% pectin–gelatin composite, which was incorporated with garlic essential oil with a concentration of 2 and 3%. This treatment provided a protective effect against changes in several physicochemical properties: inhibiting weight loss of 36.36 and 37.03%, softening of texture by 0.547 and 0.539 kg/84 mm^2, maintaining acidity of 0.0087 and 0.0081%, maintaining vitamin C content of 2.237 and 2.349 mg/gr, anti-oxidant activity (IC_{50}) 546.587 and 524.907; it also provided a protective effect on chili colors changing to red, and retains better total dissolved solid values.

Keywords: edible coating; fish gelatin; red chili; pectin; garlic essential oil

Citation: Heristika, W.; Ningrum, A.; Supriyadi; Munawaroh, H.S.H.; Show, P.L. Development of Composite Edible Coating from Gelatin-Pectin Incorporated Garlic Essential Oil on Physicochemical Characteristics of Red Chili (*Capsicum annnum* L.). *Gels* **2023**, *9*, 49. https://doi.org/10.3390/gels9010049

Academic Editors: Baskaran Stephen Inbaraj, Kandi Sridhar and Minaxi Sharma

Received: 15 December 2022
Revised: 3 January 2023
Accepted: 4 January 2023
Published: 6 January 2023

1. Introduction

Chili is a fruit plant that is very common in everyday life, and in Indonesia, almost everyone uses it in food and as a spice in daily cooking [1]. Chili can be harvested when it is still green and when it is ripe [1]. Because chili is a climacteric fruit, the physicochemical properties of chili can be changed during the ripening stage due to respiration, production of ethylene, and other physiological reactions. Because of their color, pungency, flavor, and perfume, chilies (*Capsicum* sp.) are employed as a spice in many national cuisines [2]. A good quality chili or pepper has several characteristics such as firmness, fresh calyx and pedicel, and lack of bruises, abrasions, and disease. Shrivel and wilting both have a significant impact on the visual quality of chilies [3]. Damage to chili during storage, especially at room temperature, includes physical damage such as injury, loss of weight, decreased water content, firmness, freshness, and vitamin C content [3].

Chili can be marketed in modern markets or traditional markets. In modern markets, chilies are occasionally packaged in plastic wrap and stored at a cooler temperature. This is in contrast to sales in traditional markets, in which chilies are usually stacked in an open room. When the chili arrives in the hands of consumers, it is sometimes only stored at room temperature at home. Therefore, it is necessary to find a method to be able to extend its shelf life, especially at room temperature [4].

Currently, the use of natural preservatives is increasingly in demand and has garnered special attraction. One method that can be applied to maintain quality and extend the shelf life of fruits and vegetables is to provide an edible coating [5–8]. Chili with a thin edible coating can be consumed without the need to be separated or thrown away because the edible coating is not dangerous when ingested in the body [9,10].

Edible coatings can be made from hydrocolloids, lipids, or composites of the two. In addition, antimicrobial agents can also be added to inhibit microbial growth of the product being coated [11–14]. Another consideration that is also important in choosing this coating solution material is raw materials that are sustainable, environmentally friendly practical, and economical [15–17]. The use of materials according to these criteria will certainly increase stability and protection during the product's shelf life and is expected to reduce total operational costs while reducing waste generated [13,14,18,19]. One of the most widely applied biopolymers for various types of food is gelatin. Gelatin has good mechanical properties, good optical properties, and a barrier effect on the gas flow [20,21]. An alternative source of gelatin that can be utilized is gelatin from fish industry waste [20,22]. The use of by-products or agro-industrial and aquatic waste such as tilapia fish skin by-products will add economic value and have a positive impact from an environmental and social perspective [21,23]. Tilapia skin gelatin is a species of warm water fish with a greater bloom value than cold water fish with gelling and melting values similar to mammalian gelatin, making it suitable for use as an edible coating [10].

The use of gelatin has a limitation—namely, it has a fairly high value of water vapor permeability (WVP) due to the hydrophilic nature of gelatin; therefore, it is necessary to improve this property [24,25]. One of the ways to improve the properties of gelatin is to composite it with polysaccharides such as pectin, which can correct these deficiencies. Pectin has good mechanical properties and is a good barrier to oxygen and oil [26].

The usage of gelatin in food has been widely employed, although edible coatings have been widely used on chili peppers themselves, including coating with chitosan, which has been proven to retain quality and decrease damage to green pepper [7]. In addition to composites, we can also improve the mechanical and also functional properties of edible coatings by adding antimicrobial agents such as essential oils to slow down spoilage. Essential oils that have been widely used, including rosemary in gelatin from fish, have been shown to reduce WVP by up to 54% because essential oils are hydrophobic so they can improve the hydrophilic properties of gelatin. In addition, garlic essential oil is an antifungal agent, and mycotoxins in corn and edible films are made of protein [27].

Several studies on the use of edible coatings with composite materials on fruits and vegetables—especially chili commodities—have been carried out [28]. The application of chitosan–gelatin composite material as an edible coating on post-harvest red bell pepper demonstrated excellent firmness retention compared to single coatings (noncomposite coating) and untreated fruit [10]. Few reports are available on the use of the edible coating gelatin–pectin composite with incorporated garlic essential oil in red chili to extend the period of usability of red chili; however, to the best of our knowledge, no data is yet available on the combined influence of this composite, or on expanding the shelf-life and maintaining red chili quality traits. As a result, the purpose of this experiment was to investigate the effects of the application of the edible coating gelatin–pectin composite with incorporated garlic essential oil on the shelf-life, weight loss, physicochemical parameters (such as color, firmness, vitamin C, titrable acidity) of red chili, and disease incidence during their postharvest storage at room temperature.

2. Results and Discussion

2.1. FTIR

The FTIR (Fourier Transform Infrared) spectra of a gelatin–pectin composite coating solution containing garlic essential oil and a gelatin–pectin coating solution without garlic oil are shown in Figure 1. In general, both coating solutions showed similar main peaks due to the predominant gelatin–pectin composition compared to the essential oil concentration.

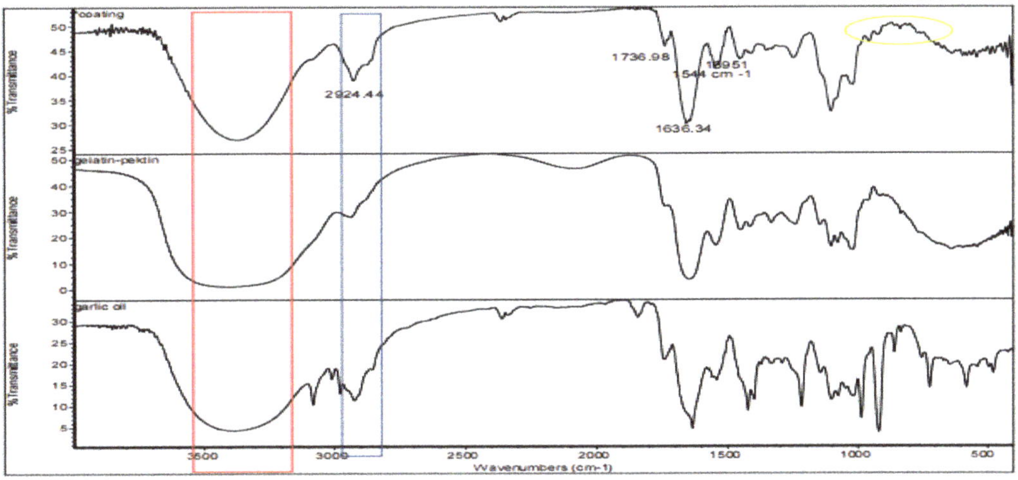

Figure 1. FTIR of an edible coating of Tilapia gelatin composite gelatin–pectin with incorporated garlic essential oil.

The presence of polyhydroxy chemicals such as flavonoids, nonflavonoids, and saponins is indicated by the existence of a wide peak at wave numbers 3373.49 and 3404.91 cm^{-1} on samples with coating with essential oil and without essential oil, respectively. The coating solution containing the essential oil has a higher hydrophobicity than the coating solution without essential oil, as evidenced by the highest peak amplitude at wave number 2924.44 cm^{-1}. This is following the hydrophobic moieties CH, CH_2, CH_3 as aromatic compounds. Peak is also seen at wave number 2363 cm^{-1} like the peak that also appears in the spectrum of garlic essential oil. Strong peaks were also found at about 1630 cm^{-1} for coating with essential oils, showing the existence of carbonyl and carboxylic group stretching. The peak at 1395 cm^{-1} reveals the presence of flavonoids, tannins, saponins, and glycosides by indicating the O-H bend of carboxylic acids. The peak at 1036 cm^{-1} indicates the organosulfur group including alliin, allicin, and diallyl disulfide [29].

The peak at approximately 720 cm^{-1} indicates the C-S bond strain in garlic oil. In the standard spectrum of garlic oil, high-intensity peaks appear in the range of 990 cm^{-1} to 900 cm^{-1}. These peaks are caused by the $=CH_2$ deformation of the vinyl groups found in sulfides and dithiins. This peak was also seen in the coating samples, but it was not as strong as the standard essential oil. Thus, it can be concluded that essential oils can provide a cross-linking effect with the coating solution.

2.2. Weight Loss

Figure 2 indicates that applying an edible gelatin–pectin coating supplemented with garlic essential oil to red chili reduces the percentage of weight loss in red chili held at room temperature for 14 days more efficiently than the uncoated sample ($p < 0.05$). The proportion of weight loss in untreated samples (control) was much greater than in samples coated with an edible coating (Figure 2). Weight loss is commonly attributed entirely to water loss; however, loss of other components may also contribute to this problem. Nonetheless, other than water loss, the contribution from other components is regarded as minimal. This water loss reduces the turgor and hardness of the fruits. It could cause acceleration in the surface depression and deformation of produce. Water loss is associated with several other changes occurring in fruits and can act as a trigger to initiate these changes [30]. The edible coatings can cover the surface layer of the fruit, preventing respiration, transpiration, and syneresis [8,15,16]. This data supports the advantages of putting the edible coating on fresh-cut red chili pieces, owing to the establishment of a polymeric barrier, which has been shown to prevent water loss from fresh-cut samples in other fruits [17,18].

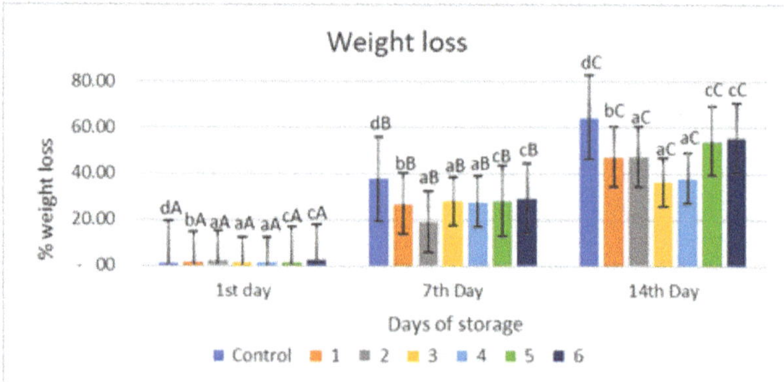

Figure 2. Effect of an edible coating of Tilapia gelatin–pectin composite with incorporated garlic essential oil on weight loss of red chili during storage The value shown is the average of the three experimental replications. Control: Uncoated Sample; 1: Gelatin75%-Pectin25%-Garlic Essential Oil2%; 2: Gelatin75%-Pectin25%-Garlic Essential Oil Oil3%; 3: Gelatin50%-Pectin50%-Garlic Essential Oil Oil2%; 4: Gelatin50%-Pectin50%-Garlic Essential Oil Oil3%; 5: Gelatin25%-Pectin75%-Garlic Essential Oil Oil2%; 6: Gelatin25%-Pectin75%-Garlic Essential Oil Oil3%. $^{a–d}$ non-capital letters show the difference between the samples ($p < 0.05$). $^{A–C}$ capital letters show the difference between the day of storage ($p < 0.05$).

Figure 2 depicts the significant weight decrease of the treated and untreated fruits ($p < 0.05$). This weight reduction in the items might be attributed to moisture loss, transpiration, and respiration, which gave the weight loss of the red chili. Water loss in the fruit occurs mostly through the fruit cuticle, the fruit's physical properties, or both. The loss of a carbon atom in each cycle of respiration may result in weight loss as well. A similar result has been made about bell peppers and green chilies. As a result, an appropriate concentration of applications of edible coating might prevent weight loss [17,18].

Overall, based on Figure 2, the percentage of weight loss increased with storage time, the percentage of weight loss in the control sample on the 1st, 7th and 14th observation days was 1.12, 37.76, and 64.71%, respectively, while in the coated samples the highest weight loss was in samples 5 and 6 at 54.48 and 55.57%, respectively, on the 14th day of storage, and the lowest weight loss was in samples 3 and 4 at 36.36 and 38.03%, respectively, on the 14th day of storage. Based on Figure 2, treatment and storage time had a significant effect ($p < 0.05$) on weight loss.

2.3. *Total Vitamin C*

Figure 3 shows the results of an evaluation of the vitamin C levels in red chili with an edible coating of a gelatin composite supplemented with garlic essential oil. Figure 3 shows that there was a significant reduction in the control sample and the sample with the edible coating treated with the addition of essential oil ($p < 0.05$). The decrease in vitamin C levels in red chili during storage was caused by the oxidation process. Vitamin C oxidizes rapidly to L-dehydroascorbic acid and then to L-dicotigulonic acid [19]. Changes in vitamin C content are caused by the unstable nature of vitamin C, which is easily oxidized when exposed to oxygen. Vitamin C contains a hydroxyl functional group (OH) that is highly reactive in the presence of an oxidizing hydroxyl group and will be oxidized to a carbonyl group. This is in line with research conducted that showed that there was a decrease in vitamin C levels in samples during storage [31]. Ascorbic acid (vitamin C) is a hydrophilic antioxidant that scavenges damaging harmful additional ROS and free radicals. In the present research, in all treatments, the ascorbic acid content declined continuously throughout the storage period, and on day 14 of storage, its content in the control samples (untreated) decreased higher than in the coated sample. The use of an edible coating gelatin–pectin composite containing garlic essential oil decreased the loss of ascorbic acid in red chili.

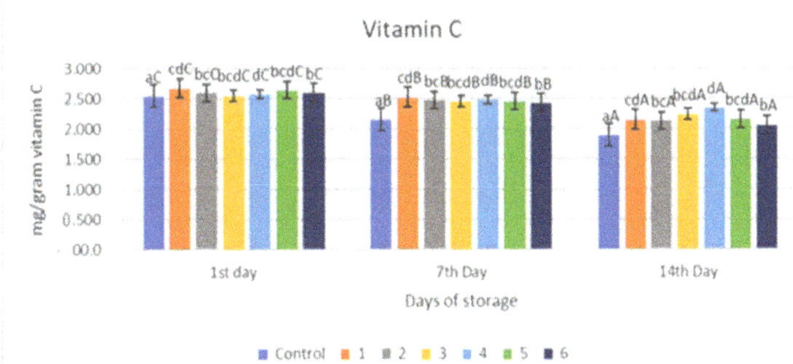

Figure 3. Total Vitamin C of red chili during storage. $^{a–d}$ non-capital letters show the difference between the sample ($p < 0.05$). $^{A–C}$ capital letters show the difference between the day of storage ($p < 0.05$). Control: Uncoated Sample; 1: Gelatin75%-Pectin25%-Garlic Essential Oil2%; 2: Gelatin75%-Pectin25%-Garlic Essential Oil Oil3%; 3: Gelatin50%-Pectin50%-Garlic Essential Oil Oil2%; 4: Gelatin50%-Pectin50%-Garlic Essential Oil Oil3%; 5: Gelatin25%-Pectin75%-Garlic Essential Oil Oil2%; 6: Gelatin25%-Pectin75%-Garlic Essential Oil Oil3%.

The decline in vitamin C in samples without treatment (control) was substantially greater ($p < 0.05$) than in samples treated with an edible coating (Figure 3). This is because the application of an edible coating can inhibit the diffusion of O_2 into the fruit tissue. Therefore, the oxidation reaction that causes damage to vitamin C can also be inhibited. Meanwhile, in untreated samples (control), the diffusion of O_2 was not inhibited as a result, and vitamin C degradation continued. Our findings demonstrated that the edible coating gelatin–pectin composite incorporated with garlic essential oil had a substantial effect on retaining the ascorbic acid content of red chili after 14 days of storage at room temperature. The capacity of the edible coating composite film to improve the inner atmosphere status and postpone the ripening process of red chili was linked to the gradual drop in ascorbic acid level in coated samples.

Naturally, ascorbate oxidase is the primary enzyme responsible for the degradation of ascorbic acid. Ascorbate oxidase levels increase under stress. Ascorbic acid oxidation can result in a loss of nutritional value. Essential oils have antioxidant properties and can help to minimize oxidative stress. As a result, it appears that an edible coating composite of gelatin–pectin incorporated with garlic essential oil improves the antioxidant activity of the investigated oil over time by protecting it from degradation by oxygen, light, and temperature, resulting in less ascorbic acid degradation in the treated sample.

The value of vitamin C from the chili samples treated with coating and the control samples during the storage period can be seen in Figure 3. From Figure 3, it can be seen that there was a decrease in the content of vitamin C ($p < 0.05$) during the storage period in all samples. Samples with a composition of gelatin:pectin 50:50 (samples 3 and 4) were able to maintain the highest vitamin C content, namely 2.237 and 2.349 mg/g; as well as for other samples treated, the greater the concentration of essential oil given, the better it was in maintaining the vitamin C content. The uncoated control sample had the lowest vitamin C content on day 14 of storage, namely 1.898 mg/g, and was significantly different ($p < 0.05$) compared to all samples treated on all observation days.

2.4. Firmness

The results of the firmness analysis of red chilies coated with an edible coating of a gelatin–pectin composite incorporated with garlic essential oil can be seen in Figure 4. Firmness is a very important parameter in determining the quality of chili that can be accepted by consumers, and this hardness value will decrease linearly during the storage period even though the samples have been treated. The F_{max} value indicates the hardness level of the sample; higher F_{max} values indicate that the sample tested has a harder texture. Figure 4 shows similar results in that the hardness value of all samples decreased with increasing storage time. The largest significant decrease in hardness value ($p < 0.05$) was in the control sample on all days of storage, from 0.64 kg/84 mm^2 on the first day to 0.449 kg/84 mm^2 on the 14th day. Meanwhile, the hardness value for each sample differed significantly from day to day. The best hardness results during storage occurred in samples with coating 3, namely from 0.63 kg/84 mm^2 on the first day to 0.547 kg/84 mm^2 on the 14th day. Therefore, based on the results coating 3 gives the best results.

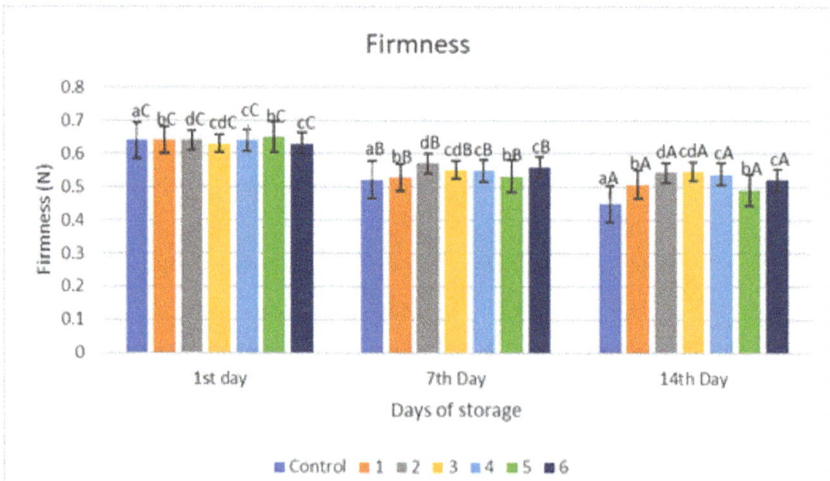

Figure 4. Firmness of red chili during storage. $^{a–d}$ non-capital letters show the difference between the sample ($p < 0.05$). $^{A–C}$ capital letters show the difference between the day of storage ($p < 0.05$). Control: Uncoated Sample; 1: Gelatin75%-Pectin25%-Garlic Essential Oil2%; 2: Gelatin75%-Pectin25%-Garlic Essential Oil Oil3%; 3: Gelatin50%-Pectin50%-Garlic Essential Oil Oil2%; 4: Gelatin50%-Pectin50%-Garlic Essential Oil Oil3%; 5: Gelatin25%-Pectin75%-Garlic Essential Oil Oil2%; 6: Gelatin25%-Pectin75%-Garlic Essential Oil Oil3%.

The results of the firmness of red chili during storage revealed that the gelatin–pectin composite edible coating mixed with garlic essential oil greatly delayed the decline of the firmness in the coated sample when compared to red chili control samples. The firmness of the room control samples decreased more throughout the 14-day storage period. This might be attributed to an increased rate of transpiration by the fruit tissue, which leads to a loss of cell turgor and, as a result, tissue stiffness. Firmness retention was greater in treated samples compared to controls under both storage conditions [17,18]. The results of this investigation demonstrated that a gelatin–pectin composite edible coating infused with garlic essential oil considerably delayed firmness loss.

Several mechanisms—including lipid oxidation, water transpiration, and hydrolysis of pectin components—contribute to the loss of firmness in fruits and vegetables during storage. Figure 4 depicts the findings of the hardness of control and treated red chili. According to Figure 4, the longer the storage period, the lower the F_{max} value, although the edible coating treatment can mitigate the Firmness decline. One of the most significant quality characteristics of all fruits and vegetables, including red chili, is firmness. Tissue softening is therefore a serious problem that has an impact on visual quality.

Based on Figure 4, the loss of firmness in coated chili fruits was reduced due to barrier qualities against gases and water loss transpiration, hence keeping cell turgor and firmness. During storage, firmness in all groups showed a downward trend, and firmness in the coated group of each variety was significantly higher than its control group, which can be explained by the coating reducing the respiration rate and ethylene production, thereby reducing the activity of cell-wall-degrading enzymes. The findings are consistent with prior research on the application, which looked at the influence of the edible coating on the firmness of fresh fruit and vegetables during storage. Previous research has also shown a similar result, in which the application of an edible coating of biopolymers influenced firmness through the application of an edible coating in cherry fruits during storage in order to reduce the firmness due to water loss transpiration [5]. In addition to the transpiration, this is a result of protopectin decomposing into soluble pectin under the action of cell wall degrading enzymes such as pectin galactosidase, polygalacturonase, and pectin methylesterase, which reduce adhesion between cells and the mechanical strength of the cell wall, leading to the softening of fruit tissues and decreased firmness [5].

2.5. Color

2.5.1. Lightness

Color and visual appearance are significant quality criteria that have a direct impact on the customer's sense of quality. It is one of the most important and distinctive characteristics of red chili. The L* or lightness value on red chili samples that were treated with a coating treatment of a pectin–gelatin composite solution incorporated with garlic essential oil and samples that were not given any treatment (control) can be seen in Table 1. The brightness (L*) of samples treated with the coating and the control samples decreased with storage time. The lowest brightness value was in the control sample, namely 20.4 $\pm$ 0.12 on the 14th day of storage. This decrease in the brightness value is due to the activity of the PPO (polyphenol oxidase) enzyme which causes a decrease in brightness to browning in fruits. Browning reactions can occur due to the reaction of oxygen with polyphenolic compounds catalyzed by polyphenol oxidase enzymes to form brown melanin compounds. PPO (Polyphenol Oxidase) enzyme activity produces a loss in brightness on the surface of the fruit flesh and leads it to brown [21]. Oxygen can react directly with polyphenol compounds if some cells or tissues are exposed due to injury. Enzymatic browning degrades the aesthetic look of fresh/fresh-cut fruits and vegetables, induces undesired taste changes, and promotes nutritional loss. Such modifications make the goods less appealing to customers. In any case, browning damages the original color of the product [22]. However, the drop in brightness value is larger in the control sample than in the sample treated with the edible coating because the edible coating can minimize the interaction between oxygen and polyphenol chemicals. Gelatin–pectin combined with garlic essential oil as a component of edible coatings has high oxygen barrier properties [23,24].

Table 1. Color Parameter during Storage.

Sample	Color Parameter								
	L*			a*			b*		
	1st Day	7th Day	14th Day	1st Day	7th Day	14th Day	1st Day	7th Day	14th Day
Control	27.42 ± 1.16	24.43 ^{Aa} ± 0.24	20.4 ^{Aa} ± 0.12	38.94 ^{Bb} ± 1.14	30.38 ^{Aa} ± 1.33	28.02 ^{Aa} ± 0.53	24.69 ^{BCb} ± 1.94	15.71 ^{Aa} ± 0.80	14.59 ^{Aa} ± 0.89
1	28.22 ± 0.57	25.87 ^{Bb} ± 1.07	25.42 ^{Cb} ± 1.27	34.77 ^{Aa} ± 1.61	32.83 ^{Ba} ± 4.06	30.41 ^{Ba} ± 3.73	21.70 ^{Ba} ± 0.86	19.71 ^{Ba} ± 3.78	16.71 ^{Ba} ± 4.39
2	28.16 ± 0.60	25.99 ^{Bb} ± 0.53	24.29 ^{Cb} ± 0.72	37.93 ^{Ab} ± 1.62	38.50 ^{Bb} ± 1.15	35.70 ^{Bb} ± 1.19	25.84 ^{Bb} ± 1.86	25.84 ^{Bb} ± 0.98	24.48 ^{Bb} ± 1.34
3	26.84 ± 1.00	26.46 ^{Bb} ± 1.16	26.11 ^{Db} ± 1.38	33.68 ^{Aa} ± 1.20	32.69 ^{Ba} ± 2.60	31.34 ^{Ba} ± 3.09	17.91 A^{a} ± 2.16	17.87 ^{Ba} ± 3.21	18.62 ^{Ba} ± 1.23
4	28.49 ± 0.57	27.48 ^{Bb} ± 0.89	27.54 ^{Db} ± 0.80	39.47 ^{Ab} ± 1.10	37.74 ^{Ab} ± 2.58	33.63 ^{Bb} ± 1.29	24.18 A^{b} ± 0.72	25.22 ^{Bb} ± 2.59	20.60 ^{Bb} ± 0.30
5	28.37 ± 0.66	25.50 ^{Bb} ± 0.26	22.20 ^{Bb} ± 0.95	39.73 ^{Ba} ± 1.30	31.20 ^{Ba} ± 2.21	28.63 ^{Ba} ± 0.90	25.92 ^{Ca} ± 2.41	17.94 ^{Ba} ± 1.55	15.44 ^{Ba} ± 1.29
6	29.40 ± 0.87	26.08 ^{Bb} ± 0.73	23.50 ^{Bb} ± 1.28	39.74 ^{Bb} ± 1.70	38.92 ^{Bb} ± 2.45	33.73 ^{Bb} ± 1.17	28.08 C^{b} ± 2.20	27.82 ^{Bb} ± 2.83	26.68 ^{Bb} ± 5.10

$^{a,\,b}$ non-capital letter show the difference between the sample ($p < 0.05$). $^{A–C}$ capital letter show the difference between the day of storage ($p < 0.05$). Control: Uncoated Sample; 1: Gelatin75%-Pectin25%-Garlic Essential Oil2%; 2: Gelatin75%-Pectin25%-Garlic Essential Oil Oil3%; 3: Gelatin50%-Pectin50%-Garlic Essential Oil Oil2%; 4: Gelatin50%-Pectin50%-Garlic Essential Oil Oil3%; 5: Gelatin25%-Pectin75%-Garlic Essential Oil Oil2%; 6: Gelatin25%-Pectin75%-Garlic Essential Oil Oil3%.

The brightness of the samples treated with either 2% or 3% garlic essential oil did not have many different results ($p > 0.05$). In samples 5 and 6, the brightness value on the 14th day of storage is close to the brightness value of the control sample. The decrease in brightness in controls tends to be greater than that of samples treated with edible coatings. Therefore, it can be concluded that the provision of edible coatings can maintain better brightness due to the presence of a layer as a barrier that reduces contact between oxygen and polyphenolic compounds. Previous research also reported the same thing, namely that the brightness value (L) of bell pepper decreased significantly during the storage period at room temperature and cold temperature [32,33]. This could be explained as coating formulations being more permeable to O2 due to the presence of a plasticizer, which decreases the intermolecular interactions among adjacent polymeric chains and facilitates gas mobility and permeability, which are functions that could cause changes in color due to oxidation damages that will influence the brightness value. In addition, the main function of the effect of coatings and edible films is to preserve changes in physicochemical parameters and biochemical processes such as enzymatic oxidation caused by polyphenol oxidase (PPO), which is responsible for the browning. This is in contrast with previously published works indicating the effectiveness of coatings [34].

2.5.2. a (Degree of Redness)

The results of the examination of the reddish (a*) color of red chili coated with an edible gelatin composite coating are shown in Table 1. According to Table 1, the red parameter (a) of all red chili samples decreased significantly ($p < 0.05$) during storage. Because some carotenoid pigments in red chili can function as antioxidants, it is sensitive to light and oxygen during storage [25]. Samples with coatings of gelatin–pectin composite incorporated with garlic essential oil tend to protect against the reduction of a value during storage. The decrease in the number suggests that coatings can keep color more effectively. This might be due to the altered environment in the fruit caused by the edible covering, which influences the respiration rate, postponing color difference among samples more than the storage duration itself [26].

On the 14th day, it was seen that the control sample had a lower redness value than the treated sample. A high redness value (a*) of chilies during storage indicates the red color of chilies is approaching bright red, and the lower a* value means that the red color of the chilies is closer to dark red. This change can be caused by the oxidation of carotene and xanthophyll pigments, which are sensitive and easily damaged if exposed to light and oxygen [1]. Treatment with an edible pectin–gelatin composite coating incorporated with garlic essential oil could maintain a better red color value in the sample during storage at room temperature because the respiration rate slowed down.

2.5.3. b (Degree of Yellowness)

The results of the examination of the yellowish parameter (b*) on red chili covered with an edible coating of gelatin–pectin mixed with garlic essential oil are shown in Table 1. On the first day, the b* value ranged from 17–28; on the 14th day, it was seen that the control sample had a lower value than the treated sample. According to Table 1, the yellow parameter of all red chili samples tends to fade with time. The decrease in the b value also suggests that coatings can hold color more effectively. This can also be attributed to the protective effect of edible coating [8].

2.6. Titratable Acidity

Titratable acidity (TA) is a measure of the amount of acid in a solution and is believed to be an indicator of chili maturity [26,27]. It demonstrates that the titrable acidity of all samples declined as storage time increased. There is a significant difference ($p < 0.05$) in TA changes over the storage period (Figure 5). Throughout the storage time, the value of acidity (TA) decreased for all treatments. The decrease in TA in all treatments over the storage period is consistent with the usual process of fruit ripening. The coating treatment considerably ($p < 0.05$) increased the TA during storage durations. This might be because the coating treatment makes less O_2 accessible for the respiratory process, thereby delaying the usage of organic acids. As a result, the decrease in TA during storage might be caused by a variety of reasons, including organic acid breakdown. Organic acids, such as citric acid, are important substrates for respiration. Acidity drop and pH increase are likely to occur in rapidly respiring fruits. External application or coating decreases respiration rates and inhibits organic acid activity. The gelatin–pectin composite with garlic essential oil treatment performed better in sustaining TA and pH at room temperature and under storage conditions than control samples, most likely due to its buffering function for organic acid component degradation. The findings are consistent with prior studies of edible coating for green chilies on the TA and pH of green chilies while they were stored [35]. The results showed that using a chitosan–pullulan composite edible film kept the acidity and pH of bell peppers stable throughout storage [33]. This might be related to the activation of defense enzymes slowing down the respiration rate, metabolic activity, and regulation of enzymatic activity and anthracnose.

TA values in all samples decreased significantly ($p < 0.05$) with increasing storage time (Figure 5). The biggest decrease in TA values was in the control sample, namely 0.0112, 0.0078, and 0.0066% on storage days 1, 7, and 14, respectively. The gelatin–pectin composite incorporated with garlic essential oil proved to be better at retaining TA under room temperature storage conditions (±29 °C, RH ± 69%) compared to the control sample. This study observed that TA contents of red chili in all groups showed a downward trend and that coating treatment inhibited the decomposition of TA due to the oxygen barrier effect of the coating reducing the oxygen concentration inside the fruit, which also reduced the intensity of respiration and related enzyme activities, thereby reducing the consumption of TA.

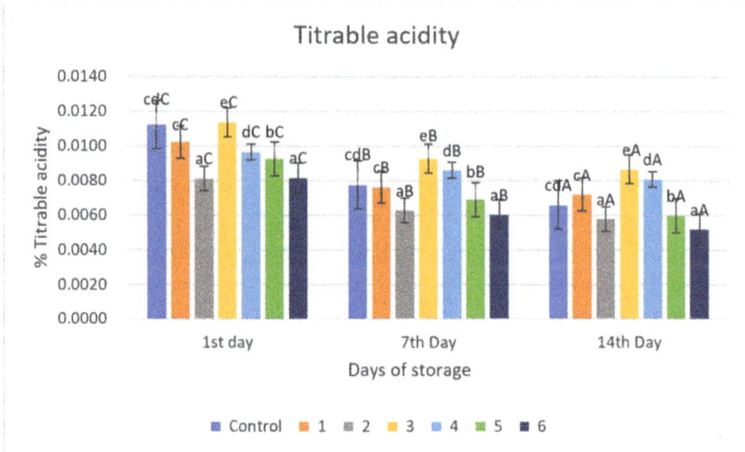

Figure 5. Titrable acidity of red chili during storage. [a–e] non-capital letters show the difference between the day of storage ($p < 0.05$). [A–C] capital letters show the difference between the day of storage ($p < 0.05$). Control: Uncoated Sample; 1: Gelatin75%-Pectin25%-Garlic Essential Oil2%; 2: Gelatin75%-Pectin25%-Garlic Essential Oil Oil3%; 3: Gelatin50%-Pectin50%-Garlic Essential Oil Oil2%; 4: Gelatin50%-Pectin50%-Garlic Essential Oil Oil3%; 5: Gelatin25%-Pectin75%-Garlic Essential Oil Oil2%; 6: Gelatin25%-Pectin75%-Garlic Essential Oil Oil3%.

2.7. *Total Soluble Solids*

The total soluble solids (TSS) of fruits and vegetables are known to be an essential determinant in consumer acceptability. Because of the hydrolytic conversion of complex polysaccharides into simpler sugars and the transformation of pectic components, the TSS behavior of red chili steadily rises with the progression of time storage.

The analysis of total soluble solids in red chilies coated with an edible coating of gelatin–pectin composite incorporated with garlic essential oil shown in Figure 6 can be taken as an example of the quantity of sugar in a material. Total soluble solids tend to increase during storage. The increase in the total value of soluble solids during storage is caused by the accumulation of glucose as a result of the hydrolysis of carbohydrates, which is faster than the process of converting glucose into energy and H_2O [36]. During the ripening process, the carbohydrate content in red chili can also be hydrolyzed into glucose, fructose, and sucrose. Reducing sugar levels can change following the fruit respiration pattern [37]. The appearances of the coated and uncoated red chili samples are shown in Figure 6. The change of physicochemical characteristics including TSS will also give effect to the appearance of the red chili during storage.

The value of the increase in the total dissolved solids value of the control sample was significantly different ($p < 0.005$) compared to the treated sample. On the first day, the control sample had a value of 14.9 brix and increased to 18.8 brix on day 14 (Figure 6). There was a lower increase in value in the treated samples 1, 2, 3, and 4, which were around 16 brix on day 14. Samples 5 and 6 had higher brix values, namely 17.72 and 17.56 brix, and it can be seen that a decrease in gelatin content increased the total dissolved solids, which is because the gelatin layer is proven to be a good barrier thereby delaying the ripening process so that the total dissolved solids value becomes lower.

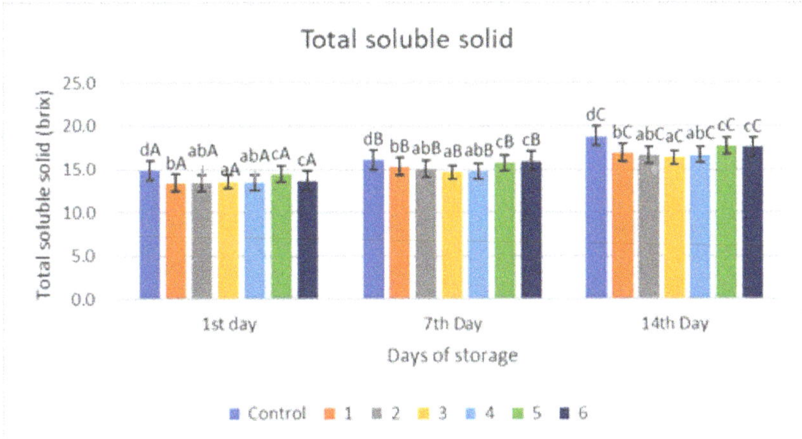

Figure 6. Total Soluble Solid of red chili during storage. [a–d] non-capital letters show the difference between the day of storage ($p < 0.05$). [A–C] capital letters show the difference between the day of storage ($p < 0.05$). Control: Uncoated Sample; 1: Gelatin75%-Pectin25%-Garlic Essential Oil2%; 2: Gelatin75%-Pectin25%-Garlic Essential Oil Oil3%; 3: Gelatin50%-Pectin50%-Garlic Essential Oil Oil2%; 4: Gelatin50%-Pectin50%-Garlic Essential Oil Oil3%; 5: Gelatin25%-Pectin75%-Garlic Essential Oil Oil2%; 6: Gelatin25%-Pectin75%-Garlic Essential Oil Oil3%.

2.8. Antioxidant Activity

Figure 7 depicts the total antioxidant (IC_{50}) analysis of red chili covered with an edible coating of a gelatin–pectin composite supplemented with garlic essential oil. It can be seen in Figure 7 that there is a significant difference in the control sample compared to the treatment sample ($p < 0.005$), especially on the seventh day, but for all coating treatments on the same day, there is no significant difference. On the first day of storage, a low IC_{50} value was present at coating 1–coating 4; coating 5 and 6 were almost the same as the control, but were still below.

There was a significant increase in all treatments on day 7 and some coatings still increase in IC_{50} value until day 14 and some decreased. This fluctuating result is due to the level of maturity of chili itself; the growth conditions of the plant can be different, thus it will affect the antioxidant capacity of the fruit tissue. On the last day of storage, it was seen that the addition of pectin and garlic essential oil to the coating solution affected the IC_{50} value. The greater the concentration of pectin, the lower the IC_{50} value.

The addition of garlic oil will affect its antioxidant activity, and there was an increase in the percentage of inhibition along with the addition of the concentration of garlic oil used. This is because garlic contains the main compounds callicin, alliin, allyl cysteine, and allyl disulphide, which are active against free radical damage [38]. In Figure 7, it can be seen that the control sample has a high IC_{50} value compared to all coating samples on all observation days, which is in line with the results of vitamin C analysis in this study where coating samples had the highest levels of vitamin C compared to the control. Vitamin C has an important role in antioxidant capacity, and it is an organic acid needed by fruit as a substrate in the respiration process. The treatment given is expected to inhibit the respiration process so that the vitamin C content is maintained and, of course, it will also maintain the antioxidant capacity of the fruit [39].

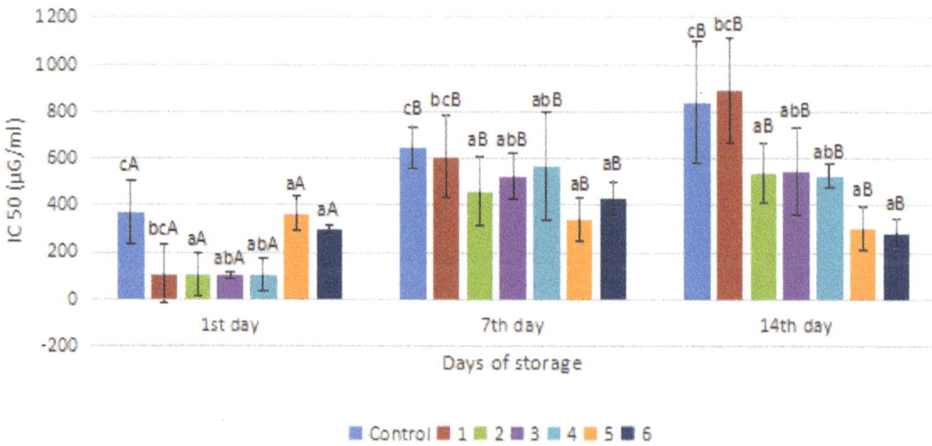

Figure 7. Total antioxidant of red chili during storage. [a–c] non-capital letters show the difference between the day of storage ($p < 0.05$). [A, B] capital letters show the difference between the day of storage ($p < 0.05$). Control: Uncoated Sample; 1: Gelatin75%-Pectin25%-Garlic Essential Oil2%; 2: Gelatin75%-Pectin25%-Garlic Essential Oil Oil3%; 3: Gelatin50%-Pectin50%-Garlic Essential Oil Oil2%; 4: Gelatin50%-Pectin50%-Garlic Essential Oil Oil3%; 5: Gelatin25%-Pectin75%-Garlic Essential Oil Oil2%; 6: Gelatin25%-Pectin75%-Garlic Essential Oil Oil3%.

Based on Figure 7, on the first day of storage, the lowest IC_{50} value was in samples 1 to 4 and the highest was in the control sample, which was 368.76. On the seventh day of storage, there was a significant increase in all treatments; later, almost all samples still experienced an increase in IC_{50} values until day 14, and some also decreased. This fluctuating yield is due to several factors such as the maturity level of the chili itself and different plant growth conditions that can affect the antioxidant capacity of the fruit tissue [40]. In general, on the 14th day, it was seen that the addition of pectin and garlic essential oil to the coating solution affected the IC_{50} value. The greater the pectin concentration, the lower the IC_{50} value, namely in samples 5 and 6. Among all the samples, the best was the one containing higher essential oils. The addition of garlic essential oil will thus affect its antioxidant activity. There was an increase in the percentage of inhibition along with the addition of the concentration of garlic essential oil used. This is because garlic contains the main compounds callicin, alliin, allyl cysteine, and allyl disulphide, which are active against damage caused by free radicals [27]. Coating chili peppers has been shown to provide a better IC_{50} value, and the total radical-scavenging activity of chili peppers is also influenced by the synergism between the total antioxidants in the sample, such as vitamins C and E, and the content of carotenoid pigments [41]. In this study, the treated samples were able to retain their vitamin C content better than the control, where vitamin C is an organic acid needed as a substrate in the respiration process. Coating can inhibit this process so that the vitamin C content is better maintained and, of course, it will also be able to maintain the antioxidant capacity of the fruit [13].

2.9. Decay Incidence

One of the causes of rotting in red chili besides anthracnose is the damage caused by putrefactive fungi such as *Aspergillus* sp. and *Fusarium* sp. The percentage of damage (decay incidence) can be seen in Table 2. A significant difference ($p < 0.05$) was seen in the damage between the control sample and the treated sample. On the first day, almost all of the treated samples were not damaged, while the control samples on the first day were damaged by 22.22%. Along with the storage time, the percentage of damage caused by these microorganisms increased, and on the 7th day, the treated samples began to experience damage. The lowest damage was in coated samples 6 and 3 and the highest was in the control sample. At the end of storage on the fourteenth day, the lowest damage was in sample 3 with a 50:50 composite composition and the highest damage was in the control sample of 68.89%. In general, the data shows that the addition of more garlic essential oil (3%) can slightly reduce the damage caused by these microorganisms. This may be due to the addition of 2% essential oil which is good enough to prevent damage caused by microorganisms in red chili. However, the addition of essential oils to the coating solution still shows significant potential for antimicrobial activity, and the addition of essential oil to chitosan can extend the shelf life by inhibiting the growth of unwanted microorganisms [9,42].

Table 2. Decay incidence of red chilli during storage.

Sample	Decay Incidence (%)		
	1st Day	7th Day	14th Day
Control	22.22 ^{cA} ± 12.02	42.22 ^{cB} ± 6.67	68.89 ^{cC} ± 14.53
1	0.00 ^{bA} ± 0.00	33.33 ^{bB} ± 10.00	48.89 ^{bC} ± 10.54
2	0.00 ^{abA} ± 0.00	22.22 ^{abB} ±12.02	40.00 ^{abC} ± 10.00
3	2.22 ^{aA} ± 6.67	17.78 ^{aB} ± 15.63	28.89 ^{aC} ± 14.53
4	0.00 ^{bA} ± 0.00	35.56 ^{bB} ± 13.33	42.22 ^{bC} ± 6.67
5	0.00 ^{bA} ± 0.00	33.33 ^{bB} ± 12.02	44.44 ^{bC} ± 18.56
6	4.44 ^{aA} ±8.82	11.11 ^{aB} ± 8.82	33.33 ^{aC} ± 16.67

$^{a-c}$ non-capital letter show the difference between the sample ($p < 0.05$). $^{A-C}$ capital letter show the difference between the day of storage ($p < 0.05$). Control: Uncoated Sample; 1: Gelatin75%-Pectin25%-Garlic Essential Oil2%; 2: Gelatin75%-Pectin25%-Garlic Essential Oil Oil3%; 3: Gelatin50%-Pectin50%-Garlic Essential Oil Oil2%; 4: Gelatin50%-Pectin50%-Garlic Essential Oil Oil3%; 5: Gelatin25%-Pectin75%-Garlic Essential Oil Oil2%; 6: Gelatin25%-Pectin75%-Garlic Essential Oil Oil3%.

Furthermore, the coating helps delay senescence, which makes the commodity more susceptible to pathogen infection as a result of loss of cellular or tissue integrity. A similar study was also conducted on bell peppers, in which the fruit coated with chitosan and chitosan-gelatin composite significantly inhibited microbial spoilage, namely 7.4% and 10.6%. This value is two to three times smaller than that of the uncoated peppers, which is 25.3%. The addition of gelatin to the coating does not reduce the antimicrobial effect [32]. The appearance of red chili can be seen in Figure 8.

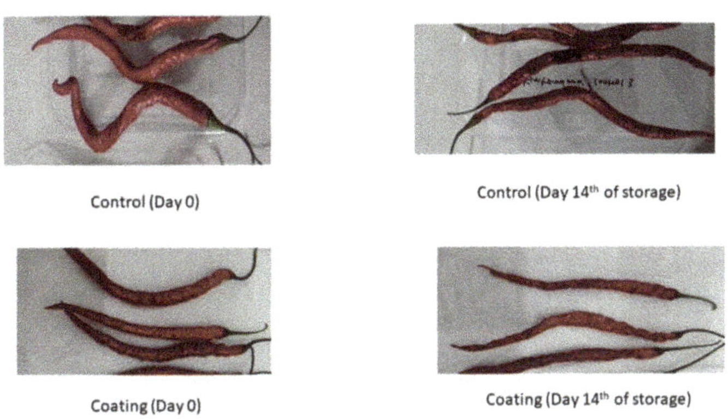

Figure 8. Appearance of Red Chili.

3. Conclusions

The treatment of edible coatings from gelatin–pectin composites, especially the concentration of 50:50, is incorporated with garlic essential oil with a concentration of 2%

(coating 3) on red chili commodities that are stored for 14 days at room temperature, which can provide a protective effect against changes in several physicochemical properties such as inhibition of weight loss by 36.36% and maintaining the decay incidence that began on the 7th day as 17.78% to 28.89% on the 14th day of storage. The edible coating treatment of the gelatin–pectin composite, especially the concentration of 50:50 that was incorporated with garlic essential oil with a concentration of 2% (coating 3), had a hardness value of 0.547 kg/84 mm^2, maintained acidity of 0.0087%, maintained vitamin C content of 2.237 mg/g, antioxidant activity (IC_{50}) 546.587, and provided a protective effect on red chili discoloration to maintain a better value of total dissolved solids. Our findings reveal that the application of the edible coating to red chilies can be an effective method to protect the quality of red chilies and has the potential to extend its storage life up to the 14th day at room temperature storage (±29 °C, RH ± 69%).

4. Materials and Methods

4.1. Materials

Red chili (*C. annum* L.) aged 1–2 days of harvesting (Pasar Jatimulyo, Lampung, Indonesia), garlic essential oil (Lansida group, Yogyakarta, Indonesia), tilapia fish gelatin (Redman fish gelatin, Ang Mo Kio, Singapore), Low Methoxyl Pectin (LMP) (Campectin 4510, Madrid, Spain), and other chemical reagents.

4.2. Preparation of Edible Coatings

First, the dissolving gelatin and pectin in the distilled water are tested separately with a concentration of 3% (*w*/*v*) at 50 °C for 120 min. Afterward, several formulas of the gelatin and pectin solutions with gelatin:pectin ratios of 75:25, 50:50, and 25:75 (*v*/*v*) were then homogenized by stirring using a magnetic stirrer at 40 °C for 3 h. Afterward, 7% glycerol was added as a plasticizer and stirred at 40 °C for 60 min. Garlic oil was added to the solution according to the concentration variations of 2 and 3% (*w*/*v*) and then added between 20 (15% *v*/*v*) while stirring continued for 90 min. The edible coating solution was then ready to use.

4.3. Application for Coating on Red Chili

The process of coating the edible coating solution on red chili (*Capsicum annum* L.) referring to the research of Bermudez-Oria et al. (2017) was carried out using the dipping technique. First, the red chili samples were dipped in the coating solution for 1 min, then drained and dried at room temperature, and then each sample was stored at room temperature for 14 days and analyzed on days 1, 7, and 14. The evaluation of the sample was carried out including physical analysis (weight loss, firmness, and color) and chemical analysis (titratable acidity, vitamin C content, and total soluble solids). The application for coating the red chili can be seen in Figure 9.

4.4. FTIR

The coating solution was applied to the KBr plate. Then, the plate was loaded in FTIR spectroscopy (Thermo, Japan) with a scan range from 400 to 4000 cm^{-1}.

4.5. Weight Loss

The calculation of weight loss is presented below [42].

$$\frac{\text{Weight loss }(\%)= \text{initial weight}(g) - \text{final weight }(g) \times 100\%}{\text{Initial weight }(g)}$$

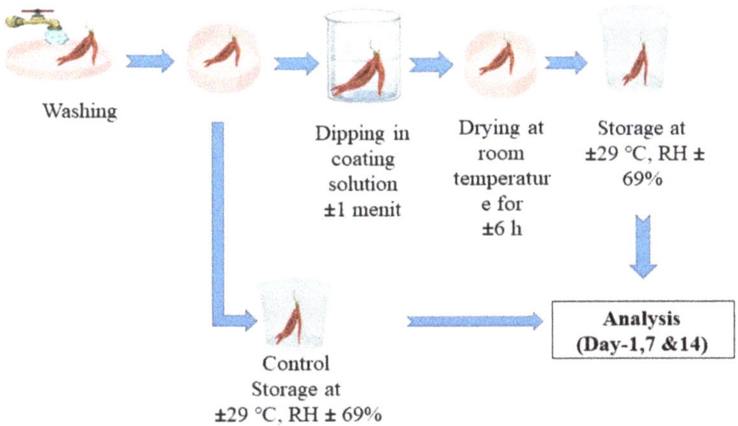

Figure 9. Application of edible coating on red chili.

4.6. Total Vitamin C

Analysis of vitamin C was carried out using modified iodine titration [43]. An amount of 2.5 g of homogenized red chili sample was weighed and put in a 100 mL volumetric flask, then 100 mL of distilled water was added into the flask. Afterward, the solution was filtered to separate the filtrate. The obtained filtrate was placed in an Erlenmeyer flask, then 2 mL of 1% starch solution was added. Afterward, titration was performed using 0.01 N Iodine solution. Vitamin C content was calculated as mg ascorbic acid/g sample (1 mL 0.01 N Iodine = 0.88 mg ascorbic acid).

4.7. Color

The Minolta CR-400 Chromameter was used to analyze color intensity. The sample is placed on top of the chromameter sensor, then light is fired at the part to be measured so that the values of L (lightness), a (green-red chromaticity), and b (yellow-blue chromaticity) will appear on the chromameter display.

4.8. Firmness

Firmness measurements were carried out using a Fruit hardness tester KM-1 (Fujiwara, Tokyo, Japan). The firmness test was measured based on the level of resistance of the fruit to the rheometer needle. The maximum load is 1 kg. Each measurement was repeated three times per test sample. The value of fruit firmness is read on a pointer scale in kg/84 mm^2 units. This value indicates the compressive force required by the needle to pierce the fruit sample.

4.9. Titratable Acidity (TA)

TA was calculated based on the AOAC method which was expressed in grams of acid/100 g of the product. The sample was titrated using 0.1 mol/L NaOH and phenolphthalein as an indicator.

4.10. Total Soluble Solid (TSS)

The Atago Master-53M refractometer was used to measure total soluble solids. One to two drops of the sample chili extract are put on the prism of the refractometer at room temperature, and the Brix % is read through the refractometer's eyepiece.

4.11. Antioxidant Activity

The DPPH technique was used to assess antioxidant activity (2,2-diphenyl-1-picrihydrazil). A UV-Vis spectrophotometer with a wavelength of 517 nm was used to measure the absorption. The antioxidant activity of the sample is expressed as a percentage of radical scavenging activity, or the sample's capacity to trap free radical molecules. The greater the number of free radical molecules that can be collected, the greater the antioxidant activity content of the sample. The mathematical results are used to create regression curves. The IC50 concentration is calculated using the resulting linear regression equation.

$$\text{DPPH radical scavenging activity } (\%) = \frac{\text{A blanko} - \text{A sample}}{\text{A blanko}} \times 100\%$$

4.12. Decay Incidence

The percentage of decay incidence of coated and uncoated (control samples) red chili were analyzed on days 1, 7, and 14 at room temperature storage (±29 °C, RH ± 69%). The decay incidence was determined by observing whether or not decay was visible on the sample surface, then the percentage of damaged samples was calculated. Samples are considered damaged if there is fungal mycelium on the surface and decay occurs. The results of these observations are expressed as the percentage of samples contaminated with fungal mycelium.

$$\text{Decay incidence } (\%) = \frac{\text{Amount of contaminated sample}}{\text{Total sample}} \times 100\%$$

4.13. Statistical Analysis

SPSS 25 was used for the statistical analysis. The analysis of variance (ANOVA) and DMRT were performed with a degree of confidence of 95%. The studies were carried out in at least triplicate, and the results were given as mean standard deviation.

Author Contributions: Conceptualization, A.N., W.H., S., H.S.H.M. and P.L.S.; methodology, A.N., W.H. and S.; software, A.N. and W.H.; validation, A.N.; formal analysis, A.N.; investigation, A.N. and W.H.; resources, A.N. and W.H.; data curation, A.N., W.H. and S.; writing—original draft preparation, A.N., W.H. and S.; writing—review and editing, A.N., W.H. and S.; visualization, A.N. and W.H.; supervision, A.N., S., H.S.H.M. and P.L.S.; project administration, A.N. and W.H.; funding acquisition, A.N. and W.H. All authors have read and agreed to the published version of the manuscript.

Funding: The authors thanks to RTA (Rekognisi Tugas Akhir) 2022(3550/UN1.P.III/Dit-Lit/PT.01.05/2022), Universitas Gadjah Mada for supporting the research.

Acknowledgments: The author wishes to thank LPDP for providing Scholarship and also RTA 2022(3550/UN1.P.III/Dit-Lit/PT.01.05/2022), Universitas Gadjah Mada for supporting the research.

Conflicts of Interest: The authors declare no conflict of interest.

References

1. Muthmainnah, N.; Suratman; Solichatun. Postharvest application of an edible coating based on chitosan and gum Arabic for controlling respiration rate and vitamin C content of chilli (*Capsicum frustecens* L.). *IOP Conf. Ser. Mater. Sci. Eng.* **2019**, *633*, 012028. [CrossRef]
2. Manjunatha, M.; Anurag, R.K. Effect of modified atmosphere packaging and storage conditions on quality characteristics of cucumber. *J. Food Sci. Technol.* **2014**, *51*, 3470–3475. [CrossRef] [PubMed]
3. Jansasithorn, R.; East, A.R.; Hewett, E.W.; Heyes, J.A. Skin cracking and postharvest water loss of Jalapeño chilli. *Sci. Hortic.* **2014**, *175*, 201–207. [CrossRef]
4. Wibowo, C.; Haryanti, P.; Wicaksono, R. Effect of edible coating application by spraying method on the quality of red chili during storage. *IOP Conf. Ser. Earth Environ. Sci.* **2021**, *746*, 012004. [CrossRef]
5. Zhang, Y.L.; Cui, Q.L.; Wang, Y.; Shi, F.; Fan, H.; Zhang, Y.Q.; Lai, S.T.; Li, Z.H.; Li, L.; Sun, Y.K. Effect of edible carboxymethyl chitosan-gelatin based coating on the quality and nutritional properties of different sweet cherry cultivars during postharvest storage. *Coatings* **2021**, *11*, 396. [CrossRef]
6. Christopoulos, M.V.; Gkatzos, D.; Kafkaletou, M.; Bai, J.; Fanourakis, D.; Tsaniklidis, G.; Tsantili, E. Edible Coatings from *Opuntia ficus-indica* Cladodes Alongside Chitosan on Quality and Antioxidants in Cherries during Storage. *Foods* **2022**, *11*, 699. [CrossRef]
7. Figs, F.; Pingo, L.; Vieira, T.M. Composite Coatings of Chitosan and Alginate Emulsions with Olive Oil to Enhance Postharvest Quality and Shelf Life of. *Foods* **2021**, *10*, 718.

8. Shiekh, K.A.; Ngiwngam, K.; Tongdeesoontorn, W. Polysaccharide-Based Active Coatings Incorporated with Bioactive Compounds for Reducing Postharvest Losses of Fresh Fruits. *Coatings* **2022**, *12*, 8. [CrossRef]
9. Khan, M.R.; Di Giuseppe, F.A.; Torrieri, E.; Sadiq, M.B. Recent advances in biopolymeric antioxidant films and coatings for preservation of nutritional quality of minimally processed fruits and vegetables. *Food Packag. Shelf Life* **2021**, *30*, 100752. [CrossRef]
10. Li, H.; Tang, R.; Mustapha, W.A.W.; Liu, J.; Faridul Hasan, K.M.; Li, X.; Huang, M. Application of Gelatin Composite Coating in Pork Quality Preservation during Storage and Mechanism of Gelatin Composite Coating on Pork Flavor. *Gels* **2022**, *8*, 21. [CrossRef]
11. De Elche, M.; Grimal, A.; Baraldi, E. Composite Edible Coatings and Modified Atmosphere. *Coatings* **2021**, *11*, 308.
12. Mannozzi, C.; Glicerina, V.; Tylewicz, U.; Castagnini, J.M.; Canali, G.; Rosa, M.D.; Romani, S. Influence of two different coating application methods on the maintenance of the nutritional quality of fresh-cut melon during storage. *Appl. Sci.* **2021**, *11*, 8510. [CrossRef]
13. Kehila, S.; Alkalai-Tuvia, S.; Chalupovicz, D.; Poverenov, E.; Fallik, E. Can edible coatings maintain sweet pepper quality after prolonged storage at sub-optimal temperatures? *Horticulturae* **2021**, *7*, 387. [CrossRef]
14. Al-Hilifi, S.A.; Al-Ali, R.M.; Al-Ibresam, O.T.; Kumar, N.; Paidari, S.; Trajkovska Petkoska, A.; Agarwal, V. Physicochemical, Morphological, and Functional Characterization of Edible Anthocyanin-Enriched Aloe vera Coatings on Fresh Figs (*Ficus carica* L.). *Gels* **2022**, *8*, 645. [CrossRef]
15. Cheng, Y.J.; Wu, Y.J.; Lee, F.W.; Ou, L.Y.; Chen, C.N.; Chu, Y.Y.; Kuan, Y.C. Impact of Storage Condition on Chemical Composition and Antifungal Activity of Pomelo Extract against *Colletotrichum gloeosporioides* and Anthracnose in Post-harvest Mango. *Plants* **2022**, *11*, 2064. [CrossRef]
16. Gallego, M.; Arnal, M.; Talens, P.; Toldrá, F.; Mora, L. Effect of gelatin coating enriched with antioxidant tomato by-products on the quality of pork meat. *Polymers* **2020**, *12*, 1032. [CrossRef]
17. Petkoska, A.T.; Daniloski, D.; D'Cunha, N.M.; Naumovski, N.; Broach, A.T. Edible packaging: Sustainable solutions and novel trends in food packaging. *Food Res. Int.* **2021**, *140*, 109981. [CrossRef]
18. Ghosh, M.; Singh, A.K. Potential of engineered nanostructured biopolymer based coatings for perishable fruits with Coronavirus safety perspectives. *Prog. Org. Coatings* **2022**, *163*, 106632. [CrossRef]
19. Badawy, M.E.I.; Rabea, E.I.; El-Nouby, M.A.M.; Ismail, R.I.A.; Taktak, N.E.M. Strawberry Shelf Life, Composition, and Enzymes Activity in Response to Edible Chitosan Coatings. *Int. J. Fruit Sci.* **2017**, *17*, 117–136. [CrossRef]
20. Salsabiela, S.; Sekarina, A.S.; Bagus, H.; Audiensi, A.; Azizah, F.; Heristika, W.; Susanto, E.; Siti, H.; Munawaroh, H.; Show, P.L.; et al. Development of Edible Coating from Gelatin Composites with the Addition of Black Tea Extract (*Camellia sinensis*) on Minimally Processed Watermelon (*Citrullus lanatus*). *Polymers* **2022**, *14*, 2628. [CrossRef]
21. Ningrum, A.; Perdani, A.W.; Supriyadi; Munawaroh, H.S.H.; Aisyah, S.; Susanto, E. Characterization of Tuna Skin Gelatin Edible Films with Various Plasticizers-Essential Oils and Their Effect on Beef Appearance. *J. Food Process. Preserv.* **2021**, *45*, e15701. [CrossRef]
22. Sutrisno, E.; Ningrum, A.; Supriyadi; Munawaroh, H.S.H.; Aisyah, S.; Susanto, E. Characterization of tuna (*Thunnus albacares*) skin gelatin edible film incorporated with clove and ginger essential oils and different surfactants. *Food Res.* **2021**, *5*, 440–450. [CrossRef] [PubMed]
23. Toniciolli Rigueto, C.V.; Rosseto, M.; Alessandretti, I.; de Oliveira, R.; Wohlmuth, D.A.R.; Ferreira Menezes, J.; Loss, R.A.; Dettmer, A.; Pizzutti, I.R. Gelatin films from wastes: A review of production, characterization, and application trends in food preservation and agriculture. *Food Res. Int.* **2022**, *162*, 112114. [CrossRef] [PubMed]
24. Ningrum, A.; Hapsari, M.W.; Nisa, A.A.; Munawaroh, H.S.H. Edible film characteristic from yellowfin skin tuna (*Thunnus albacares*) gelatin enriched with cinnamon (*Cinnamomum zeylanicum*) and roselle (*Hibiscus sabdariffa*). *Food Res.* **2020**, *4*, 1646–1652. [CrossRef] [PubMed]
25. Vijayakumar, A.; Badwaik, L.S. Recent advancement in improvement of properties of polysaccharides and proteins based packaging film with added nanoparticles: A review. *Int. J. Biol. Macromol.* **2022**, *203*, 515–525. [CrossRef]
26. Yeddes, W.; Djebali, K.; Aidi Wannes, W.; Horchani-Naifer, K.; Hammami, M.; Younes, I.; Saidani Tounsi, M. Gelatin-chitosan-pectin films incorporated with rosemary essential oil: Optimized formulation using mixture design and response surface methodology. *Int. J. Biol. Macromol.* **2020**, *154*, 92–103. [CrossRef]
27. Handayasari, F.; Suyatma, N.E.; Nurjanah, S. Physiochemical and antibacterial analysis of gelatin–chitosan edible film with the addition of nitrite and garlic essential oil by response surface methodology. *J. Food Process. Preserv.* **2019**, *43*, e14265. [CrossRef]
28. Kulawik, P.; Jamróz, E.; Janik, M.; Tkaczewska, J.; Krzyściak, P.; Skóra, M.; Guzik, P.; Milosavljević, V.; Tadele, W. Biological activity of biopolymer edible furcellaran-chitosan coatings enhanced with bioactive peptides. *Food Control* **2022**, *137*, 108933. [CrossRef]
29. Tongnuanchan, P.; Benjakul, S.; Prodpran, T.; Pisuchpen, S.; Osako, K. Mechanical, thermal and heat sealing properties of fish skin gelatin film containing palm oil and basil essential oil with different surfactants. *Food Hydrocoll.* **2016**, *56*, 93–107. [CrossRef]
30. Yousuf, B.; Wu, S.; Siddiqui, M.W. Incorporating essential oils or compounds derived thereof into edible coatings: Effect on quality and shelf life of fresh/fresh-cut produce. *Trends Food Sci. Technol.* **2021**, *108*, 245–257. [CrossRef]
31. Ganduri, V.S.R. Evaluation of pullulan-based edible active coating methods on Rastali and Chakkarakeli bananas and their shelf-life extension parameters studies. *J. Food Process. Preserv.* **2020**, *44*, e14378. [CrossRef]

32. Poverenov, E.; Zaitsev, Y.; Arnon, H.; Granit, R.; Alkalai-Tuvia, S.; Perzelan, Y.; Weinberg, T.; Fallik, E. Effects of a composite chitosan-gelatin edible coating on postharvest quality and storability of red bell peppers. *Postharvest Biol. Technol.* **2014**, *96*, 106–109. [CrossRef]
33. Kumar, N.; Pratibha; Neeraj; Ojha, A.; Upadhyay, A.; Singh, R.; Kumar, S. Effect of active chitosan-pullulan composite edible coating enrich with pomegranate peel extract on the storage quality of green bell pepper. *LWT* **2021**, *138*, 110435. [CrossRef]
34. Bari, A.; Giannouli, P. Evaluation of Biodegradable Gelatin and Gelatin–Rice Starch Coatings to Fresh Cut Zucchini Slices. *Horticulturae* **2022**, *8*, 1031. [CrossRef]
35. Ali, A.; Maqbool, M.; Alderson, P.G.; Zahid, N. Effect of gum arabic as an edible coating on antioxidant capacity of tomato (*Solanum lycopersicum* L.) fruit during storage. *Postharvest Biol. Technol.* **2013**, *76*, 119–124. [CrossRef]
36. Aguilar-Veloz, L.M.; Calderón-Santoyo, M.; Carvajal-Millan, E.; Martínez-Robinson, K.; Ragazzo-Sánchez, J.A. *Artocarpus heterophyllus* Lam. Leaf extracts added to pectin-based edible coating for *Alternaria* sp. control in tomato. *LWT* **2022**, *156*, 113022. [CrossRef]
37. Sucheta; Chaturvedi, K.; Sharma, N.; Yadav, S.K. Composite edible coatings from commercial pectin, corn flour and beetroot powder minimize post-harvest decay, reduces ripening and improves sensory liking of tomatoes. *Int. J. Biol. Macromol.* **2019**, *133*, 284–293. [CrossRef]
38. Safari, Z.S.; Ding, P.; Nakasha, J.J.; Yusoff, S.F. Combining chitosan and vanillin to retain postharvest quality of tomato fruit during ambient temperature storage. *Coatings* **2020**, *10*, 1222. [CrossRef]
39. Escamilla-García, M.; Rodríguez-Hernández, M.J.; Hernández-Hernández, H.M.; Delgado-Sánchez, L.F.; García-Almendárez, B.E.; Amaro-Reyes, A.; Regalado-González, C. Effect of an edible coating based on chitosan and oxidized starch on shelf life of *Carica papaya* L., and its physicochemical and antimicrobial properties. *Coatings* **2018**, *8*, 318. [CrossRef]
40. Glowacz, M.; Rees, D. Exposure to ozone reduces postharvest quality loss in red and green chilli peppers. *Food Chem.* **2016**, *210*, 305–310. [CrossRef]
41. Ali, A.; Bordoh, P.K.; Singh, A.; Siddiqui, Y.; Droby, S. Post-harvest development of anthracnose in pepper (*Capsicum* spp): Etiology and management strategies. *Crop Prot.* **2016**, *90*, 132–141. [CrossRef]
42. Mohammadi, A.; Hashemi, M.; Hosseini, S.M. Postharvest treatment of nanochitosan-based coating loaded with *Zataria multiflora* essential oil improves antioxidant activity and extends shelf-life of cucumber. *Innov. Food Sci. Emerg. Technol.* **2016**, *33*, 580–588. [CrossRef]
43. Sapei, L.; Hwa, L. Study on the Kinetics of Vitamin C Degradation in Fresh Strawberry Juices. *Procedia Chem.* **2014**, *9*, 62–68. [CrossRef]

gels

MDPI

Article

Developing a Prolamin-Based Gel for Food Packaging: In-Vitro Assessment of Cytocompatibility

Franziska Trodtfeld [1,2,*], Tina Tölke [2] and Cornelia Wiegand [1,*]

1 Department of Dermatology, Jena University Hospital, Friedrich Schiller University Jena, Am Klinikum 1, D-07747 Jena, Germany
2 INNOVENT e.V., Prüssingstraße 27 B, D-07745 Jena, Germany; tt@innovent-jena.de
* Correspondence: franziska.trodtfeld@med.uni-jena.de (F.T.); c.wiegand@med.uni-jena.de (C.W.)

Abstract: Growing environmental concerns drive efforts to reduce packaging waste by adopting biodegradable polymers, coatings, and films. However, biodegradable materials used in packaging face challenges related to barrier properties, mechanical strength, and processing compatibility. A composite gel was developed using biodegradable compounds (prolamin, d-mannose, citric acid), as a coating to increase the oxygen barrier of food packaging materials. To improve gel stability and mechanical properties, the gels were physically cross-linked with particles synthesized from tetraethyl orthosilicate and tetramethyl orthosilicate precursors. Additionally, biocompatibility assessments were performed on human keratinocytes and fibroblasts, demonstrating the safety of the gels for consumer contact. The gel properties were characterized, including molecular structure, morphology, and topography. Biocompatibility of the gels was assessed using bioluminescent ATP assay to detect cell viability, lactate dehydrogenase assay to determine cell cytotoxicity, and a leukocyte stimulation test to detect inflammatory potential. A composite gel with strong oxygen barrier properties in low-humidity environments was prepared. Increasing the silane precursor to 50 wt% during gel preparation slowed degradation in water. The addition of citric acid decreased gel solubility. However, higher precursor amounts increased surface roughness, making the gel more brittle yet mechanically resistant. The increase of precursor in the gel also increased gel viscosity. Importantly, the gels showed no cytotoxicity on human keratinocytes or fibroblasts and had no inflammatory effects on leukocytes. This composite gel holds promise for oxygen barrier food packaging and is safe for consumer contact. Further research should focus on optimizing the stability of the oxygen barrier in humid environments and investigate the potential sensitizing effects of biodegradable materials on consumers.

Keywords: food packaging; composite gels; biodegradable polymers; cytocompatibility; in vitro testing

Citation: Trodtfeld, F.; Tölke, T.; Wiegand, C. Developing a Prolamin-Based Gel for Food Packaging: In-Vitro Assessment of Cytocompatibility. *Gels* **2023**, *8*, 740. https://doi.org/10.3390/gels9090740

Academic Editor: Baskaran Stephen Inbaraj

Received: 17 August 2023
Revised: 8 September 2023
Accepted: 10 September 2023
Published: 12 September 2023

1. Introduction

The food packaging industry is continuously looking for ways to improve packaging materials to ensure optimal durability and strength for food products' shelf lives. The food packaging industry's primary concerns are to ensure the safety of food contents and to prolong their shelf life by protecting them from spoilage, oxidation, and external contaminants. But in response to growing environmental concerns, there is a concerted effort to additionally reduce packaging waste through the increased use of biodegradable polymers, coatings and films [1,2]. Degradable materials used in packaging face challenges in terms of their barrier properties, mechanical strength, and processing compatibility [3,4]. Key considerations in the development and introduction of these materials are: (i) ensuring adequate barrier properties to protect the packaged products from oxygen and the following degrading oxidation reactions; (ii) achieving sufficient mechanical strength for durability; and (iii) finding processing methods that are compatible and efficient. Nanocomposite films and coatings containing inorganic particles have been investigated for their potential applications in food packaging [5–8]. These nanocomposites have exhibited satisfactory barrier properties, attributed to the particles effectively impeding the diffusion of molecules [9–11]. In addition to their exceptional barrier performance, the incorporation of nanomaterials can enhance stability, mechanical strength, and overall durability in degradable food packaging coatings [12,13]. Additionally, the incorporation of nanomaterials serves to enhance the water resistance of the inherently hydrophilic degradable polymers. This aspect holds particular significance, considering that the majority of food products characteristically contain higher moisture levels [14,15]. A composite gel with an oxygen barrier for food packaging was developed using in-situ sol-gel polymerization in a matrix of biodegradable polypeptides, polysaccharides, and organic acids. Controlled sol-gel conditions (time, temperature, pH) created a semi-solid gel with a continuous network of nanoparticles and pores [16,17]. Polar groups in proteins enhanced the oxygen barrier by slowing down oxygen diffusion [2,18]. D-mannose served as a carrier, and citric acid as a catalyst for the silane precursor reaction [19,20]. Physical cross-linking with particles synthesized from tetraethyl orthosilicate and tetramethyl orthosilicate improved gel stability [21,22]. Comprehensive analysis included molecular structure, morphology, and topography investigations.

At the same time, new materials for food packaging potentially coming into contact with the skin need to be assessed for biocompatibility and eliciting inflammatory reactions and skin irritation in humans [23]. Natural polymers, derived from sources such as plants or animals, may contain sensitizing or irritating components that can initiate inflammatory reactions in susceptible individuals [24–26]. These reactions can range from mild symptoms, such as skin irritation, to more severe consequences, including tissue damage and health complications [27–29]. Identifying and evaluating skin irritants in natural polymers and their derivatives is necessary for consumer safety. To develop safe materials, the synthesized gels underwent a series of biological assays. Cytotoxicity was assessed using in vitro assays with human keratinocytes and fibroblasts exposed to various extraction mediums that mimic environmental conditions. Additionally, an initial evaluation of inflammatory potential was conducted by detecting sulfidoleukotrienes following leukocyte stimulation [30]. This preliminary assessment allows to gain insights into the gels initial response to immune cells and potential initiates of inflammation [31].

This study examines the physical and chemical properties of a novel prolamine silica composite gel designed as a coating to enhance oxygen barrier properties in food packaging materials. Additionally, it proposes a testing method to evaluate cytocompatibility for biodegradable materials, thus ensuring the safety by utilization.

2. Results and Discussion

Sol-gel coatings (ProMa) with different amounts precursor consisting of tetramethoxysilane (TMOS) and tetraethoxysilane (TEOS) were obtained by in situ polymerization in a 70% ethanol solution containing extracted wheat gluten prolamins, d-mannose and citric acid.

2.1. IR Spectroscopy Studies

Prolamins extracted in 70% EtOH/H_2O v/v % from gluten, the storage protein unit of wheat and the coating with the polymer matrix and 50 wt% precursor were analyzed with FT-IR spectroscopy to investigate molecular structure of the gel.

The proposed network structure of the material consists of different types of chemical bonding, including peptide bonds within the polypeptides, glycoside bonds of d-mannose, and intra- and intermolecular hydrogen bonds. Additionally, ionic bonds between citric acid and amide groups, as well as potential hydrogen bonding, contribute to cross-linking. It is also hypothesized that the silica particles may possess a negatively charged surface, allowing for the adsorption of polymers due to favorable surface area and energy ratios [32,33]. These various bonding interactions result in a network structure, providing mechanical stability and reinforcing the material's overall properties. However, further research is needed to confirm and fully understand the significance of these bonds in the network.

In Figure 1 the presence of the amide bonds I and II in the polypeptides within the polypeptide structure of the coating is clear from the distinct peaks observed at 1650 cm^{-1} and 1550 cm^{-1}, respectively [34].

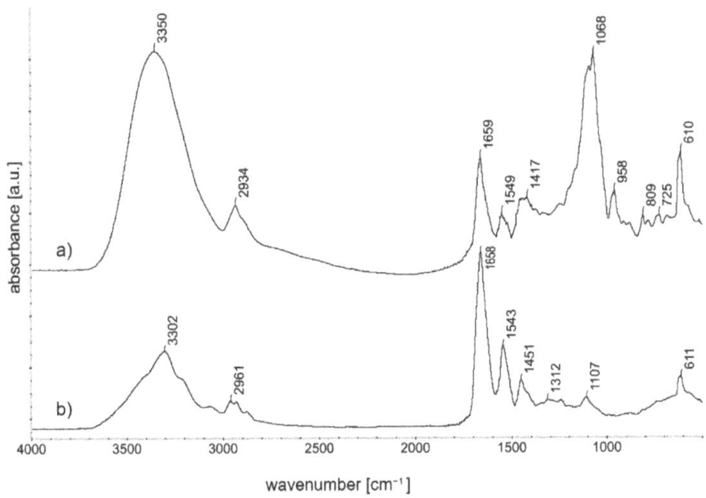

Figure 1. FT-IR spectra (4000–500 cm^{-1} spectrum): (**a**) ProMa 50 wt% precursor gel on silicon wafer (**b**) Prolamins extracted from Gluten Powder (>75%) on silicon wafer.

The peak at ~950 cm^{-1} and the peak at 1000–1200 cm^{-1} (C-O-C stretching) can be attributed to the glycosidic linkage in a polysaccharide. The peaks 809 cm^{-1} and 1068 cm^{-1} are characteristic IR signals of the d-mannose [35]. Silica bonds are superimposed by the matrix bonds in the spectrum, but can be determined at 1096 cm^{-1} and 960 cm^{-1} when the polymer matrix is subtracted from the spectra [36]. The peak around ~3300 cm^{-1} is characteristic for –OH groups and the ~2900 cm^{-1} for saturated hydrocarbons [37,38].

2.2. *Tensile Strength Examinations of Composite Gels*

Texture analysis was performed to evaluate additional material properties of structure and composition, which are relevant for processability, coating stability and material strength. The strain-stress curve (Figure 2) of a material can show the mechanical properties such as elasticity and strength. In this study, the gels were cast dried and analyzed as thin foils to determine differences between the coating compositions. The results are not directly transferable to conclude the material behavior if the gels are coated onto different materials but allow a basic comparison between the gel compositions tested.

The strain stress curve shows that the polymer matrix ProMa 0 wt% precursor is an elastic material that exhibits a long necking phase before the material breaks [39]. As the amount of silica precursor for this material is increased in ProMa 10 wt% precursor and ProMa 20 wt%, the elasticity decreases while the yield strength of the material increases. From this, it can be concluded that the material becomes harder and more resistant to mechanical forces [40]. On the other hand, the fracture point is reached at a lower load. This means that the material becomes more brittle and less ductile [41,42].

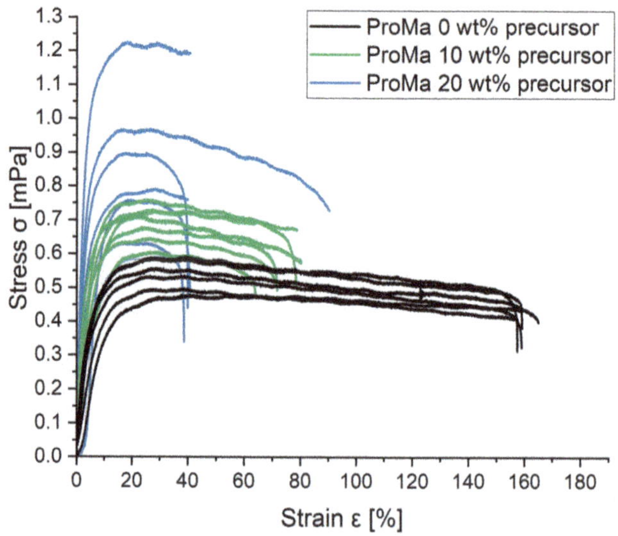

Figure 2. Stress σ [mPa] as a function of Strain ε [%] for ProMa with 0/10/20 wt% precursor.

The Young modulus was fitted to the strain stress curves and the average of six curves was taken (Table 1). The Young modulus is a material parameter that describes the proportional relationship between stress and strain during the deformation of a solid body with linear-elastic behavior and serves as an indicator of its stiffness and strength.

Table 1. Young modulus [MPa] of ProMa with 0/10/20 wt% precursor determines from stress strain curve.

	ProMa 0 wt% p.	**ProMa 10 wt% p.**	**ProMa 20 wt% p.**
Young Modulus MPa	10.01	17.12	21.07
SD	4.96	10.07	7.28

A higher Young modulus as seen in ProMa 20 wt% precursor compared to ProMa 0 wt% precursor, signifies a greater ability to resist deformation under external pressure, indicating increased stiffness and strength. It reflects the sample capacity to withstand stress without undergoing permanent deformation. This correlation supports that the incorporation of silica precursor contributes to the reinforcement and improved mechanical resistance of the polymer matrix [17,18].

This data emphasizes the observed trend that as the silane precursor concentration within the gel rises, intermolecular chain entanglement becomes more pronounced. The rigidity of the gel increases and the movement of molecular chains of the degradable matrix decreases [43]. This heightened entanglement is manifested in reinforced gel interactions, as evidenced by an increase in tensile strength and a corresponding decrease in elongation at break [14,44]. Hence, it becomes imperative to determine the ideal precursor quantity that can enhance the mechanical properties, specifically elevating tensile strength while keeping material brittleness within acceptable limits.

2.3. Atomic Force Microscopy Analysis

The analysis of the surface topography and structure of the gels was conducted by atomic force microscopy (AFM). Silica particles have a rough, irregular surface structure, and when dispersed in a polymer matrix [45], they produce a roughened surface (Figure 3).

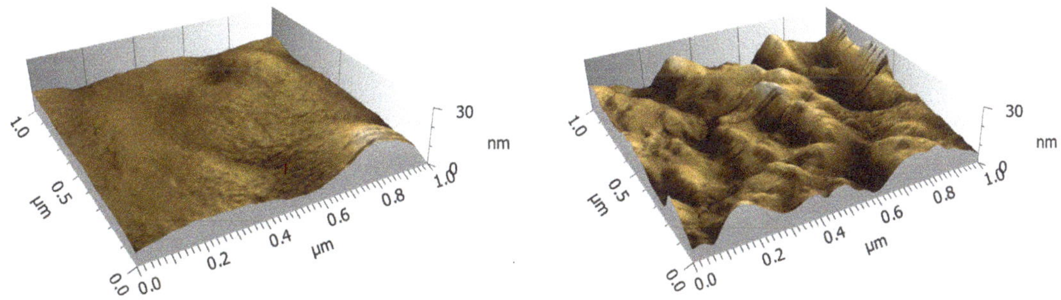

(a) (b)

Figure 3. AFM picture of (**a**) ProMa 0 wt% precursor (**b**) Proma 50 wt% precursor.

The arithmetic mean of the surface level (Sa) increased from 3.18 nm for ProMa with 0 wt% of precursor to 9.48 nm for ProMa 50 wt% precursor (measured over a 10 µm × 10 µm area) [46]. Increased particle aggregation, which is present in samples with high silane precursor can lead to an increase in surface roughness, which result in the increased material brittleness observed in texture analysis (Figure 2) [47,48]. This supports the restricting of molecular chain movement and increased entanglement. The roughness increases the surface area of the material, which promotes the adhesion of additional coating. At the same time, the material becomes more hydrophilic, meaning it attracts more water. This could be beneficial for distribution and adhesion of additional coating, such as a water vapor barrier [49].

2.4. Scanning Electronic Microscopy Analysis

The material surface and morphology of cross sections were determined using scanning electronic microscopy (SEM). The distribution of particles within the sample appears to be homogeneous through the polymer matrix, indicating an even dispersion throughout the material. Figure 4 provides insight into the size estimation of the silica particle agglomerates, which is estimated to be around 100 nm. This finding suggests that the silica particles have formed relatively uniform agglomerates within the material.

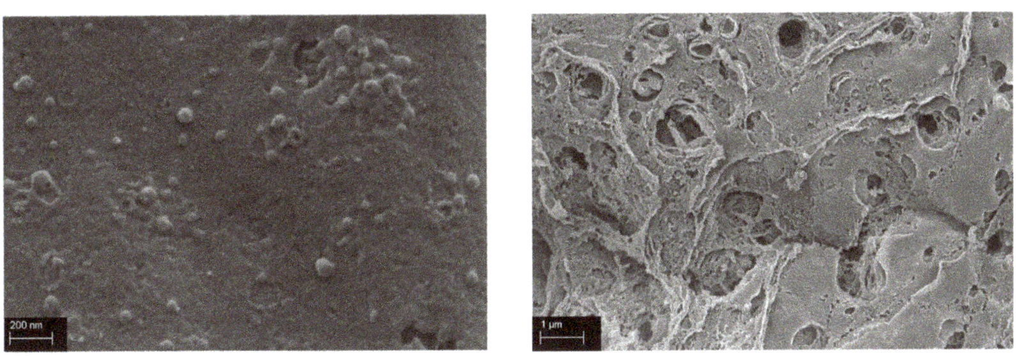

(a) (b)

Figure 4. SEM images of ProMa + 20 wt% precursor (**a**) surface and (**b**) fracture edge, dried with CO_2 critical control point method and gold sputtered.

2.5. Rheological Investigations of Composite Gels

The rheological properties were analyzed to obtain data for viscosity, flow behavior and altering of the gel, to understand the processability and applications possibility. By examining the data from viscosity measurements, changes in the material's flow behavior over time can be observed, reflecting the altering rheological properties [50]. This information is crucial to understanding the material's process ability and gelling time. The rheological properties and viscosity of the polymer matrix vary depending on the amount of silica precursor present during the modification process [51,52]. The viscosity of all samples decreased with increasing shear rate, which demonstrates the shear thinning effect [53,54]. This phenomenon can be attributed to the entanglement of polymers within the network and their subsequent detanglement in response to applied shear stress. The viscosity for the gel ProMa 0 wt% precursor does decrease from 9.4 mPa·s after 1 d to 6 mPa·s at 7 d whereas for the ProMa 50 wt% precursor the viscosity increases from 8.76 mPa·s after 1 d up to 28.3 mPa·s at 7 d. Figure 5 provides a visual representation of the viscosity immediately after synthesis and the median viscosity measured over a one-week period after synthesis.

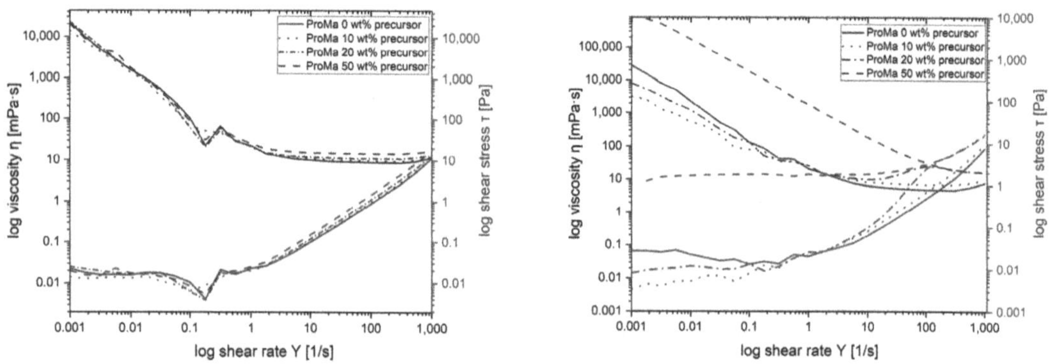

(a) (b)

Figure 5. Viscosity curve of ProMa + 0/10/20/50 wt% precursor (**a**) directly after synthesis (**b**) median of measuring 1 d, 3 d and 7 d after synthesis.

At very low shear rates, the viscosity tends to infinity. After the shear-thinning valley, a slight shear thickening effect is observed at very high shear rates. This shear thickening phenomenon is attributed to the increased frequency of particle collisions within the solution, even in samples without addition of silica precursor, suggesting that it is not solely caused by formed particles.

Furthermore, the addition of up to 50 wt% silica precursors to the solution results in a slightly increased viscosity trend, in general, the materials behave in the same ways as a thin liquid and are very easy to process. During one week of gel altering, the viscosity curve intercepts at a log shear rate of 100. If the flow rate and viscosity overlap, it means that at low shear stress (low flow rate) the viscosity of the material is high indicating that it is thick and viscous. However, with increasing shear stress (higher flow rate), the viscosity decreases, and the material behaves more like a thin fluid [53]. The decline of polymer entanglement in ProMa with 50 wt% precursor indicates a shift towards silica network formation, which will increase the amount of chemical bonding in the gel. This leads to increased viscosity, which is crucial for coating processes [54]. Additionally, silica particles aggregate, strengthening the gel and further elevating viscosity [33]. This viscosity change highlights distinct processing times, depending on precursor concentration. Lower concentrations of precursor such as ProMa 10 wt% and ProMa 20 wt% are not showing a strong viscosity increase, this could be due to reduced aggregation and therefore less crowding, improved electrostatic repulsions and increases solvent interactions [55,56]. This results in longer shelf life compared to ProMa 50 wt% precursor, underscoring viscosity's role in solution stability.

2.6. Water Stability

The water stability of the gels was determined to analyze if the stability of the degradable gel coating could be improved upon gel alteration by incorporation of silica. The amount of silica precursor also directly affects the coating's ability to dissolve or disperse in water [57]. Higher concentrations of silica precursor as in inorganic material generally result in reduced water solubility, while lower concentrations may lead to increased solubility and influenced functionality, as it is demonstrated in Figure 6 [58,59]. The decline in solubility of the gels containing 50 wt% precursor can be attributed to a substantial increase in hydrogen bonding. These interactions occur between the SiO_2 particles and the degradable matrix, leading to a reduction in both the mobility of the polymer chains and the accessibility of hydroxyl groups. This decrease in solubility is a result of these intricate hydrogen bonds forming, which restrict the overall movement of the polymer chains and limit the availability of hydroxyl groups within the system [60]. Integration of silica precursor for 10 wt% and for 20 wt% did not show a decrease in solubility of the gel matrix, concluding that the amount of silica particles formed is not sufficient for available hydroxyl groups of the polymer chains. The incorporation of citric acid into the polymer matrix plays a significant role in enhancing the water stability of the material. Observation shows that with citric acid incorporation decreases the weight loss further from around 60% to around 50% weight loss. Citric acid is used to lower the gel's pH to 3. In this pH range, silica particles typically reach their point of zero charge (PZC), which typically falls within the pH range of 2 to 4 [61]. This results in a neutrally charged surface that demonstrates a strong affinity for adsorbing anionic polymers, such as prolamins. However, it's important to note that prolamins, which generally carry a net negative charge, may be less likely to adsorb onto the silica particle surface in a pH 3 environment. This is due to the fact that at pH 3, the carboxyl groups of the amino acids in prolamins are likely to be protonated, resulting in a reduction or neutralization of their negative charge. As a result, the electrostatic attraction between prolamins and the silica surface may be diminished, making adsorption less likely. In conclusion citric acid's hydrophilic properties introduce more hydroxyl groups, forming hydrogen bonds with neutral silica particle surfaces forming a sturdy three-dimensional network in the matrix that prevents water-induced dissolution [62,63].

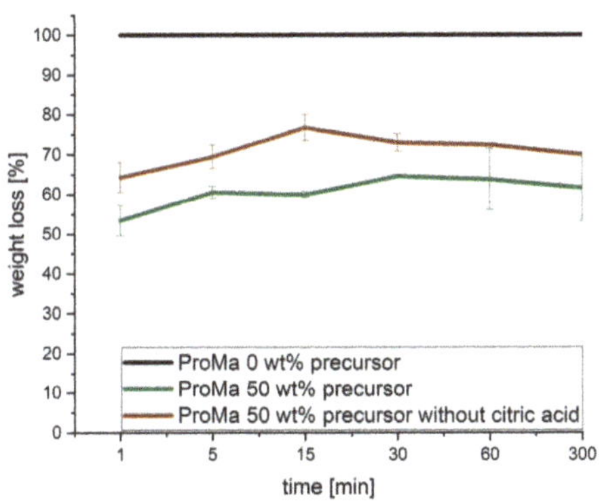

Figure 6. Weight loss [%] over time [min] of ProMa 0/50 wt% precursor and ProMa 50 wt% precursor without adding citric acid during synthesis. A cross-linking of the components chemically and not just physically could increase mechanical stability and decrease solubility of the gel in water. Chemical cross-linking can also make the material less susceptible to natural degradation processes. Microorganisms that aid in the biodegradation of polymers might find it more challenging to break down the material due to the increased cross-linking, leading to reduced biodegradability and change the environmental impact of the material.

2.7. Oxygen Permeability of Composite Gels on PET

The oxygen permeability was tested of the composite gel coated onto a matrix (PET), to determine if the gels increase the oxygen barrier properties of the matrix material at different relative humidities (r.H.) (Figure 7).

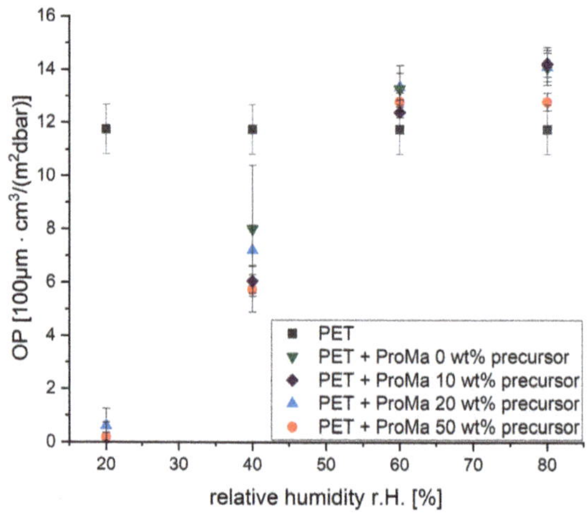

Figure 7. Oxygen Permeability (OP) for PET+ ProMa 0/10/20/50 wt% precursor at different relative humidities 20–80 r.H. [%].

The gels increase the oxygen barrier properties of PET [64,65] in low r.H. environments significantly, but the additional barrier properties cannot be sustained in higher r.H. environments over 60% r.H. When r.H. is increased, there are more water molecules in the surrounding environment. In the case of a polymer matrix, water molecules are attracted to the polar groups within the matrix due to the presence of polar or hydrophilic regions. This attraction leads to the adsorption of water molecules onto the polymer matrix. As water molecules adsorb onto the polymer matrix, they can create gaps or voids within the structure. These gaps act as channels through which other molecules, such as oxygen molecules, can enter. Oxygen permeation through the polymer matrix can be problematic in certain applications [1,66]. The higher the relative humidity and the more water molecules adsorbed, the greater the potential for gaps and increased permeation of oxygen through the matrix [67]. The addition of silica to the polymer matrix plays a vital role in improving the water stability of the material. A subtle reduction in oxygen permeability, noticeable at 80% relative humidity, is evident in gels comprising 50 wt% precursor, while no discernible difference is observed in gels with precursor concentrations ranging from 0 wt% to 20 wt%. Silica particles, known for their hydrophilic properties, effectively act as a barrier against water permeability [32] when incorporated into the polymer matrix. Consequently, the incorporation of silica greatly improves the durability and longevity of the polymer matrix in moisture-sensitive conditions. As can be seen in Figure 7, the oxygen permeability decreases only slightly compared to the polymer coating without precursors. The incorporation of silica precursors leads to an improvement in water stability but does not prevent the formation of voids in the matrix, even though silica incorporation has been reported to improve the oxygen barrier significantly [68,69]. In order to increase the stability of the oxygen barrier in humid environments, it would be useful to increase the crystallinity of the material and decrease the amount of polar groups so that the oxygen barrier remains intact but the attraction to water might decrease [70]. Tailoring particle shape during synthesis could greatly enhance oxygen barrier performance. A lamellar morphology proves more effective in improving barrier properties than spherical particle [71,72]. This also suggests a promising approach for enhancing oxygen barrier stability in high-humidity environments [73]. In summary, higher r.H. does lead to the adsorption of water molecules to the polymer matrix, creating gaps within the structure. These gaps allow oxygen molecules to permeate through the matrix, reducing the material's overall effectiveness as an oxygen barrier. Therefore, multilayer oxygen barriers and water vapor barriers would be the best option for sufficient food packaging with this gel coating [49].

Cytocompatibility

2.8. In-Vitro Cytotoxicity Testing

The material extracts were subjected to testing in accordance with standard conditions specified in DIN EN ISO 10993-5 [74], which is primarily intended for medical products and demands an incubation between 24 h and 72 h [75]. However, since food packaging materials typically have shorter contact times compared to medical products, the materials were incubated with cell medium for a 24-h period to allow interaction with the cells. This modified incubation period ensures that the materials have sufficient exposure within a timeframe that more closely matches their contact time in food packaging applications. Subsequently, the extracts were assessed for cellular triphosphate (ATP) content and lactate dehydrogenase release to determine effects on cell viability as well as cell cytotoxicity (Figure 8).

Remarkably, none of the tested materials exhibited any cytotoxic effects, which was confirmed by the lactate dehydrogenase assay. However, it is worth noting that the ATP concentration in fibroblasts decreased slightly, which may be due to reduced fibroblast proliferation associated with the presence of citric acids in the extract. It is known that fibroblasts are sensitive to specific organic acids [76]. However, this sensitivity is concentration and acid dependent. The pH of the cell medium solution after extraction was measured for control and was determined at pH ~8.0 for fibroblasts. Since the citric acid only slightly affected cell proliferation and no direct cytotoxic effect was observed, the results were considered acceptable to continue further investigations.

Since the material may be exposed to skin cells under various conditions, it was important to investigate the effects of human sweat and saliva on the extraction of specific compounds. To achieve this, extracts of the material were prepared using artificial saliva (SV) and artificial sweat solutions (SSL). The same tests that were conducted with the cell medium were then performed with these extracts (Figure 9). Subjecting the materials to artificial saliva and artificial sweat solutions aimed to simulate more realistic conditions for use and evaluate any potential cytotoxicity. These additional tests provide a comprehensive assessment of the material's cell compatibility, considering the potential interactions with skin cells in the presence of specific compounds found in human sweat and saliva. The results of these tests will contribute to a deeper understanding of the gels performance and safety in real-life scenarios.

No cytotoxic effects for the material extracts were detected in either the sweat solution or the artificial saliva, which indicates their cell compatibility. However, a slight decrease in fibroblasts was observed when exposed to the sweat solution extracts, which may be due to the presence of citric acids in combination with the lower pH of the sweat solution (pH~5.5) that inhibits cell proliferation [77]. There can be several reasons for this, as the pH affects various cell components (i.e., the enzymatic activity, the DNA stability and replication, the protein function, and the cellular metabolism) and can also cause cell cycle arrest [78].

Interestingly, increased cell proliferation was observed in the artificial saliva (pH 6.8). Exposure to saliva can result in improved cell growth and proliferation, due to growth factors, protease inhibitors and a humid environment [79,80]. Since artificial saliva is not a complete replica of human saliva, the constituents are mainly sodium chloride, potassium chloride, potassium thiocyanate, potassium dihydrogen phosphate and urea [81–84]. These electrolytes play an important role in maintaining the balance of fluids in the body and are involved in various bodily functions. It appears that the saliva acts as a buffer to neutralize organic acids. When acids are present in the mouth, saliva helps maintain the pH balance by neutralizing them [82]. Since the citric acid is the component that lowers the pH and can hinder cell proliferation, buffering this component in addition with a dense nutrient supply for the cells may result in increased cell proliferation. In addition, the artificial saliva incorporates urea, which can disrupt protein structures. This leads to differences in the materials degradation depended on the extraction medium, shown in Figure 10. This supports the hypothesis that the conditions for testing must be adapted to the intended purpose, to enable accurate safety assessment of the material. Protein structure disruption can produce free amino acids and smaller peptides, which can influence cell proliferation. Especially glutamine is a crucial amino acid for cell metabolism and is required for the synthesis of nucleotides, proteins, and other cellular components. It plays a role in supporting the energy needs of rapidly proliferating cells [85,86].

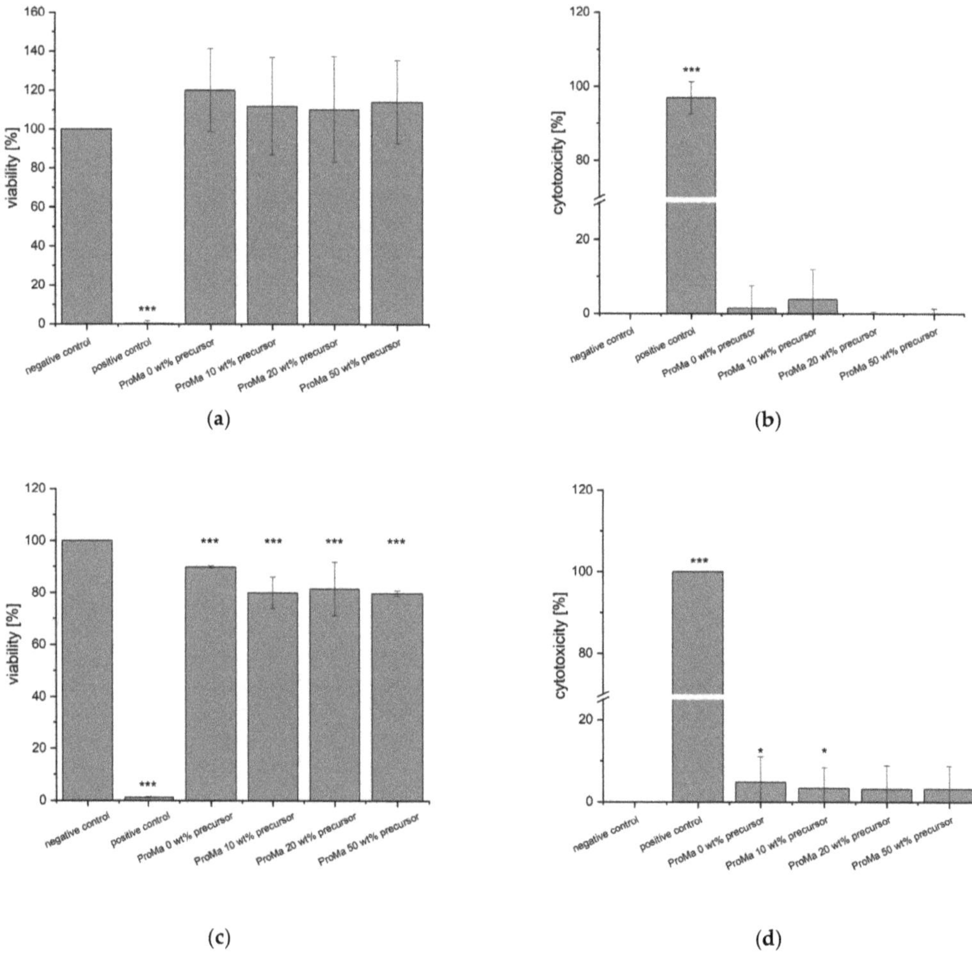

Figure 8. (**a**) cell viability [%] of HaCaT cells exposed to sample extracts in DMEM for 24 h determined with bioluminescent ATP assay (**b**) cytotoxicity [%] of HaCaT cells exposed to sample extracts in DMEM for 24 h, determined with LDH-Assay (**c**) cell viability [%] of Fibroblasts exposed to sample extracts in DMEM for 24 h, determined with bioluminescent ATP assay (**d**) cytotoxicity [%] of Fibroblasts exposed to sample extracts in DMEM for 24 h, determined with LDH-Assay. Asterisks indicate significant deviations from the negative control (* $p < 0.05$; *** $p < 0.001$).

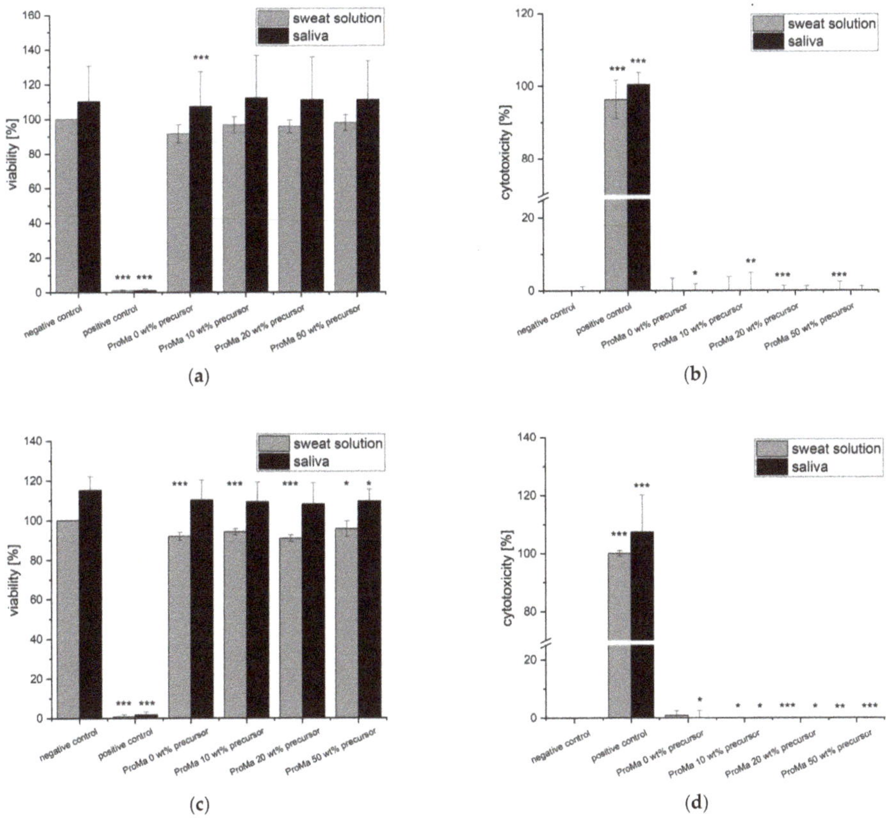

Figure 9. (**a**) cell viability [%] of HaCaT cells exposed to sample extracts in artificial sweat solution (SSL) and artificial saliva (SV) for 24 h, determined with bioluminescent ATP Assay (**b**) cytotoxicity [%] of HaCaT cells exposed to sample extracts in SSL and SV for 24 h, determined with LDH Assay (**c**) cell viability [%] of Fibroblasts exposed to sample extracts in SSL and SV for 24 h, determined with bioluminescent ATP Assay (**d**) cytotoxicity [%] of Fibroblasts exposed to sample extracts in SSL and SV for 24 h, determined with LDH Assay. Asterisks indicate significant deviations from the negative control (* $p < 0.05$; ** $p < 0.01$; *** $p < 0.001$).

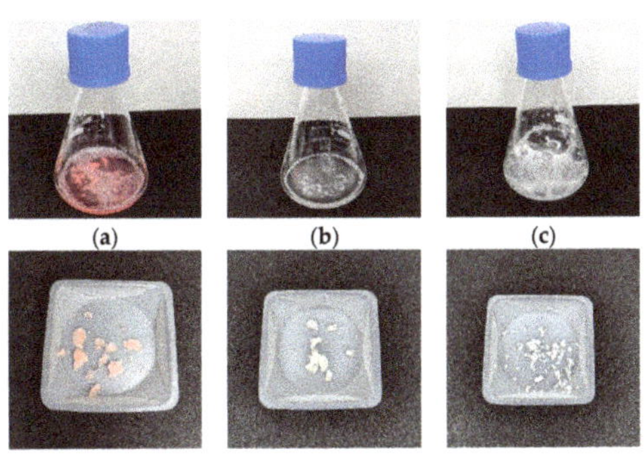

Figure 10. Extract solutions (**a**) DMEM, (**b**) artifical sweat solution, and (**c**) artificial saliva after 24 h, 37 °C and the sample after supernatant removal.

2.9. In-Vitro Inflammation Potential Testing

Another important factor for asserting consumer safety of new materials, especially those which are based on natural polymers, is the inflammatory potential [29,87]. Inflammation is a natural defense mechanism, but excessive or prolonged inflammation can lead to tissue damage and other health complications. Testing materials for their potential to induce inflammation helps to ensure that they do not cause harmful effects when exposed to living tissues or cells.

During inflammation, various cells, particularly leukocytes such as mast cells and eosinophils, produce and release leukotrienes [88]. Upon stimulation by inflammatory signals, the cell membrane phospholipids release arachidonic acid. The arachidonic acids activates the 5-lipoxygenase, which converts arachidonic acid into a reactive intermediate, being then further processed to leukotriene B4 (LTB4) and the cysteinyl leukotrienes (LTC4, LTD4, and LTE4) [89,90]. The produced leukotrienes contribute to the inflammatory response by promoting vasodilation and increasing vascular permeability, as well as attracting and activating immune cells [91]. They act as signaling molecules, homing other leukocytes to the site of inflammation and promoting the release of other inflammatory mediators, such as cytokines [92]. The levels of leukotrienes (LTC4, LTD4, and LTE4) in the affected tissues or body fluids can be measured to assess the extent and severity of inflammation. An immunoassay technique [31] was used to measure the release of sulfidoleukotrienes (sLTs) from isolated leucocytes after incubation with the respective material extracts (Figure 11).

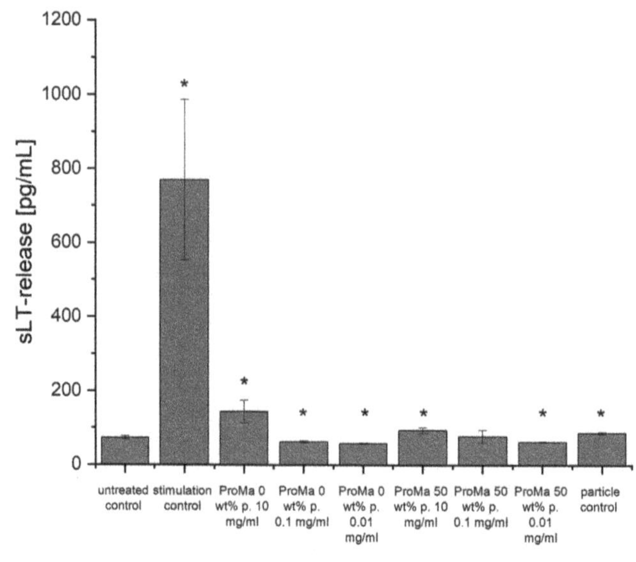

Figure 11. Concentrations of sulfidoleukotrienes (sLT) in (pg/mL). For the leukocyte stimulation assay, the measurements were performed in triplicate (n = 3) for each individual and in duplicate (n = 2) for ELISA. Asterisks indicate significant deviations from the untreated control (* $p < 0.05$).

The concentrations of sLTs (expressed in picograms per milliliter, pg/mL) were determined. A reading of 100 pg/mL above the untreated control indicates a significant inflammatory stimulation [93,94]. However, none of the tested materials resulted in a peak concentration that high. The polymer matrix ProMa 0 10 mg/mL led to a slight increase in the sLT production, which could be an effect of the sample's degradation into the buffer and the cell interaction with polypeptides and organic acids. As the assay components probably stabilize the pH sufficiently, the slight increase in sLT release might be caused by degradation of the sample so that the components would be free to interact with the leukocytes. The samples with incorporated silica demonstrated greater withstanding to dissolution. However, they also release ions into the buffer that reduce the pH. The materials were milled to yield small particles and formed an emulsion with the buffer. Since particles can lead to an increase in sLT production due to physical disruption of the cell membrane, a particle control was also prepared. There is a correlation between the leukocyte stimulation assay and skin sensitization testing [95]. By adopting it to safety evaluations, an effective screening procedure for testing of the inflammatory potential of materials can be established, thereby promoting consumer safety. The comprehensive evaluation of materials for their inflammatory potential and skin sensitization is crucial in ensuring the safety and efficacy of materials used in the food industry. This testing approach helps identify potential risks, allowing for the development of safer and more biocompatible materials and products. As a result, it plays a pivotal role in enhancing consumer trust and ensuring the overall quality of food-related materials and applications.

3. Conclusions

This study presents the development of a composite gel, consisting of a biodegradable polymer matrix based on prolamins with increased incorporation of a silica precursor. The functional oxygen barrier of the gel in food packaging materials was effectively demonstrated when subjected to low relative humidity conditions, as evidenced by the oxygen transmission rate measurements. However, when precursor amounts were increased to 50 wt%, a slight decrease in the gel's oxygen permeability was observed under higher relative humidity conditions. This can be attributed to the inherent property of silica particles to absorb moisture and subsequently safeguard the integrity of the gel matrix. The addition of precursor amounts did not enhance the gel's barrier properties in low humidity environments, nor did it diminish the gel's oxygen barrier properties in such conditions. Significantly, the incorporation of the silane precursor greatly improved water stability. The resulting silica particles interacted with hydroxyl groups in polysaccharides, polypeptides and citric acid reducing their interactions with water and restricting polymer chain movement. It was observed that higher precursor amounts improved mechanical resistance but also increased brittleness, as molecular movement of the polymer chains became hindered, and entanglements increased. This was accompanied by an increase in surface roughness due to particle agglomeration, which further supported the entanglement of the gels. Future research endeavors should prioritize the optimization of water stability and hygroscopic properties to achieve a reduced oxygen transmission rate, particularly in elevated humidity environments.

The cell compatibility of the gels was assessed through cytotoxicity testing using human keratinocytes and fibroblasts, according to DIN EN ISO 10993-5. Encouragingly, no cytotoxic effects were detected, indicating the safety of the gels for consumer contact. Furthermore, a leukocyte stimulation test was performed to evaluate the gel's potential for causing inflammatory effects in end consumer contact scenarios. From the results obtained, it can be concluded that the gels are safe, based on the parameters of the selected bioassays. However, it is necessary to recognize that the leukocyte testing alone might not be sufficient to comprehensively evaluate the interaction of the gels with human skin. To ensure more comprehensive testing, for further research additional in vitro assays for possible skin sensitization and evaluation of allergenic potential is recommended. This step is crucial in assessing any potential risks related to sensitization and ensuring the overall safety and efficacy of the gels for consumer use.

Overall, this research contributes valuable insight to the development of composite gels for potential food packaging applications, demonstrating promising oxygen barrier properties and cytocompatibility. The study emphasizes the significance of thorough testing to guarantee the safety and suitability of these materials for human contact.

4. Materials and Methods

4.1. Materials

Wheat Gluten (CAS No. 8002-80-0) for prolamin extraction was purchased from Carl Roth (Karlsruhe, Germany) with a >75% protein content purity, d-Mannose > 99% (CAS No. 3458-28-4) was bought from Thermo Fisher Scientific (Waltham, MA, USA). The precursor tetraethyl orthosilicate (CAS No.78-10-4) and tetramethyl orthosilicate (CAS No. 681-84-5) were purchased with 98% purity from Sigma Aldrich (St. Louis, MO, USA). The citric acid (99.9%, CAS No. 77-92-9) was acquired from VWR International (Dresden, Germany). For the cell medium, the Dulbecco's Modified Eagle Medium DMEM was bought from BioConcept Ltd. (Allschwil, Switzerland). The antibiotic-antimycotic solution (10,000 U/mL penicillin, 10,000 µm/mL streptomycin, 25 µg/mL amphotericin) was bought from PromoCell (Heidelberg, Germany), the fetal calf serum from PAN Biotech (Aidenbach, Germany) and the insulin (5 mg/mL) from PELOBiotech (Planegg, Germany). The Recombinant Human Fibroblast Growth Factor-2 (rhFGF2) was bought from Sartorius CellGenix GmbH (Freiburg, Germany), trypsin ethylenediaminetetraacetic acid and gentamycin (10 mg/mL) from Gibco, Life Technologies Limited (Thermo Fisher Scientific, Waltham, MA, USA). The bioassay ATPLite M Assay was purchased from PerkinElmer (Waltham, MA, USA) the cytotoxicity detection kit (LDH) from Roche (Basel, Switzerland) and the CAST® ELISA for the sulfidoleukotrienes detection from Bühlmann Laboratories (Schönenbuch, Switzerland). The dextran solution was also obtained from Bühlmann Laboratories (Schönenbuch, Switzerland). For the preparation of the artificial sweat solution the natriumchlorid NaCl (>99.5%, CAS No. 7647-14-5), the hydrochloric acid 37% HCL (CAS No. 7647-01-0) and the sodium hydroxide NaOH (>99%, CAS No. 1310-73-2) were all purchased from Carl Roth (Karlsruhe, Germany) and the L-Histidine $C_6H_9O_2N_3 . H_2O$ (>98.5%, CAS No. 5934-29-2) from AppliChem GmbH (Darmstadt, Germany). The sodium dihydrogen phosphate dehydrate $NaH_2PO_4 \cdot 2H_2O$ (> 98%, CAS No. 13472-35-0) and artificial saliva for pharmaceuticals acquired from Sigma Aldrich (St. Louis, MO, USA). For extract sterile filtration a 0.2 µm filter from Sarstedt AG & Co.KG (Nümbrecht, Germany) and for blood storage EDTA monovettes were used (Sarstedt AG & Co.KG).

4.2. Preparation of Composite Gel

In a beaker, 20 g of wheat gluten were extracted with 200 mL of 70% ethanol. After stirring for 1 h the mixture was centrifuged with 4000 rpm for 10 min. The supernatant was used for further synthesis. 1 g of d-mannose was solved in 10 mL supernatant of the prolamin extraction. The precursors tetraethoxysilane and tetramethoxysilane were mixed in the volume ratio 30:70% *v*/*v*. Then 0.14 mL, 0.28 mL and 0.68 mL of this precursor was pipetted in the pre solution of prolamin and d-mannose. Anhydrous citric acid was added; the solution was stirred for 1 h and altered for 12 h at room temperature.

For the texture analysis, the sol-gels (ProMa) were weighed in at 6 g in a 70 × 40 mm Teflon cast and dried for 24 h in a desiccator filled with dry silica. For the water stability testing a volume of 1 mL of the solution was carefully pipetted into 25 mm^2 Teflon casts and allowed to dry for a duration of 24 h.

4.3. IR Spectroscopy Studies

The chemical structure and composition of the gels were studied using the IR MB 3000, ABB Automation products GmbH (Alzenau, Germany). The gels were applied on double polishes silicon wafers, and dried at 80 °C in the oven for 30 min. The prepared samples were then analyzed with FT-IR spectroscopy, absorbance spectrum 4000–500 cm^{-1}.

4.4. Tensile Strength Examinations of Composite Gels

The tensile strength of the gels was determined using the Texture Analyzer TA. XT2i, Stable Micro Systems Ltd. The films were punched with a stamping press into samples of 35 mm length, 2 mm width. For the sample with 50 wt% precursor it was not possible to solvent cast a stable film, the thickness was too high and the material to brittle. The samples were clamped into two brackets of the texture analyzer with gap distance of 20 mm. The thickness of each sample was measured before the experiment using a caliper gauge and multiplied with the height of the experiment design to determine the cross-section area of the samples. The samples were pulled apart with a speed of 1 mm/s until they break. The tensile strength stress σ and strain ε can be derived, this was determined from measured load and deflection the material sample, cross sectional area S0 and length l0. The stress strain curve was determined following Equations (1) and (2) [41].

$$\sigma = \frac{F}{S_0} \tag{1}$$

$$\varepsilon = \frac{dl}{l0} \tag{2}$$

4.5. Atomic Force Microscopy Analysis

The surface analysis of the gels was obtained with the MFP 3D-Classic, Asylum Research. The gels ProMa 0 wt% precursor and ProMa 50 wt% precursor were diluted 1:1 V% with 70% ethanol and applied in a thin layer on a polished silicon wafer, dried in an oven at 80 °C for 30 min. After drying observations were carried out using an MFP 3D-Classic Atomic Force Microscope (Asylum Research, Santa Barbara, CA, USA), operating in air under ambient humidity and room temperature conditions. Gels samples were examined in tapping mode with a resolution of 256 × 256. The drive frequency was 273.9 kHz. Areas of 1 × 1 micrometer and 10 × 10 micrometer were acquired, with an aluminum coated Micro Cantilever of 160 µm in length (160AC-NA, OPUS by MikroMasch, Sofia, Bulgaria) with a force constant 26 N/m (nominal value). The images were analyzed by MarSurf MfM Premium (Version 7.4.8737) according to ISO 25178 [96], including the calculation of roughness parameters.

4.6. Scanning Electron Microscopy Analysis

The morphology of the samples was analyzed on the surface and the cross section of dried gel samples. The gel samples were dispersed in 80%, 90%, 95% and 99% ethanol for each 5 min and dried using CPD 7501 Critical Point Drying Apparatus (Quorum Technologies, East Sussex, UK). The samples were gold sputtered with a turbomolecular-pumped sputter coater Quorum Q150V ES plus (Quorum Technologies, East Sussex, UK) and analyzed using Field Emission SEM Zeiss Gemini SEM 460 (Carl Zeiss AG, Oberkochen, Germany).

4.7. Rheological Investigations of Composite Gel

The viscosity of gel solutions was analyzed at different altering states with an Rheometer MCR 301 (Anton Paar GmbH, Graz, Austria) at T = +25 °C with a cone plate system (25 mm, 1 °C angle) following the measuring methodology of Mezger [97]. The measurement is shearing rate controlled with logarithmically increasing steps. With increasing shear rate, the measuring point duration was steadily shortened, starting at the shear rate of 0.001 Hz with 1000 s per measuring point and at the end of the experiment at the shear rate of 1000 Hz with only 1 s per measuring point. For the solutions altered for 1 d, 3 d, 7 d the measuring point duration was steadily shortened, starting at the shear rate of 0.001 1/s with 350 s per measuring point and at the end of the experiment at the shear rate of 1000 1/s with 10 s per measuring point.

4.8. Water Solubility of Composite Gels

The water solubility was determined in accordance to the protocol developed by Pearson [98]. After the drying process, the samples were weighed and subsequently immersed in 10 mL of deionized water for specified time intervals (1 min, 5 min, 15 min, 30 min, 60 min, 300 min). Following the immersion period, the water was filtered using a paper filter (Grade 392, Satorius AG, Göttingen, Germany), dried at room temperature for 24 h, and weighed again. The weight loss percentage was calculated by comparing the initial weight of the samples before immersion with the final weight after filtration and drying.

4.9. Gas Permeability of Composite Gels on PET

The gels were applied as a 4 μm wet film (dried film thickness 900 nm) onto a polyethylene foil (PET) with a thickness of 12 μm. Subsequently, the samples underwent a drying process at 80 °C for a duration of 30 min. After drying, the samples were carefully inserted into a Gas-Transmission Tester, manufactured by Brugger Feinmechanik GmbH. The testing procedure employed oxygen as the test gas following the ISO 15105-1 [99] guidelines and demonstrated from Startek et al. [64]. The chamber was maintained at a temperature of 23 °C, and the relative humidity was adjusted to 20%, 40%, 60%, and 80% relative humidity (r.H.).

4.10. In-Vitro Cytotoxicity Testing

Cell compatibility was assessed by preparing sample extracts following the DIN EN ISO 10993-12 [100] guidelines as previously reported [77]. To accomplish this, 200 mg samples (n = 3) of each material were mixed with either 10 mL of DMEM, 10 mL of a controlled sweat solution (5 g NaCl, 2.2 g $NaH_2PO_4 \cdot 2H_2O$, 0.5g $C_6H_9O_2N_3 \cdot H_2O$ + HCL, 0.1 N NaOH—pH 5.5) or artificial saliva for pharmaceuticals (250 mL, Sigma-Aldrich) in Erlenmeyer flasks. The flasks were then shaken in a ThermoBath (GFL, Deutschland) for 24 h at 37 °C. Following the incubation period, any residual material was removed, and the extracts were sterilized using filtration through a 0.2 μm filter (Sarstedt AG & Co.KG).

Human keratinocytes HaCaT (provided by Prof. Fusenig, Heidelberg) were cultured in DMEM supplemented with 1% antibiotic-antimycotic solution and 10% fetal calf serum in 75 cm^2 cell culture flasks. Normal human fibroblasts (PELOBiotech GmbH, Planegg, Germany) were cultured in DMEM supplemented with 500 μL insulin, 2.5 mL gentamycin, 250 μL rhFGF2 and 2% fetal calf serum. The cultures were maintained at 37 °C in a humidified atmosphere with 5% CO_2 for seven days.

Afterwards, the cells were detached using trypsin ethylenediaminetetraacetic acid and seeded into 96-well plates at a cell density of 2×10^5 cells/mL human keratinocytes and 1×10^5 cells/mL fibroblasts. The plates were then incubated for 48 h. Subsequently, the medium was replaced with fresh medium, a controlled sweat solution or artificial saliva (negative control), material extracts, and Triton-X 100 (cytotoxicity control). The cells were incubated for 24 h.

Cell viability and proliferation were assessed using the ATPLite M Assay according to the manufacturer's instructions. The cell count was calculated as a percentage of the negative control at each respective time point. The cytotoxic effects of the material were determined using the cytotoxicity detection kit according to the manufacturer's instructions. Optical density was measured at 490 nm using a SPECTROstar-Omega plate reader (BGM Labtech). Cytotoxicity was calculated using Equation (3).

$$\text{cytotoxicity } [\%] = \frac{\text{n} - \text{positive control}}{\text{negative control} - \text{positive control} \cdot 100} \quad (3)$$

4.11. In-Vitro Inflammation Potential Testing

The inflammation potential was assessed using an enzyme immunoassay for the quantitative determination of sulfidoleukotrienes produced by isolated leukocytes.

Materials were dried and milled in fine particles and suspensions of 10 mg/mL, 0.1 mg/mL and 0.01 mg/mL were prepared in stimulation buffer. For particle control pure SiO_2 particles were prepared following the same process as for the gels, a dispersion of 10 mg/mL was prepared. Particles were washed 5 times with distilled H_2O to remove possible contaminants.

For the study, EDTA whole blood samples from two patients were collected [101] and stored in EDTA monovettes. The blood was transferred into 50 mL tubes, and 15 mL of the blood sample was mixed with 4 mL of dextran solution. The mixture was inverted and allowed to sediment for 30 min. After sedimentation, the supernatant was carefully collected using a Pasteur pipette and transferred into new 50 mL tubes. These tubes were then centrifuged at 160× g for 10 min to remove any remaining supernatant. The discarded supernatant was replaced with 15 mL of stimulation buffer, which was added to the isolated leukocytes. A volume of 200 µL of the leukocyte stimulation buffer solution was subsequently added to the samples and controls, with each determination performed in triplicate. The samples were sealed and incubated at 37 °C for 50 min. Following incubation, the samples were vortexed and centrifuged at 4 °C, 1000× g for 6 min. The supernatants from the triplicate determinations were combined (pooled). These pooled supernatants were then frozen and stored at −20 °C for further analysis.

The concentration of sulfidoleukotrienes (sLTC4, sLTD4, sLTE4) was determined using the CAST assay kit. The release of sulfidoleukotrienes (sLT) from the stimulation control and allergen was calculated using the following Equation (4):

$$\text{sLT} = \text{sLT}\ [\alpha]\left(\frac{\text{pg}}{\text{mL}}\right) - \text{sLT}\ [\text{backround}]\left(\frac{\text{pg}}{\text{mL}}\right) \quad (4)$$

Here, "α" represents the control and samples.

The color absorbance was measured spectrophotometrically at a wavelength of 405 nm. For the quantification of sLTD4, a four-parameter fit regression model was used, with a calibration range of 50–3200 pg/mL. To determine a positive stimulation response, a threshold of ≥100 pg sLT/mL was considered after subtracting the value obtained from the negative control.

4.12. Statistical Analysis

IBM SPSS Software 29.0.0.0 was used to test the biological data for normal distribution, subsequently a Mann–Whitney-U-Test was carried out to determine the central tendencies of two independent random samples. Asterisks indicate significant deviations from the negative control (* $p < 0.05$; ** $p < 0.01$; *** $p < 0.001$).

Author Contributions: Conceptualization, F.T., C.W. and T.T.; Methodology, F.T. and C.W.; Software, F.T.; Validation, F.T.; Formal analysis, F.T.; Investigation, F.T.; Resources, C.W. and T.T.; Data curation, F.T.; Writing—original draft preparation, F.T.; Writing—review and editing, T.T. and C.W.; Supervision, C.W. All authors have read and agreed to the published version of the manuscript.

Funding: This research was partly funded by the German federal ministry for economics and climate action (project number 49MF200077).

Institutional Review Board Statement: Leukocyte isolation from the blood of healthy volunteers was conducted according to the guidelines of the Declaration of Helsinki and approved by the ethics committee of the medical faculty of the Friedrich Schiller University Jena, Jena, Germany (2022-2690_1). Blood donors gave their written informed consent prior to the voluntary blood donation.

Informed Consent Statement: Not applicable.

Data Availability Statement: Data available on request.

Acknowledgments: We would like to thank Torsten Walter for so kindly sharing his professional expertise with us, as well as Jun Xu, Denise Rietz, Petra Koecher, and Martina Schweder for the technical support.

Conflicts of Interest: The authors declare no conflict of interest.

References

1. R, R.; Philip, E.; Madhavan, A.; Sindhu, R.; Pugazhendhi, A.; Binod, P.; Sirohi, R.; Awasthi, M.K.; Tarafdar, A.; Pandey, A. Advanced Biomaterials for Sustainable Applications in the Food Industry: Updates and Challenges. *Environ. Pollut.* **2021**, *283*, 117071. [CrossRef]
2. Calva-Estrada, S.J.; Jiménez-Fernández, M.; Lugo-Cervantes, E. Protein-Based Films: Advances in the Development of Biomaterials Applicable to Food Packaging. *Food Eng. Rev.* **2019**, *11*, 78–92. [CrossRef]

3. Niranjana Prabhu, T.; Prashantha, K. A Review on Present Status and Future Challenges of Starch Based Polymer Films and Their Composites in Food Packaging Applications. *Polym. Compos.* **2018**, *39*, 2499–2522. [CrossRef]
4. Taherimehr, M.; YousefniaPasha, H.; Tabatabaeekoloor, R.; Pesaranhajiabbas, E. Trends and Challenges of Biopolymer-based Nanocomposites in Food Packaging. *Compr. Rev. Food Sci. Food Saf.* **2021**, *20*, 5321–5344. [CrossRef]
5. Luo, Q.; Hossen, M.A.; Zeng, Y.; Dai, J.; Li, S.; Qin, W.; Liu, Y. Gelatin-Based Composite Films and Their Application in Food Packaging: A Review. *J. Food Eng.* **2022**, *313*, 110762. [CrossRef]
6. Basumatary, I.B.; Mukherjee, A.; Kumar, S. Chitosan-Based Composite Films Containing Eugenol Nanoemulsion, ZnO Nanoparticles and Aloe Vera Gel for Active Food Packaging. *Int. J. Biol. Macromol.* **2023**, *242*, 124826. [CrossRef] [PubMed]
7. Venkatesan, R.; Rajeswari, N. ZnO/PBAT Nanocomposite Films: Investigation on the Mechanical and Biological Activity for Food Packaging: ZnO/PBAT Nanocomposite Films. *Polym. Adv. Technol.* **2017**, *28*, 20–27. [CrossRef]
8. Zhang, W.; Ahari, H.; Zhang, Z.; Jafari, S.M. Role of Silica (SiO2) Nano/Micro-Particles in the Functionality of Degradable Packaging Films/Coatings and Their Application in Food Preservation. *Trends Food Sci. Technol.* **2023**, *133*, 75–86. [CrossRef]
9. Schilling, T.; Habel, C.; Rosenfeldt, S.; Röhrl, M.; Breu, J. Impact of Ultraconfinement on Composite Barriers. *ACS Appl. Polym. Mater.* **2020**, *2*, 3010–3015. [CrossRef]
10. Pinto, L.; Bonifacio, M.A.; De Giglio, E.; Santovito, E.; Cometa, S.; Bevilacqua, A.; Baruzzi, F. Biopolymer Hybrid Materials: Development, Characterization, and Food Packaging Applications. *Food Packag. Shelf Life* **2021**, *28*, 100676. [CrossRef]
11. Sun, M.; Liu, N.; Ni, S.; Bian, H.; Fu, Y.; Chen, X. Poplar Hot Water Extract Enhances Barrier and Antioxidant Properties of Chitosan/Bentonite Composite Film for Packaging Applications. *Polymers* **2019**, *11*, 1614. [CrossRef] [PubMed]
12. Chen, J.; Luo, L.; Cen, C.; Liu, Y.; Li, H.; Wang, Y. The Nano Antibacterial Composite Film Carboxymethyl Chitosan/Gelatin/Nano ZnO Improves the Mechanical Strength of Food Packaging. *Int. J. Biol. Macromol.* **2022**, *220*, 462–471. [CrossRef] [PubMed]
13. Kumari, S.V.G.; Pakshirajan, K.; Pugazhenthi, G. Recent Advances and Future Prospects of Cellulose, Starch, Chitosan, Polylactic Acid and Polyhydroxyalkanoates for Sustainable Food Packaging Applications. *Int. J. Biol. Macromol.* **2022**, *221*, 163–182. [CrossRef] [PubMed]
14. Zhang, W.; Rhim, J.-W. Titanium Dioxide (TiO2) for the Manufacture of Multifunctional Active Food Packaging Films. *Food Packag. Shelf Life* **2022**, *31*, 100806. [CrossRef]
15. Mesgari, M.; Aalami, A.H.; Sathyapalan, T.; Sahebkar, A. A Comprehensive Review of the Development of Carbohydrate Macromolecules and Copper Oxide Nanocomposite Films in Food Nanopackaging. *Bioinorg. Chem. Appl.* **2022**, *2022*, 7557825. [CrossRef]
16. Rovera, C.; Türe, H.; Hedenqvist, M.S.; Farris, S. Water Vapor Barrier Properties of Wheat Gluten/Silica Hybrid Coatings on Paperboard for Food Packaging Applications. *Food Packag. Shelf Life* **2020**, *26*, 100561. [CrossRef]
17. Videira-Quintela, D.; Martin, O.; Montalvo, G. Emerging Opportunities of Silica-Based Materials within the Food Industry. *Microchem. J.* **2021**, *167*, 106318. [CrossRef]
18. Schmid, M.; Dallmann, K.; Bugnicourt, E.; Cordoni, D.; Wild, F.; Lazzeri, A.; Noller, K. Properties of Whey-Protein-Coated Films and Laminates as Novel Recyclable Food Packaging Materials with Excellent Barrier Properties. *Int. J. Polym. Sci.* **2012**, *2012*, e562381. [CrossRef]
19. Liu, J.; Huang, W.; Xing, Y.; Li, R.; Dai, J. Preparation of Durable Superhydrophobic Surface by Sol–Gel Method with Water Glass and Citric Acid. *J. Sol-Gel Sci. Technol.* **2011**, *58*, 18–23. [CrossRef]
20. Lei, B.; Chen, X.; Koh, Y.-H. Effects of Acidic Catalysts on the Microstructure and Biological Property of Sol–Gel Bioactive Glass Microspheres. *J. Sol-Gel Sci. Technol.* **2011**, *58*, 656–663. [CrossRef]
21. Coradin, T.; Allouche, J.; Boissiere, M.; Livage, J. Sol-Gel Biopolymer/Silica Nanocomposites in Biotechnology. *Curr. Nanosci.* **2006**, *2*, 219–230. [CrossRef]
22. Mesa, M.; Becerra, N.Y. Silica/Protein and Silica/Polysaccharide Interactions and Their Contributions to the Functional Properties of Derived Hybrid Wound Dressing Hydrogels. *Int. J. Biomater.* **2021**, *2021*, 1–13. [CrossRef] [PubMed]
23. Fluhr, J.W.; Darlenski, R.; Angelova-Fischer, I.; Tsankov, N.; Basketter, D. Skin Irritation and Sensitization: Mechanisms and New Approaches for Risk Assessment. 1. Skin Irritation. *Skin Pharmacol. Physiol.* **2008**, *21*, 124–135. [CrossRef] [PubMed]
24. Esser, P.R.; Wölfle, U.; Dürr, C.; von Loewenich, F.D.; Schempp, C.M.; Freudenberg, M.A.; Jakob, T.; Martin, S.F. Contact Sensitizers Induce Skin Inflammation via ROS Production and Hyaluronic Acid Degradation. *PLoS ONE* **2012**, *7*, e41340. [CrossRef] [PubMed]
25. Watanabe, H.; Unger, M.; Tuvel, B.; Wang, B.; Sauder, D.N. Review: Contact Hypersensitivity: The Mechanism of Immune Responses and T Cell Balance. *J. Interferon Cytokine Res.* **2002**, *22*, 407–412. [CrossRef]
26. Bonneville, M.; Chavagnac, C.; Vocanson, M.; Rozieres, A.; Benetiere, J.; Pernet, I.; Denis, A.; Nicolas, J.-F.; Hennino, A. Skin Contact Irritation Conditions the Development and Severity of Allergic Contact Dermatitis. *J. Investig. Dermatol.* **2007**, *127*, 1430–1435. [CrossRef]
27. Avonto, C.; Chittiboyina, A.G.; Wang, M.; Vasquez, Y.; Rua, D.; Khan, I.A. In Chemico Evaluation of Tea Tree Essential Oils as Skin Sensitizers: Impact of the Chemical Composition on Aging and Generation of Reactive Species. *Chem. Res. Toxicol.* **2016**, *29*, 1108–1117. [CrossRef]
28. Li, Y.; Damodaran, S. In Vitro Digestibility and IgE Reactivity of Enzymatically Cross-Linked Heterologous Protein Polymers. *Food Chem.* **2017**, *221*, 1151–1157. [CrossRef]
29. Mortimer, S.; Reeder, M. Botanicals in Dermatology: Essential Oils, Botanical Allergens, and Current Regulatory Practices. *Dermatitis* **2016**, *27*, 317–324. [CrossRef]

30. Barber, D.; Arias, J.; Boquete, M.; Cardona, V.; Carrillo, T.; Gala, G.; Gamboa, P.; García-Robaina, J.C.; Hernández, D.; Sanz, M.L.; et al. Analysis of Mite Allergic Patients in a Diverse Territory by Improved Diagnostic Tools. *Clin. Exp. Allergy* **2012**, *42*, 1129–1138. [CrossRef]
31. de Weck, A.L.; Sanz, M.L. Cellular Allergen Stimulation Test (CAST) 2003, a Review. *J. Investig. Allergol. Clin. Immunol.* **2004**, *14*, 253–273. [PubMed]
32. Shin, Y.; Lee, D.; Lee, K.; Ahn, K.H.; Kim, B. Surface Properties of Silica Nanoparticles Modified with Polymers for Polymer Nanocomposite Applications. *J. Ind. Eng. Chem.* **2008**, *14*, 515–519. [CrossRef]
33. Gaharwar, A.K.; Rivera, C.P.; Wu, C.-J.; Schmidt, G. Transparent, Elastomeric and Tough Hydrogels from Poly(Ethylene Glycol) and Silicate Nanoparticles. *Acta Biomater.* **2011**, *7*, 4139–4148. [CrossRef]
34. Ji, Y.; Yang, X.; Ji, Z.; Zhu, L.; Ma, N.; Chen, D.; Jia, X.; Tang, J.; Cao, Y. DFT-Calculated IR Spectrum Amide I, II, and III Band Contributions of *N* -Methylacetamide Fine Components. *ACS Omega* **2020**, *5*, 8572–8578. [CrossRef] [PubMed]
35. Kanou, M.; Nakanishi, K.; Hashimoto, A.; Kameoka, T. Influences of Monosaccharides and Its Glycosidic Linkage on Infrared Spectral Characteristics of Disaccharides in Aqueous Solutions. *Appl. Spectrosc.* **2005**, *59*, 885–892. [CrossRef]
36. Rubio, F.; Rubio, J.; Oteo, J.L. A FT-IR Study of the Hydrolysis of Tetraethylorthosilicate (TEOS). *Spectrosc. Lett.* **1998**, *31*, 199–219. [CrossRef]
37. Freda, M.; Piluso, A.; Santucci, A.; Sassi, P. Transmittance Fourier Transform Infrared Spectra of Liquid Water in the Whole Mid-Infrared Region: Temperature Dependence and Structural Analysis. *Appl. Spectrosc.* **2005**, *59*, 1155–1159. [CrossRef]
38. Wexler, A.S. Infrared Determination of Structural Units in Organic Compounds by Integrated Intensity Measurements: Alkanes, Alkenes and Monosubstituted Alkyl Benzenes. *Spectrochim. Acta* **1965**, *21*, 1725–1742. [CrossRef]
39. Riahi, Z.; Khan, A.; Rhim, J.-W.; Shin, G.H.; Kim, J.T. Gelatin/Poly(Vinyl Alcohol)-Based Dual Functional Composite Films Integrated with Metal-Organic Frameworks and Anthocyanin for Active and Intelligent Food Packaging. *Int. J. Biol. Macromol.* **2023**, *249*, 126040. [CrossRef]
40. Chu, Y.Y.; Song, X.F.; Zhao, H.X. Water-Swellable, Tough, and Stretchable Inorganic–Organic Sulfoaluminate Cement/Polyacrylamide Double-Network Hydrogel Composites. *J. Appl. Polym. Sci.* **2019**, *136*, 47905. [CrossRef]
41. Conradi, M. Nanosilica-Reinforced Polymer Composites. *Mater. Tehnol.* **2013**, *47*, 285–293.
42. Chen, L.; Chai, S.; Liu, K.; Ning, N.; Gao, J.; Liu, Q.; Chen, F.; Fu, Q. Enhanced Epoxy/Silica Composites Mechanical Properties by Introducing Graphene Oxide to the Interface. *ACS Appl. Mater. Interfaces* **2012**, *4*, 4398–4404. [CrossRef] [PubMed]
43. Hashemi Tabatabaei, R.; Jafari, S.M.; Mirzaei, H.; Mohammadi Nafchi, A.; Dehnad, D. Preparation and Characterization of Nano-SiO2 Reinforced Gelatin-k-Carrageenan Biocomposites. *Int. J. Biol. Macromol.* **2018**, *111*, 1091–1099. [CrossRef] [PubMed]
44. Zhang, W.; Zhang, Y.; Cao, J.; Jiang, W. Improving the Performance of Edible Food Packaging Films by Using Nanocellulose as an Additive. *Int. J. Biol. Macromol.* **2021**, *166*, 288–296. [CrossRef] [PubMed]
45. Rouf, T.B.; Schmidt, G.; Cakmak, M.; Kokini, J.L. Design and Mechanistic Understanding of Graphene Oxide Reinforced Zein Nanocomposites with Improved Mechanical, Barrier and Thermal Properties. *J. Mater. Sci.* **2019**, *54*, 12533–12552. [CrossRef]
46. Zhou, H.; Tong, H.; Lu, J.; Cheng, Y.; Qian, F.; Tao, Y.; Wang, H. Preparation of Bio-Based Cellulose Acetate/Chitosan Composite Film with Oxygen and Water Resistant Properties. *Carbohydr. Polym.* **2021**, *270*, 118381. [CrossRef]
47. Beigzadeh Ghelejlu, S.; Esmaiili, M.; Almasi, H. Characterization of Chitosan–Nanoclay Bionanocomposite Active Films Containing Milk Thistle Extract. *Int. J. Biol. Macromol.* **2016**, *86*, 613–621. [CrossRef]
48. Omerović, N.; Djisalov, M.; Živojević, K.; Mladenović, M.; Vunduk, J.; Milenković, I.; Knežević, N.Ž.; Gadjanski, I.; Vidić, J. Antimicrobial Nanoparticles and Biodegradable Polymer Composites for Active Food Packaging Applications. *Compr. Rev. Food Sci. Food Saf.* **2021**, *20*, 2428–2454. [CrossRef]
49. Pasquier, E.; Mattos, B.D.; Koivula, H.; Khakalo, A.; Belgacem, M.N.; Rojas, O.J.; Bras, J. Multilayers of Renewable Nanostructured Materials with High Oxygen and Water Vapor Barriers for Food Packaging. *ACS Appl. Mater. Interfaces* **2022**, *14*, 30236–30245. [CrossRef]
50. Fan, Z.; Cheng, P.; Zhang, P.; Zhang, G.; Han, J. Rheological Insight of Polysaccharide/Protein Based Hydrogels in Recent Food and Biomedical Fields: A Review. *Int. J. Biol. Macromol.* **2022**, *222*, 1642–1664. [CrossRef]
51. Lee, H.; Rukmanikrishnan, B.; Lee, J. Rheological, Morphological, Mechanical, and Water-Barrier Properties of Agar/Gellan Gum/Montmorillonite Clay Composite Films. *Int. J. Biol. Macromol.* **2019**, *141*, 538–544. [CrossRef] [PubMed]
52. Rukmanikrishnan, B.; Ramalingam, S.; Kim, S.S.; Lee, J. Rheological and Anti-Microbial Study of Silica and Silver Nanoparticles-Reinforced k-Carrageenan/Hydroxyethyl Cellulose Composites for Food Packaging Applications. *Cellulose* **2021**, *28*, 5577–5590. [CrossRef]
53. Habib, M.A.; Khoda, B. Rheological Analysis of Bio-Ink for 3D Bio-Printing Processes. *J. Manuf. Process.* **2022**, *76*, 708–718. [CrossRef] [PubMed]
54. Kobelev, V.; Schweizer, K.S. Dynamic Yielding, Shear Thinning, and Stress Rheology of Polymer-Particle Suspensions and Gels. *J. Chem. Phys.* **2005**, *123*, 164903. [CrossRef] [PubMed]
55. Hołowacz, I.; Podbielska, H.; Bauer, J.; Ulatowska-Jar, A. Viscosity, Surface Tension and Refractive Index of Tetraethylorthosilicate-Based Sol-Gel Materials Depending on Ethanol Content. *Opt. Appl.* **2005**, *35*, 691–699.
56. Rahmani-Azad, M.; Najafi, A.; Rahmani-Azad, N.; Khalaj, G. Improvement of ZrB2 Nanopowder Synthesis by Sol-Gel Method via Zirconium Alkoxide/Boric Acid Precursors. *J. Sol-Gel Sci. Technol.* **2022**, *103*, 87–96. [CrossRef]

57. Sharifi, K.A.; Pirsa, S. Biodegradable Film of Black Mulberry Pulp Pectin/Chlorophyll of Black Mulberry Leaf Encapsulated with Carboxymethylcellulose/Silica Nanoparticles: Investigation of Physicochemical and Antimicrobial Properties. *Mater. Chem. Phys.* **2021**, *267*, 124580. [CrossRef]
58. Tunç, S.; Duman, O. Preparation and Characterization of Biodegradable Methyl Cellulose/Montmorillonite Nanocomposite Films. *Appl. Clay Sci.* **2010**, *48*, 414–424. [CrossRef]
59. Azzaoui, K.; Mejdoubi, E.; Lamhamdi, A.; Jodeh, S.; Hamed, O.; Berrabah, M.; Jerdioui, S.; Salghi, R.; Akartasse, N.; Errich, A.; et al. Preparation and Characterization of Biodegradable Nanocomposites Derived from Carboxymethyl Cellulose and Hydroxyapatite. *Carbohydr. Polym.* **2017**, *167*, 59–69. [CrossRef]
60. Yang, M.; Shi, J.; Xia, Y. Effect of SiO2, PVA and Glycerol Concentrations on Chemical and Mechanical Properties of Alginate-Based Films. *Int. J. Biol. Macromol.* **2018**, *107*, 2686–2694. [CrossRef]
61. Kosmulski, M. The pH Dependent Surface Charging and Points of Zero Charge. VII. Update. *Adv. Colloid Interface Sci.* **2018**, *251*, 115–138. [CrossRef]
62. Wu, H.; Lei, Y.; Lu, J.; Zhu, R.; Xiao, D.; Jiao, C.; Xia, R.; Zhang, Z.; Shen, G.; Liu, Y.; et al. Effect of Citric Acid Induced Crosslinking on the Structure and Properties of Potato Starch/Chitosan Composite Films. *Food Hydrocoll.* **2019**, *97*, 105208. [CrossRef]
63. Menzel, C.; Olsson, E.; Plivelic, T.S.; Andersson, R.; Johansson, C.; Kuktaite, R.; Järnström, L.; Koch, K. Molecular Structure of Citric Acid Cross-Linked Starch Films. *Carbohydr. Polym.* **2013**, *96*, 270–276. [CrossRef] [PubMed]
64. Startek, K.; Marczak, J.; Lukowiak, A. Oxygen Barrier Enhancement of Polymeric Foil by Sol-Gel-Derived Hybrid Silica Layers. *Polymer* **2020**, *195*, 122437. [CrossRef]
65. Mohamed, S.A.A.; El-Sakhawy, M.; El-Sakhawy, M.A.-M. Polysaccharides, Protein and Lipid -Based Natural Edible Films in Food Packaging: A Review. *Carbohydr. Polym.* **2020**, *238*, 116178. [CrossRef] [PubMed]
66. Wu, F.; Misra, M.; Mohanty, A.K. Challenges and New Opportunities on Barrier Performance of Biodegradable Polymers for Sustainable Packaging. *Prog. Polym. Sci.* **2021**, *117*, 101395. [CrossRef]
67. Zhuang, L.; Zhi, X.; Du, B.; Yuan, S. Preparation of Elastic and Antibacterial Chitosan–Citric Membranes with High Oxygen Barrier Ability by in Situ Cross-Linking. *ACS Omega* **2020**, *5*, 1086–1097. [CrossRef] [PubMed]
68. Xu, L.; Cao, W.; Li, R.; Zhang, H.; Xia, N.; Li, T.; Liu, X.; Zhao, X. Properties of Soy Protein Isolate/Nano-Silica Films and Their Applications in the Preservation of Flammulina Velutipes. *J. Food Process. Preserv.* **2019**, *43*. [CrossRef]
69. Cui, Y.; Cheng, M.; Han, M.; Zhang, R.; Wang, X. Characterization and Release Kinetics Study of Potato Starch Nanocomposite Films Containing Mesoporous Nano-Silica Incorporated with Thyme Essential Oil. *Int. J. Biol. Macromol.* **2021**, *184*, 566–573. [CrossRef]
70. Wang, J.; Gardner, D.J.; Stark, N.M.; Bousfield, D.W.; Tajvidi, M.; Cai, Z. Moisture and Oxygen Barrier Properties of Cellulose Nanomaterial-Based Films. *ACS Sustain. Chem. Eng.* **2018**, *6*, 49–70. [CrossRef]
71. Trinh, B.M.; Chang, B.P.; Mekonnen, T.H. The Barrier Properties of Sustainable Multiphase and Multicomponent Packaging Materials: A Review. *Prog. Mater. Sci.* **2023**, *133*, 101071. [CrossRef]
72. Huang, H.-D.; Ren, P.-G.; Zhong, G.-J.; Olah, A.; Li, Z.-M.; Baer, E.; Zhu, L. Promising Strategies and New Opportunities for High Barrier Polymer Packaging Films. *Prog. Polym. Sci.* **2023**, *144*, 101722. [CrossRef]
73. Introzzi, L.; Blomfeldt, T.O.J.; Trabattoni, S.; Tavazzi, S.; Santo, N.; Schiraldi, A.; Piergiovanni, L.; Farris, S. Ultrasound-Assisted Pullulan/Montmorillonite Bionanocomposite Coating with High Oxygen Barrier Properties. *Langmuir* **2012**, *28*, 11206–11214. [CrossRef]
74. *DIN EN ISO 10993-5:2009-10*; Biologische Beurteilung von Medizinprodukten_- Teil_5: Prüfungen Auf In-Vitro-Zytotoxizität (ISO_10993-5:2009). Deutsche Fassung EN_ISO_10993-5:2009; Beuth Verlag GmbH: Berlin, Germany, 2009.
75. Wiegand, C.; Hipler, U.-C. Evaluation of Biocompatibility and Cytotoxicity Using Keratinocyte and Fibroblast Cultures. *Skin Pharmacol. Physiol.* **2009**, *22*, 74–82. [CrossRef]
76. Liao, R.; Qi, Z.; Tang, R.; Wang, R.; Wang, Y. Methyl Ferulic Acid Attenuates Human Cardiac Fibroblasts Differentiation and Myocardial Fibrosis by Suppressing pRB-E2F1/CCNE2 and RhoA/ROCK2 Pathway. *Front. Pharmacol.* **2021**, *12*. [CrossRef] [PubMed]
77. Trodtfeld, F.; Tölke, T.; Wiegand, C. Antimicrobial Functionalization of Prolamine–Silica Hybrid Coatings with Fumaric Acid for Food Packaging Materials and Their Biocompatibility. *Antibiotics* **2022**, *11*, 1259. [CrossRef] [PubMed]
78. Sharpe, J.R.; Harris, K.L.; Jubin, K.; Bainbridge, N.J.; Jordan, N.R. The Effect of pH in Modulating Skin Cell Behaviour. *Br. J. Dermatol.* **2009**, *161*, 671–673. [CrossRef]
79. Gröschl, M.; Topf, H.G.; Bohlender, J.; Zenk, J.; Klussmann, S.; Dötsch, J.; Rascher, W.; Rauh, M. Identification of Ghrelin in Human Saliva: Production by the Salivary Glands and Potential Role in Proliferation of Oral Keratinocytes. *Clin. Chem.* **2005**, *51*, 997–1006. [CrossRef]
80. Brand, H.S.; Ligtenberg, A.J.M.; Veerman, E.C.I. Saliva and Wound Healing. *Monogr. Oral Sci.* **2014**, *24*, 52–60. [CrossRef]
81. Pytko-Polonczyk, J.; Jakubik, A.; Przeklasa-Bierowiec, A.; Muszynska, B. Artificial Saliva and Its Use in Biological Experiments. *J. Physiol. Pharmacol.* **2017**, *68*, 807–813.
82. Gal, J.Y.; Fovet, Y.; Adib-Yadzi, M. About a Synthetic Saliva for in Vitro Studies. *Talanta* **2001**, *53*, 1103–1115. [CrossRef] [PubMed]
83. Almståhl, A.; Wikström, M. Electrolytes in Stimulated Whole Saliva in Individuals with Hyposalivation of Different Origins. *Arch. Oral Biol.* **2003**, *48*, 337–344. [CrossRef] [PubMed]

84. Hegde, M.N.; Attavar, S.H.; Shetty, N.; Hegde, N.D.; Hegde, N.N. Saliva as a Biomarker for Dental Caries: A Systematic Review. *J. Conserv. Dent. JCD* **2019**, *22*, 2–6.
85. Duval, D.; Demangel, C.; Munier-Jolain, K.; Miossec, S.; Geahel, I. Factors Controlling Cell Proliferation and Antibody Production in Mouse Hybridoma Cells: I. Influence of the Amino Acid Supply. *Biotechnol. Bioeng.* **1991**, *38*, 561–570. [CrossRef] [PubMed]
86. Meijer, A.J.; Dubbelhuis, P.F. Amino Acid Signalling and the Integration of Metabolism. *Biochem. Biophys. Res. Commun.* **2004**, *313*, 397–403. [CrossRef]
87. Ramos, L.R.; Santos, J.S.; Daguer, H.; Valese, A.C.; Cruz, A.G.; Granato, D. Analytical Optimization of a Phenolic-Rich Herbal Extract and Supplementation in Fermented Milk Containing Sweet Potato Pulp. *Food Chem.* **2017**, *221*, 950–958. [CrossRef] [PubMed]
88. Leukotrienes Orchestrating Allergic Skin Inflammation—Sadik—2013—Experimental Dermatology—Wiley Online Library. Available online: https://onlinelibrary.wiley.com/doi/full/10.1111/exd.12239 (accessed on 1 August 2023).
89. 5-Lipoxygenase, a Key Enzyme for Leukotriene Biosynthesis in Health and Disease—ScienceDirect. Available online: https://www.sciencedirect.com/science/article/pii/S138819811400167X (accessed on 1 August 2023).
90. Peters-Golden, M.; Henderson, W.R. Leukotrienes. *N. Engl. J. Med.* **2007**, *357*, 1841–1854. [CrossRef]
91. Sasaki, F.; Yokomizo, T. The Leukotriene Receptors as Therapeutic Targets of Inflammatory Diseases. *Int. Immunol.* **2019**, *31*, 607–615. [CrossRef]
92. Honda, T.; Kabashima, K. Prostanoids and Leukotrienes in the Pathophysiology of Atopic Dermatitis and Psoriasis. *Int. Immunol.* **2019**, *31*, 589–595. [CrossRef]
93. Sanz, M.L.; Gamboa, P.M.; Antépara, I.; Uasuf, C.; Vila, L.; Garcia-Avilés, C.; Chazot, M.; De Weck, A.L. Flow Cytometric Basophil Activation Test by Detection of CD63 Expression in Patients with Immediate-Type Reactions to Betalactam Antibiotics. *Clin. Exp. Allergy* **2002**, *32*, 277–286. [CrossRef]
94. Gamboa, P.M.; Sanz, M.L.; Caballero, M.R.; Antepara, I.; Urrutia, I.; Jauregui, I.; Gonzalez, G.; Dieguez, I.; De Weck, A.L. Use of CD63 Expression as a Marker of in Vitro Basophil Activation and Leukotriene Determination in Metamizol Allergic Patients. *Allergy* **2003**, *58*, 312–317. [CrossRef] [PubMed]
95. Ferrer, M.; Sanz, M.L.; Prieto, I.; Oehling, A. In Vitro Antigen-Specific Sulphidoleukotriene Production in Patients Allergic to Dermatophagoides Pteronyssinus. *Clin. Exp. Allergy J. Br. Soc. Allergy Clin. Immunol.* **1998**, *28*, 709–714. [CrossRef] [PubMed]
96. *DIN EN ISO 25178-1:2016-12*; Geometrische Produktspezifikation_(GPS)_-Oberflächenbeschaffenheit: Flächenhaft_- Teil_1: Angabe von Oberflächenbeschaffenheit (ISO_25178-1:2016). Deutsche Fassung EN_ISO_25178-1:2016; Beuth Verlag GmbH: Berlin, Germany, 2016.
97. Mezger, T. *The Rheology Handbook: For Users of Rotational and Oscillatory Rheometers*; Vincentz Network: Hannover, Germany, 2020; ISBN 978-3-86630-532-8.
98. Pearson, G.J. Long Term Water Sorption and Solubility of Composite Filling Materials. *J. Dent.* **1979**, *7*, 64–68. [CrossRef] [PubMed]
99. ISO 15105-1:2007—Plastics—Film and Sheeting—Determination of Gas-Transmission Rate—Part 1: Differential-Pressure Methods. Available online: https://www.iso.org/standard/41677.html (accessed on 9 September 2023).
100. *DIN EN ISO 10993-12:2021-08*; Biologische Beurteilung von Medizinprodukten_- Teil_12: Probenvorbereitung Und Referenzmaterialien (ISO_10993-12:2021). Deutsche Fassung EN_ISO_10993-12:2021; Beuth Verlag GmbH: Berlin, Germany, 2021.
101. Riedl, R.; Wallert, M.; Lorkowski, S.; Wiegand, C. Effects of Histamine and the α-Tocopherol Metabolite α-13′-COOH in an Atopic Dermatitis Full-Thickness Skin Model. *Molecules* **2023**, *28*, 440. [CrossRef] [PubMed]

gels

MDPI

Article

Evaluation of a Fish Gelatin-Based Edible Film Incorporated with *Ficus carica* L. Leaf Extract as Active Packaging

Hanan Rizqy Fauzan, Andriati Ningrum * and Supriyadi Supriyadi

Department of Food and Agricultural Product Technology, Faculty of Agricultural Technology, Universitas Gadjah Mada, Flora Street No. 1, Bulaksumur, Yogyakarta 55281, Indonesia
* Correspondence: andriati_ningrum@ugm.ac.id

Abstract: The significant concerns associated with the widespread use of petroleum-based plastic materials have prompted substantial research on and development of active food packaging materials. Even though fish gelatin-based films are appealing as active food packaging materials, they present practical production challenges. Therefore, this study aimed to develop an edible film using *Ficus carica* L. leaf extract (FLE), as it is affordable, accessible, and has superoxide anion radical scavenging action. This edible film was produced by adding FLE to mackerel skin gelatin at varied concentrations (2.5–10% *w*/*w*). The results showed that adding FLE to gelatin films significantly affected the tensile strength (TS), elongation at break (EAB), transmittance and transparency, solubility, water vapor permeability (WVP), antioxidant activity, and antibacterial activity. Among all the samples, the most promising result was obtained for the edible film with FLE 10%, resulting in TS, EAB, solubility, WVP, antioxidant activity, and antibacterial activity against *S. aureus* and *E. coli* results of 2.74 MPa, 372.82%, 36.20%, 3.96×10^{-11} g/msPa, 45.49%, 27.27 mm, and 25.10 mm, respectively. The study's overall findings showed that fish gelatin-based films incorporated with FLE are promising eco-friendly, biodegradable, and sustainable active packaging materials.

Keywords: edible film; fish gelatin; sustainable packaging; *Ficus carica* L.

Citation: Fauzan, H.R.; Ningrum, A.; Supriyadi, S. Evaluation of a Fish Gelatin-Based Edible Film Incorporated with *Ficus carica* L. Leaf Extract as Active Packaging. *Gels* **2023**, *8*, 918. https://doi.org/10.3390/gels9110918

Academic Editors: Baskaran Stephen Inbaraj, Kandi Sridhar, Minaxi Sharma and Aris Giannakas

Received: 10 August 2023
Revised: 2 October 2023
Accepted: 9 October 2023
Published: 20 November 2023

1. Introduction

Packaging is essential for maintaining the quality of food products by protecting them from all forms of physical, chemical, and microbiological damage [1]. The most common material used in food packaging comes from petroleum. The packaging industry prefers these materials because they form highly transparent films and have mechanical strength, water vapor resistance, and gas resistance [2]. However, if not treated properly in the process, these films can cause environmental pollution. Most petroleum-based plastics are not biodegradable and can break down into dangerous microplastics, threatening ecosystems and wildlife. Furthermore, improper disposal of plastic packaging can result in litter accumulating in natural environments and contributing to plastic pollution in oceans and rivers. Therefore, it is necessary to develop an alternative food packaging film made from natural, biodegradable, and sustainable materials [3,4].

The use of biodegradable biopolymers for food packaging has gained interest, given the inconvenience to sustainability of conventional food packaging [5]. Biodegradable and renewable polymers, including proteins, can replace non-environmentally friendly synthetic packaging. Proteins have the potential to form transparent films with specific mechanical properties, excellent oxygen barrier capabilities, and the ability to form tissues and induce plasticity and elasticity [6,7]. Gelatin is one form of protein that may be utilized to generate edible films. Gelatin is a denatured protein with exceptional gelling characteristics and is commonly used in the food industry. Gelatin-based edible films have potent gas and lipid barrier properties and can incorporate antibacterial or antioxidant compounds [6,7].

Globally, most gelatin comes from the skin and bones of land mammals [8]. However, gelatin derived from land mammals can cause bovine spongiform encephalopathy (BSE) and foot and mouth disease (FMD), so its safety is questionable. Even foods and medicines with a gelatin composition derived from land mammals have been banned by several European countries [9]. Gelatin derived from the skin of narrow-barred Spanish mackerel (*Scomberomorus commerson*) can be an alternative because the gelatin content is relatively high, at around 18% [10]. In Indonesia, mackerel production in 2020 reached 435,835.39 tons. This vast amount of production will undoubtedly result in a substantial amount of fish skin by-products produced by the fish processing industry [11].

Gelatin-based films can be combined with plant extracts with bioactive compounds that enhance the functional properties of the films. Many edible films have been developed by incorporating plant extracts, such as green tea extract (GTE) and *Lepidium sativum* extract (LSE), into gelatin. A higher GTE content integrated into gelatin films confers significantly higher tensile strength and lower elongation at break, water solubility, and water vapor permeability. Moreover, there is no notable difference in the mechanical parameters of tensile strength and elongation at break in films enriched with LSE [12,13].

Ficus carica L., commonly known as fig, originated in western Asia (Turkey), and its cultivation spread to other parts of the world through Mediterranean countries. This plant belongs to the Moraceae family and is believed to have been used as a medicinal herb for centuries. The fruit, roots, and leaves of the *Ficus carica* L. plant contain phytochemicals and various bioactive compounds. Specifically, the leaves are rich in polyphenols with antioxidant, anticancer, antidiabetic, and anti-inflammatory properties and hepatoprotective activity that could benefit human health [14–17]. Therefore, the leaf extracts of this plant are suitable for incorporation into gelatin to form an edible film. Fig leaf extract (FLE) contains flavonoids, terpenoids, and tannins that can enhance the physical properties of gelatin edible films and increase their antioxidant activity [18,19]. Nevertheless, no study has reported the incorporation of FLE into fish gelatin as an active food packaging material. Hence, this study was carried out to assess the physical properties, antioxidant activity, and antimicrobial activity of mackerel skin gelatin-based films enriched with fig leaf extract (FLE) as a bioactive component in these films to create eco-friendly, biodegradable, and sustainable active packaging.

2. Results and Discussion

2.1. Fig Leaf Extract (FLE) Antioxidant Activity

Before being used for film preparation, fig leaf extract was tested for its antioxidant activity using the DPPH method. From the tests that were conducted, fig leaf extract requires a concentration of 27.18 $\pm$ 0.44 ppm to inhibit 50% of DPPH free radicals. In comparison, BHT requires 19.75 $\pm$ 0.23 ppm. These findings demonstrated that fig leaf flavonoids had high antioxidant activity. Flavonoids and polyphenols have a higher concentration of phenolic hydroxyl groups, which exert antioxidant action by lowering hydroxyl groups. They can help stabilize free radicals and act as antioxidants by supplying hydrogen ions [20].

2.2. Color

As FLE was incorporated into fish gelatin films, the color of the films changed from greenish to yellowish, and this finding was validated by color parameter measurements. As shown in Table 1, increasing FLE incorporation into fish gelatin films raised the b and ΔE values of the films while decreasing the L and a values. The ΔE measurement quantifies the difference between the displayed color and the original color standard of the input content. The results were consistent with a prior study, which found an increase in yellowness after incorporating plant extract into gelatin-based films [7,21]. Previous research also showed that increasing the concentration of fig leaf extract causes a decrease in brightness and redness but an increase in yellowness and ΔE in chitosan-based films [7,22]. The observed color variations in the film were generated by naturally colored pigments in the *Ficus carica* L. extract (Figure 1).

Table 1. Color properties and transparency at 600 nm of the films.

Film Samples	L*	a*	b*	ΔE*	Transparency (600 nm)
FLE 0%	86.19 ± 0.93 [e]	−1.52 ± 0.08 [e]	3.82 ± 0.21 [a]	-	0.61 ± 0.02 [a]
FLE 2.5%	82.21 ± 0.89 [d]	−2.48 ± 0.22 [d]	7.56 ± 0.12 [b]	5.57 ± 0.66 [a]	1.95 ± 0.24 [b]
FLE 5%	80.27 ± 0.66 [c]	−4.37 ± 0.29 [c]	12.89 ± 0.59 [c]	11.20 ± 0.87 [b]	3.68 ± 0.01 [c]
FLE 7.5%	77.47 ± 0.43 [b]	−7.22 ± 0.26 [b]	18.15 ± 0.27 [d]	17.72 ± 0.32 [c]	6.12 ± 0.28 [d]
FLE 10%	71.14 ± 0.78 [a]	−9.78 ± 0.18 [a]	22.85 ± 0.22 [e]	25.64 ± 0.34 [d]	7.44 ± 0.23 [e]

FLE: *Ficus carica* L. leaf extract; L*: brightness; a*: redness; b*: yellowness; ΔE*: total color difference. Significant differences ($p < 0.05$) are shown by different superscript letters in the columns.

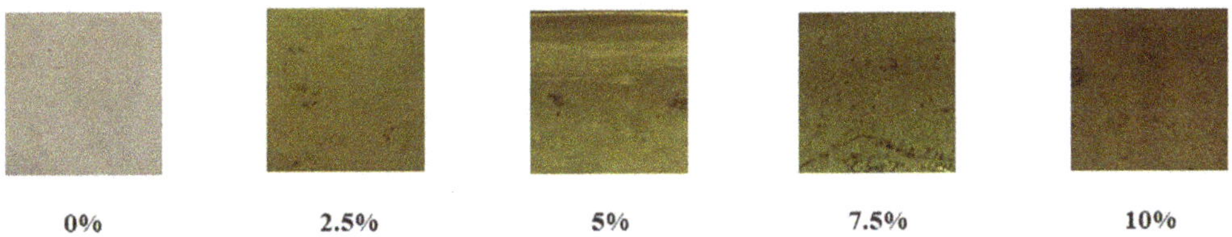

Figure 1. Development of an edible film enriched with *Ficus carica* L. leaf extract.

2.3. Transmittance and Transparency

Transmittance refers to the ability of an edible film to allow the passage of light or other electromagnetic radiation through it. Films with high transmittance are more transparent, allowing more light to pass through, while those with low transmittance are less transparent or even opaque [7]. The transmittance of fish gelatin edible films incorporated with FLE at 200–800 nm wavelength is shown in Figure 2. The percentage of light that could pass through the edible film decreased as the concentration of FLE increased.

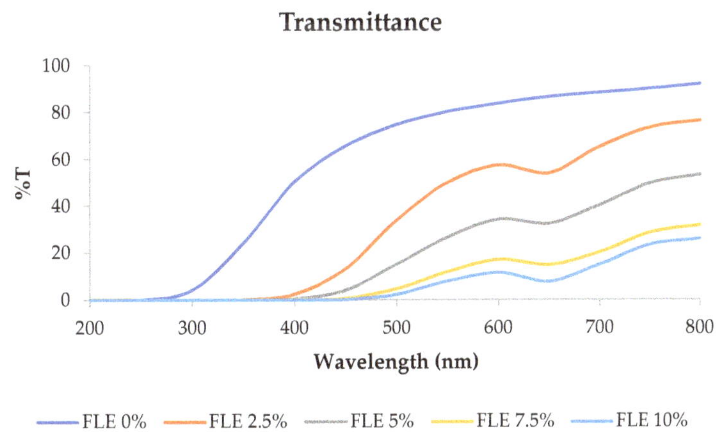

Figure 2. Transmittance of fish gelatin edible films incorporated with *Ficus carica* L. leaf extract.

These results determined the transparency value of the film, where the transparency value was directly proportional to the film's transmittance. The transparency values are presented in Table 1. Increasing the concentration of FLE significantly affected the transparency of the film samples. Higher transparency values correspond to a higher visible light absorbance, which indicates a lower transparency of the film [23–33]. Therefore, the edible film with 10% FLE had the best UV barrier but had the lowest transparency.

This finding was in line with previous studies showing that phenolic compounds distributed within a film matrix affect the morphology and light transmission of films [34–36]. The edible film based on fish skin gelatin incorporated with FLE exhibited good UV barrier properties, making it suitable for application in products vulnerable to UV light-induced damage, especially foods susceptible to UV light-induced lipid oxidation [37,38].

2.4. Thickness

Table 2 shows the thickness of the gelatin films containing various concentrations of FLE. The addition of *Ficus carica* L. leaf extract to gelatin films resulted in no significant variation in film thickness. FLE was equally distributed in the gap between the gelatin film matrix because the hydroxyl groups of polyphenols in FLE generated intermolecular hydrogen interactions with the amino/hydroxyl groups in gelatin. As a consequence, the integration of FLE had no significant effect on the thickness of the gelatin films [23].

Table 2. Physical properties of the films.

Film Samples	Thickness (mm)	Tensile Strength (MPa)	Elongation at Break (%)	Elastic Modulus (MPa)
FLE 0%	0.126 ± 0.01^{a}	1.13 ± 0.20^{a}	321.58 ± 15.33^{ab}	0.35 ± 0.05^{a}
FLE 2.5%	0.124 ± 0.01^{a}	1.21 ± 0.21^{a}	302.37 ± 17.97^{a}	0.40 ± 0.09^{ab}
FLE 5%	0.127 ± 0.01^{a}	1.68 ± 0.21^{b}	325.03 ± 25.58^{ab}	0.52 ± 0.07^{b}
FLE 7.5%	0.126 ± 0.01^{a}	1.84 ± 0.25^{b}	348.23 ± 9.28^{bc}	0.53 ± 0.06^{b}
FLE 10%	0.127 ± 0.01^{a}	2.47 ± 0.28^{c}	372.82 ± 14.89^{c}	0.66 ± 0.05^{c}

FLE: *Ficus carica* L. leaf extract. Significant differences ($p < 0.05$) are shown by different superscript letters in the columns.

2.5. Tensile Strength, Elongation at Break, and Elastic Modulus

The effect of FLE inclusion in gelatin films on tensile strength (MPa), elongation at break (%), and elastic modulus was studied. Tensile strength (TS) is directly related to polymer chain cohesion, whereas elongation at break (EAB) is related to flexibility and extensibility before breaking. Moreover, the elastic modulus is a quantity that measures an object or substance's resistance to being deformed elastically when a stress is applied to it. The interaction of the gelatin polymer chain with other components, including antioxidant extracts, water, and plasticizers, can have a substantial impact on the mechanical characteristics of the resultant mix films. Furthermore, the combination of the biopolymer matrix, plasticizer, and active agent might affect the mechanical characteristics of the resultant active blend films [24].

Table 2 shows the tensile strength (MPa), elongation at break (%), and elastic modulus (MPa) data. The TS of the gelatin-based films rose as the FLE concentration increased. This was due to the ability of hydroxyl groups in polyphenolic chemicals to form hydrogen bonds with hydrogen acceptors in gelatin molecules. The formation of intermolecular hydrogen bonds between FLE and gelatin rendered the films more compact and resistant to tensile stress [25]. A previous study discovered that including green tea extract raised the TS of gelatin films, which was induced by interactions between the phenolic compounds of the extract and the amino acids of the gelatin [26]. The addition of FLE, on the other hand, enhanced the EAB of the gelatin-based films. The slightly higher EAB in these gelatin-based films was most likely due to polyphenols acting as plasticizers, increasing the flexibility of FG-HBE films [27]. When plant extract was introduced into gelatin films, the EAB improved similarly because the interactions between polyphenols and gelatin created a more cohesive and flexible film matrix [28]. The addition of FLE increased the rigidity of the film, thus increasing the elastic modulus. This result is in line with similar research that demonstrated an increase in Young's modulus as tensile strength values increased [29].

2.6. Solubility

Water solubility is defined as the proportion of water-soluble compounds in the film and is sometimes referred to as a film's water resistance. According to the results, the water solubility of the films decreased dramatically in proportion to the FLE content (Table 3). These findings are consistent with prior studies in which green tea extract was combined with gelatin films [22]. With the addition of FLE, the gelatin films' solubility decreased due to the creation of stronger cross-links between polyphenols and gelatin in the polymer matrix, which reduced the polar groups and hydrophilic qualities of the films [30].

Table 3. Physical properties of the films.

Film Samples	Solubility (%)	WVP $\times 10^{-11}$ (g/msPa)
FLE 0%	46.47 ± 0.28^{e}	5.61 ± 0.25^{c}
FLE 2.5%	42.80 ± 0.17^{d}	5.04 ± 0.61^{bc}
FLE 5%	40.86 ± 0.20^{c}	4.89 ± 0.18^{b}
FLE 7.5%	37.46 ± 0.29^{b}	4.57 ± 0.27^{ab}
FLE 10%	36.20 ± 0.09^{a}	3.96 ± 0.18^{a}

Significant differences ($p < 0.05$) are shown by different superscript letters in the columns.

2.7. Water Vapor Permeability (WVP)

WVP is a critical feature of edible films used in food packaging applications to protect food from water-induced deterioration. As demonstrated in Table 3, the WVP of FLE-gelatin films decreases steadily as the FLE content increases. The lower WVP values of all film formulations might be attributed to the formation of a film matrix via the interaction of phenolic chemicals with gelatin. The addition of fig leaf extract might limit the free volume in the polymer matrix and enhance the tortuosity of the water molecules' passage across the network, lowering the rate of water molecule diffusion through the films [31]. According to the literature, comparable behavior was reported for green tea extract-gelatin films [22].

2.8. Antibacterial Activity of the Films

The antimicrobial activities of films against Gram-negative and Gram-positive bacterial species were evaluated by determining the inhibition zones (mm) on solid medium. Table 4 shows that the inhibition zone of the film against *Staphylococcus aureus* and *Escherichia coli* increased in size as the FLE concentration increased. A larger zone of inhibition indicates greater antibacterial activity. These findings are in line with similar research describing the antibacterial ability of gelatin-based films incorporated with *Ginkgo biloba* extract [39].

Table 4. Antibacterial activity of the films.

Film Samples	Inhibition Zone (mm)	
	S. aureus	*E. coli*
FLE 0%	-	-
FLE 2.5%	18.64 ± 0.24^{a}	-
FLE 5%	23.61 ± 0.12^{b}	22.33 ± 0.22^{a}
FLE 7.5%	25.69 ± 0.20^{c}	23.17 ± 0.18^{b}
FLE 10%	27.27 ± 0.15^{d}	25.10 ± 0.28^{c}

FLE: *Ficus carica* L. leaf extract. Significant differences ($p < 0.05$) are shown by different superscript letters in the columns.

Previous studies have shown the antimicrobial activity of *Ficus carica* L. extracts against Gram-positive and Gram-negative bacteria. This antibacterial activity is due to phenolic compounds in FLE, such as flavonoids, caftaric acid, quercetin, p-hydroxybenzoic acid, caffeic acid, and gallic acid. Phenolic compounds can disrupt the integrity of bacterial walls and cell membranes, resulting in the release of intracellular components from microbial cells. This process inhibits microorganism growth by blocking electron transfer, nutrient absorption, nucleotide synthesis, and membrane ATP activity [39–41].

2.9. Antioxidant Activity of the Films

Because antioxidant films may preserve food from oxidation and deterioration, they are essential for active packaging. The DPPH radical scavenging experiment assessed the antioxidant properties of all the film samples.

As demonstrated in Figure 3, integrating FLE greatly enhanced the DPPH value of the films. The powerful antioxidant capacity of phenolic compounds in FLE was ascribed to the increased DPPH radical scavenging activity in FLE-gelatin films. The phenolic components in *Ficus carica* L. extract chemically cross-link with the gelatin compound via covalent bonds. The oxidation of phenolic compounds to radicals results in covalent bonds between phenolic chemicals and proteins. These interactions between phenolic chemicals and gelatin amino acids enhance the antioxidant properties of the films [30].

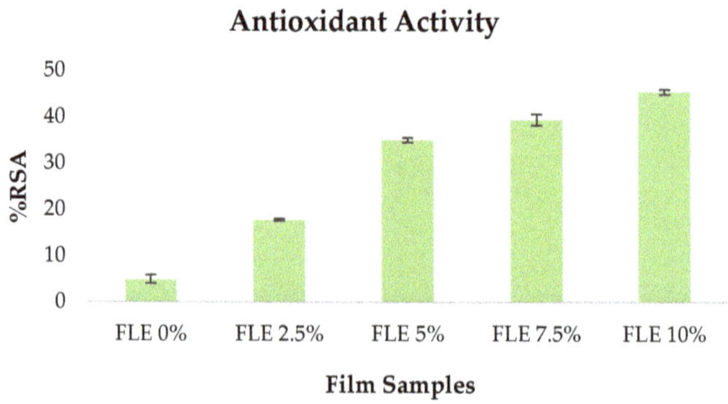

Figure 3. Antioxidant activity of fish gelatin edible films incorporated with *Ficus carica* L. leaf extract.

3. Conclusions

In this study, fig leaf extract (FLE) was incorporated into gelatin-based edible films to develop a sustainable active packaging material. The results lead to the conclusion that the incorporation of FLE improved the properties of these edible films compared to control films. The edible film enriched with 10% FLE showed the best tensile strength (2.47 MPa), elongation at break (372.82%), solubility (36.2%), WVP (3.96×10^{-11} g/msPa), antioxidant activity (45.49%), and antibacterial activity against *S. aureus* (27.27 mm) and *E. coli* (25.10 mm). These findings align with previous studies indicating that the addition of plant extracts can enhance the characteristics of gelatin-based edible films. We hope that this study promotes the development of sustainable packaging that can be applied to several food matrices. A further physicochemical investigation is required for such novel gelatin-based films to study possible relaxations of the gelatin matrix with extract molecules. This investigation represents a part of a future project involving the application of such films as active food packaging materials for preservation.

4. Materials and Methods

4.1. Materials

Fish skin gelatin was obtained from the domestic industry in Bandung, Indonesia. Fig leaves (*Ficus carica* L.) were harvested at Jogja Ara Garden in Sleman, Indonesia. CV. Nura Jaya, Surabaya, Indonesia, provided food-grade glycerol. All chemicals were analytical grade.

4.2. Preparation of Fig Leaf Extract (FLE)

Fig leaves were cleaned and dried in a cabinet dryer at 50 °C for 24 h. Dried fig leaves were crushed and passed through a sieve with a mesh size of 60. For 48 h, 75 g of fig leaf powder was macerated in 600 cc of 70% ethanol (Merck, Germany) (1:8 *w*/*v*). The solution was filtered using Whatman No. 1 filter paper and concentrated at 40 °C in a rotating vacuum evaporator.

4.3. Antioxidant Activity of FLE

The antioxidant activity of the extract was determined using a previously described DPPH radical scavenging assay with modifications [6]. A total of 4 mg of DPPH was dissolved in 100 mL of methanol to make a 0.1 mM DPPH solution. Dissolving 0.1 g of FLE in 100 mL of methanol yielded a 1000 ppm stock solution. Following that, sample test solutions with graded concentrations of 10, 20, 30, 40, 50, and 60 ppm were created. One milliliter of each sample solution concentration was combined with two milliliters of a 0.1 mM DPPH solution. The mixture was vortexed and incubated for 30 min at 27 °C. A UV-VIS spectrophotometer was used to detect absorbance at 517 nm, with methanol serving as the blank solution. Using the following equation, the antioxidant ability (%Radical Scavenging Activity/%RSA) to inhibit free radicals was obtained:

$$\%RSA = \frac{abs\ blanko - abs\ sample}{abs\ blanko} \times 100\% \quad (1)$$

The % inhibition of free radicals was plotted against the concentration of the test solution. The linear regression equation was used to get the IC_{50} value, which is the concentration of the test solution capable of blocking 50% of free radicals. Using the following formula, the computed results were inserted into the linear equation:

$$y = ax + b \quad (2)$$

where y is the % inhibition (50), a is the slope, x represents the concentration (μg/mL), and b represents the constant.

4.4. Preparation of Edible Films from Mackerel Skin Gelatin Incorporated with Fig Leaf Extract

The preparation of gelatin edible films was based on a previous method with modifications [6]. At 50 °C, gelatin powder was dissolved in distilled water at a concentration of 4% (*w*/*v*) and agitated for 30 min at 500 rpm. The gelatin solution was then mixed for 5 min with 25% (*v*/*w*) glycerol from the quantity of gelatin. Furthermore, fig leaf extract in the amounts of 0%, 2.5%, 5%, 7.5%, and 10% (*w*/*w*) of the amount of gelatin was added to the mixture and mixed for 30 min. For 5 min, the solution was homogenized with an Ultra-Turrax homogenizer at 4000 rpm. The liquid was put into the mold and dried for 20–24 h in a cabinet dryer set to 50 °C, after which the edible film sheet was removed from the mold.

4.5. Film Characterization

4.5.1. Color

The color analysis of the edible film was based on the previously reported method [6]. The color of the edible film was assessed using a digital colorimeter at six random spots on each sample with an 8 mm aperture, yielding values of L^* (brightness), a^* (red/green), and b^* (yellow/blue). The total color difference (ΔE^*) was calculated using the following equation:

$$\Delta E* = \sqrt{(L* - L_0*)^2 + (a* - a_0*)^2 + (b* - b_0*)^2} \quad (3)$$

L^*, a^*, and b^* are the color parameters of films incorporated with FLE; L_0^*, a_0^*, and b_0^* are the color parameters of the control films (100% gelatin films).

4.5.2. Transmittance and Transparency

The transmittance and transparency of the films were measured according to a previously reported method [7]. The barrier qualities of the film against ultraviolet (UV) and visible light were tested using a UV-Vis spectrophotometer at the specified wavelength range of 200–800 nm, while the transparency value of the film was computed using the following equation:

$$Transparency = 2 - log_{10}(\%T)/x \quad (4)$$

$\%T$ denotes the percent of transmittance at 600 nm, while x denotes the film thickness (mm).

4.5.3. Thickness

The determination of film thickness was based on a previously reported method [7]. Ten spots on each sample were randomly measured with a micrometer, and the average value was calculated as the edible film thickness.

4.5.4. Tensile Strength, Elongation at Break, and Elastic Modulus

Tensile strength (TS) and elongation at break (EAB) were determined using a previously reported approach and a Universal Testing Machine (UTM) [7]. Film samples were clipped to be 5 cm long and 1.5 cm wide. The film was put on the UTM stretcher and dragged by the device. With three replications, the test results yielded Fmax data, which are a measure of TS (MPa) and EAB (%). The elastic modulus (Young's modulus) was obtained from the comparison of TS and EAB [32], expressed in mega Pascals (MPa), and formulated with the following equation:

$$E = \frac{Tensile\ strength\ (\mathrm{MPa})}{Elongation\ at\ break} \quad (5)$$

4.5.5. Solubility

The solubility of the edible films was determined using a previously described procedure [7]. Briefly, 2 × 2 cm samples were dried in an oven at 105 °C for 24 h. W_1 was calculated after weighing the dry samples. Following that, the samples were immersed in 30 mL of aquades for 24 h at ambient temperature (25 °C). The samples were then dried again in a 105 °C oven for 24 h before being stored in a desiccator for 15 min. The samples were weighed, and W_2 was obtained. The percentage of sample solubility in water was determined using the following equation:

$$\%S = \frac{W_1 - W_2}{W_1} \times 100\% \quad (6)$$

S denotes the sample's solubility, W_1 the weight of the dried sample before immersion, and W_2 the weight of the dried sample after immersion.

4.5.6. Water Vapor Permeability (WVP)

The water vapor permeability of the film was calculated as previously described [7]. A glass cup-shaped container containing 6 mL of distilled water was sealed with the film at the mouth. The sample was then placed in a desiccator for 7 h at room temperature. Every hour, the sample was weighed using an analytical balance. The following equation was used to compute the WVP value:

$$WVTR = \frac{\Delta W}{A\ \Delta P} \quad (7)$$

$$WVP = \frac{WVTR \times L}{\Delta P} \quad (8)$$

where:

$WVTR$ = the water vapor transmission rate;
ΔW = the weight of water vapor passing through the film (g);
A = the surface area of the film $\pi.r^2$ (m^2);
ΔP = the partial pressure difference in water vapor (Pa);

L = the thickness of the film (mm);
WVP = the water vapor permeability (g/msPa).

4.5.7. Antibacterial Activity of the Edible Films

The antibacterial activity of the films was assessed against *Escherichia coli* (*E. coli*) and *Staphylococcus aureus* (*S. aureus*) using the method of a previous study with a slight modification [33]. The bacteria were individually cultured by transferring frozen *E. coli* and *S. aureus* into separate tubes of 10 mL of sterile nutrient broth with an inoculating loop. The organisms were incubated for 24 h at 37 °C for reactivation. Rejuvenation of the bacteria was carried out by streaking a loopful of pure culture onto a sterile nutrient agar plate and incubating at 37 °C for 24 h. The bacterial suspension density was adjusted to 0.5 M of McFarland Standard, which is equivalent to a bacterial suspension containing approximately 1×10^8 CFU/mL. A dried surface of Mueller–Hinton agar was inoculated with each bacterium with a sterile swab. A film sample with a diameter of 15 mm was placed on the agar. The plate was incubated at 37 °C for 24 h. The diameter of the inhibition zone surrounding the film disc and the contact surface of the edible film with the agar surface were observed.

4.5.8. Antioxidant Activity of the Edible Films

The measurement of the antioxidant activity of the edible films was based on a previously described method with modifications [6]. Twenty-five milligrams of film was dipped into 3 mL of methanol for 3 h to obtain a film extract solution. Then, 0.1 mL of the solution was added to 3.9 mL of 0.1 mM DPPH and incubated in the dark for 30 min before the absorbance was measured at 517 nm using a spectrophotometer. Methanol and DPPH solutions without the film extract were used as blanks, and the DPPH radical scavenging capacity was measured as follows:

$$\%RSA = \frac{abs\ blanko - abs\ sample}{abs\ blanko} \times 100\% \quad (9)$$

Author Contributions: Conceptualization, A.N. and S.S.; Methodology, H.R.F., A.N. and S.S.; Investigation, H.R.F.; Resources, A.N.; Data curation, A.N.; Writing—original draft, A.N.; Writing—review & editing, A.N.; Supervision, A.N. and S.S.; Project administration, A.N.; Funding acquisition, A.N. All authors have read and agreed to the published version of the manuscript.

Funding: The authors wish to thank Direktorat Penelitian UGM under the project Riset Tugas Akhir (RTA 2023) number 5075/UN1.P.II/Dit-Lit/PT.01.01/2023.

Acknowledgments: The authors wish to thank Direktorat Penelitian UGM under the project Riset Tugas Akhir (RTA 2023) number 5075/UN1.P.II/Dit-Lit/PT.01.01/2023.

Conflicts of Interest: The authors declare no conflict of interest.

References

1. Trajkovska Petkoska, A.; Daniloski, D.; D'Cunha, N.M.; Naumovski, N.; Broach, A.T. Edible Packaging: Sustainable Solutions and Novel Trends in Food Packaging. *Food Res. Int.* **2021**, *140*, 109981. [CrossRef]
2. Gómez-Guillén, M.C.; Pérez-Mateos, M.; Gómez-Estaca, J.; López-Caballero, E.; Giménez, B.; Montero, P. Fish Gelatin: A Renewable Material for Developing Active Biodegradable Films. *Trends Food Sci. Technol.* **2009**, *20*, 3–16. [CrossRef]
3. Gallego, M.; Arnal, M.; Talens, P.; Toldrá, F.; Mora, L. Effect of Gelatin Coating Enriched with Antioxidant Tomato By-Products on the Quality of Pork Meat. *Polymers* **2020**, *12*, 1032. [CrossRef]
4. Evode, N.; Qamar, S.A.; Bilal, M.; Barceló, D.; Iqbal, H.M.N. Plastic Waste and Its Management Strategies for Environmental Sustainability. *Case Stud. Chem. Environ. Eng.* **2021**, *4*, 100142. [CrossRef]
5. Rodríguez-Félix, F.; Corte-Tarazón, J.A.; Rochín-Wong, S.; Fernández-Quiroz, J.D.; Garzón-García, A.M.; Santos-Sauceda, I.; Plascencia-Martínez, D.F.; Chan-Chan, L.H.; Vásquez-López, C.; Barreras-Urbina, C.G.; et al. Physicochemical, Structural, Mechanical and Antioxidant Properties of Zein Films Incorporated with No-Ultrafiltered and Ultrafiltered Betalains Extract from the Beetroot (*Beta vulgaris*) Bagasse with Potential Application as Active Food Packaging. *J. Food Eng.* **2022**, *334*, 111153. [CrossRef]

6. Azizah, F.; Nursakti, H.; Ningrum, A. Supriyadi Development of Edible Composite Film from Fish Gelatin-Pectin Incorporated with Lemongrass Essential Oil and Its Application in Chicken Meat. *Polymers* **2023**, *15*, 2075. [CrossRef]
7. Ningrum, A.; Widyastuti Perdani, A.; Supriyadi; Siti Halimatul Munawaroh, H.; Aisyah, S.; Susanto, E. Characterization of Tuna Skin Gelatin Edible Films with Various Plasticizers-Essential Oils and Their Effect on Beef Appearance. *J. Food Process. Preserv.* **2021**, *45*, e15701. [CrossRef]
8. Munawaroh, H.S.H.; Gumilar, G.G.; Berliana, J.D.; Aisyah, S.; Nuraini, V.A.; Ningrum, A.; Susanto, E.; Martha, L.; Kurniawan, I.; Hidayati, N.A.; et al. In Silico Proteolysis and Molecular Interaction of Tilapia (*Oreochromis niloticus*) Skin Collagen-Derived Peptides for Environmental Remediation. *Environ. Res.* **2022**, *212*, 113002. [CrossRef]
9. Fu, Y.; Therkildsen, M.; Aluko, R.E.; Lametsch, R. Exploration of Collagen Recovered from Animal By-Products as a Precursor of Bioactive Peptides: Successes and Challenges. *Crit. Rev. Food Sci. Nutr.* **2019**, *59*, 2011–2027. [CrossRef] [PubMed]
10. Nagai, T.; Suzuki, N. Isolation of Collagen from Fish Waste Material—Skin, Bone and Fins. *Food Chem.* **2000**, *68*, 277–281. [CrossRef]
11. Ali, A.M.M.; Kishimura, H.; Benjakul, S. Physicochemical and Molecular Properties of Gelatin from Skin of Golden Carp (*Probarbus jullieni*) as Influenced by Acid Pretreatment and Prior-Ultrasonication. *Food Hydrocoll.* **2018**, *82*, 164–172. [CrossRef]
12. Salem, A.; Jridi, M.; Abdelhedi, O.; Fakhfakh, N.; Nasri, M.; Debeaufort, F.; Zouari, N. Development and Characterization of Fish Gelatin-Based Biodegradable Film Enriched with Lepidium Sativum Extract as Active Packaging for Cheese Preservation. *Heliyon* **2021**, *7*, e08099. [CrossRef]
13. Wu, J.; Chen, S.; Ge, S.; Miao, J.; Li, J.; Zhang, Q. Preparation, Properties and Antioxidant Activity of an Active Film from Silver Carp (*Hypophthalmichthys molitrix*) Skin Gelatin Incorporated with Green Tea Extract. *Food Hydrocoll.* **2013**, *32*, 42–51. [CrossRef]
14. Hussain, S.Z.; Naseer, B.; Qadri, T.; Fatima, T.; Bhat, T.A. *Fruits Grown in Highland Regions of the Himalayas*; Springer: Cham, Switzerland, 2021; ISBN 9783030755010.
15. Hajam, T.A.; Saleem, H. Phytochemistry, Biological Activities, Industrial and Traditional Uses of Fig (*Ficus carica*): A Review. *Chem. Biol. Interact.* **2022**, *368*, 110237. [CrossRef]
16. Abdel-Rahman, R.; Ghoneimy, E.; Abdel-Wahab, A.; Eldeeb, N.; Salem, M.; Salama, E.; Ahmed, T. The Therapeutic Effects of *Ficus carica* Extract as Antioxidant and Anticancer Agent. *S. Afr. J. Bot.* **2021**, *141*, 273–277. [CrossRef]
17. El Ghouizi, A.; Ousaaid, D.; Laaroussi, H.; Bakour, M.; Aboulghazi, A.; Soutien, R.S.; Hano, C.; Lyoussi, B. *Ficus carica* (Linn.) Leaf and Bud Extracts and Their Combination Attenuates Type-1 Diabetes and Its Complications via the Inhibition of Oxidative Stress. *Foods* **2023**, *12*, 759. [CrossRef]
18. Nirwana, I.; Rianti, D.; Helal Soekartono, R.; Listyorini, R.D.; Basuki, D.P. Antibacterial Activity of Fig Leaf (*Ficus carica* Linn.) Extract against Enterococcus Faecalis and Its Cytotoxicity Effects on Fibroblast Cells. *Vet. World* **2018**, *11*, 342–347. [CrossRef]
19. Yilmaz, P.; Demirhan, E.; Ozbek, B. Development of *Ficus carica* Linn Leaves Extract Incorporated Chitosan Films for Active Food Packaging Materials and Investigation of Their Properties. *Food Biosci.* **2022**, *46*, 101542. [CrossRef]
20. Wahyu, D.; Ningrum, A.; Vanidia, N.; Siti, H.; Munawaroh, H.; Susanto, E.; Show, P. Food Hydrocolloids for Health In Silico and In Vitro Assessment of Yellowfin Tuna Skin (*Thunnus albacares*) Hydrolysate Antioxidation Effect. *Food Hydrocoll. Health* **2023**, *3*, 100126. [CrossRef]
21. Jridi, M.; Abdelhedi, O.; Salem, A.; Kechaou, H.; Nasri, M.; Menchari, Y. Physicochemical, Antioxidant and Antibacterial Properties of Fish Gelatin-Based Edible Films Enriched with Orange Peel Pectin: Wrapping Application. *Food Hydrocoll.* **2020**, *103*, 105688. [CrossRef]
22. Khan, M.R.; Di Giuseppe, F.A.; Torrieri, E.; Sadiq, M.B. Recent Advances in Biopolymeric Antioxidant Films and Coatings for Preservation of Nutritional Quality of Minimally Processed Fruits and Vegetables. *Food Packag. Shelf Life* **2021**, *30*, 100752. [CrossRef]
23. Liu, Y.; Wang, S.; Lan, W.; Qin, W. Development of Ultrasound Treated Polyvinyl Alcohol/Tea Polyphenol Composite Films and Their Physicochemical Properties. *Ultrason. Sonochem.* **2019**, *51*, 386–394. [CrossRef]
24. Villasante, J.; Martin-Lujano, A.; Almajano, M.P. Characterization and Application of Gelatin Films with Pecan Walnut and Shell Extract (*Carya illinoiensis*). *Polymers* **2020**, *12*, 1424. [CrossRef] [PubMed]
25. Zhang, J.; Zou, X.; Zhai, X.; Huang, X.W.; Jiang, C.; Holmes, M. Preparation of an Intelligent PH Film Based on Biodegradable Polymers and Roselle Anthocyanins for Monitoring Pork Freshness. *Food Chem.* **2019**, *272*, 306–312. [CrossRef] [PubMed]
26. Nunes, J.C.; Melo, P.T.S.; Lorevice, M.V.; Aouada, F.A.; de Moura, M.R. Effect of Green Tea Extract on Gelatin-Based Films Incorporated with Lemon Essential Oil. *J. Food Sci. Technol.* **2021**, *58*, 1–8. [CrossRef] [PubMed]
27. Jamróz, E.; Kulawik, P.; Krzyściak, P.; Talaga-Ćwiertnia, K.; Juszczak, L. Intelligent and Active Furcellaran-Gelatin Films Containing Green or Pu-Erh Tea Extracts: Characterization, Antioxidant and Antimicrobial Potential. *Int. J. Biol. Macromol.* **2019**, *122*, 745–757. [CrossRef]
28. Liu, J.; Zhang, L.; Liu, C.; Zheng, X.; Tang, K. Tuning Structure and Properties of Gelatin Edible Films through Pullulan Dialdehyde Crosslinking. *Lwt* **2021**, *138*, 110607. [CrossRef]
29. Edhirej, A.; Sapuan, S.M.; Jawaid, M.; Zahari, N.I. Effect of Various Plasticizers and Concentration on the Physical, Thermal, Mechanical, and Structural Properties of Cassava-Starch-Based Films. *Starch Staerke* **2017**, *69*, 1500366. [CrossRef]
30. Chen, H.; Wu, D.; Ma, W.; Wu, C.; Liu, J.; Du, M. Strong Fish Gelatin Hydrogels Double Crosslinked by Transglutaminase and Carrageenan. *Food Chem.* **2022**, *376*, 131873. [CrossRef]

31. Tyuftin, A.A.; Kerry, J.P. Gelatin Films: Study Review of Barrier Properties and Implications for Future Studies Employing Biopolymer Films. *Food Packag. Shelf Life* **2021**, *29*, 100688. [CrossRef]
32. Fadilah; Susanti, A.D.; Distantina, S.; Purnamasari, D.P.; Ahmad, J.F. Mechanical Properties of Films from Carboxy Methyl Glucomannan and Carrageenan with Glycerol as Plasticizer. *AIP Conf. Proc.* **2020**, *2217*, 030022. [CrossRef]
33. Al-Hassan, A.A.; Norziah, M.H. Starch-Gelatin Edible Films: Water Vapor Permeability and Mechanical Properties as Affected by Plasticizers. *Food Hydrocoll.* **2012**, *26*, 108–117. [CrossRef]
34. Du, H.; Liu, C.; Unsalan, O.; Altunayar-Unsalan, C.; Xiong, S.; Manyande, A.; Chen, H. Development and Characterization of Fish Myofibrillar Protein/Chitosan/Rosemary Extract Composite Edible Films and the Improvement of Lipid Oxidation Stability during the Grass Carp Fillets Storage. *Int. J. Biol. Macromol.* **2021**, *184*, 463–475. [CrossRef]
35. Norajit, K.; Kim, K.M.; Ryu, G.H. Comparative Studies on the Characterization and Antioxidant Properties of Biodegradable Alginate Films Containing Ginseng Extract. *J. Food Eng.* **2010**, *98*, 377–384. [CrossRef]
36. Thivya, P.; Bhosale, Y.K.; Anandakumar, S.; Hema, V.; Sinija, V.R. Study on the Characteristics of Gluten/Alginate-Cellulose/Onion Waste Extracts Composite Film and Its Food Packaging Application. *Food Chem.* **2022**, *390*, 133221. [CrossRef]
37. Munir, S.; Hu, Y.; Liu, Y.; Xiong, S. Enhanced Properties of Silver Carp Surimi-Based Edible Films Incorporated with Pomegranate Peel and Grape Seed Extracts under Acidic Condition. *Food Packag. Shelf Life* **2019**, *19*, 114–120. [CrossRef]
38. Papadaki, A.; Manikas, A.C.; Papazoglou, E.; Kachrimanidou, V.; Lappa, I.; Galiotis, C.; Mandala, I.; Kopsahelis, N. Whey Protein Films Reinforced with Bacterial Cellulose Nanowhiskers: Improving Edible Film Properties via a Circular Economy Approach. *Food Chem.* **2022**, *385*, 132604. [CrossRef]
39. Hu, X.; Yuan, L.; Han, L.; Li, S.; Song, L. Characterization of Antioxidant and Antibacterial Gelatin Films Incorporated with: Ginkgo Biloba Extract. *RSC Adv.* **2019**, *9*, 27449–27454. [CrossRef] [PubMed]
40. El Dessouky Abdel-Aziz, M.; Darwish, M.S.; Mohamed, A.H.; El-Khateeb, A.Y.; Hamed, S.E. Potential Activity of Aqueous Fig Leaves Extract, Olive Leaves Extract and Their Mixture as Natural Preservatives to Extend the Shelf Life of Pasteurized Buffalo Milk. *Foods* **2020**, *9*, 615. [CrossRef]
41. Tkachenko, H.M.; Buyun, L.I.; Osadowski, Z.; Honcharenko, V.I.; Prokopiv, A.I. Antimicrobial Screening of the Ethanolic Leaves Extract of *Ficus carica* L. (Moraceae)—An Ancient Fruit Plant. *Introd. Plants* **2017**, *1*, 78–87. [CrossRef]

gels

MDPI

Review

A Comprehensive Review of Food Hydrogels: Principles, Formation Mechanisms, Microstructure, and Its Applications

Pinku Chandra Nath [1], Shubhankar Debnath [1], Kandi Sridhar [2], Baskaran Stephen Inbaraj [3], Prakash Kumar Nayak [4,*] and Minaxi Sharma [5,*]

[1] Department of Bio Engineering, National Institute of Technology, Agartala 799046, India
[2] Department of Food Technology, Koneru Lakshmaiah Education Foundation Deemed to be University, Vaddeswaram 522502, India
[3] Department of Food Science, Fu Jen Catholic University, New Taipei City 24205, Taiwan
[4] Department of Food Engineering and Technology, Central Institute of Technology Kokrajhar, Kokrajhar 783370, India
[5] Haute Ecole Provinciale de Hainaut-Condorcet, 11, Rue de la Sucrerie, 7800 Ath, Belgium
* Correspondence: pk.nayak@cit.ac.in (P.K.N.); minaxi86sharma@gmail.com (M.S.)

Abstract: Food hydrogels are effective materials of great interest to scientists because they are safe and beneficial to the environment. Hydrogels are widely used in the food industry due to their three-dimensional crosslinked networks. They have also attracted a considerable amount of attention because they can be used in many different ways in the food industry, for example, as fat replacers, target delivery vehicles, encapsulating agents, etc. Gels—particularly proteins and polysaccharides—have attracted the attention of food scientists due to their excellent biocompatibility, biodegradability, nutritional properties, and edibility. Thus, this review is focused on the nutritional importance, microstructure, mechanical characteristics, and food hydrogel applications of gels. This review also focuses on the structural configuration of hydrogels, which implies future potential applications in the food industry. The findings of this review confirm the application of different plant- and animal-based polysaccharide and protein sources as gelling agents. Gel network structure is improved by incorporating polysaccharides for encapsulation of bioactive compounds. Different hydrogel-based formulations are widely used for the encapsulation of bioactive compounds, food texture perception, risk monitoring, and food packaging applications.

Keywords: hydrogels; food applications; bioactive compounds; mechanical strength; microstructure

Citation: Nath, P.C.; Debnath, S.; Sridhar, K.; Inbaraj, B.S.; Nayak, P.K.; Sharma, M. A Comprehensive Review of Food Hydrogels: Principles, Formation Mechanisms, Microstructure, and Its Applications. *Gels* **2023**, *9*, 1. https://doi.org/10.3390/gels9010001

Academic Editor: Hiroyuki Takeno

Received: 18 November 2022
Revised: 10 December 2022
Accepted: 16 December 2022
Published: 20 December 2022

1. Introduction

In order to develop novel and innovative consumer-based functional foods, food processing industries are facing numerous challenges. Recent food processing trends are based on ensuring food safety and nutritional value to safeguard consumers. Furthermore, under proper storage, shipping, and delivery circumstances, the longevity of a product can be extended [1]. Researchers are creating new forms of food matrices that can be synthesized artificially [2] or with the assistance of different microbes [3], plants [4], or insects to assure quality and safety [5]. Foods with gelation characteristics have recently become much more popular in the marketplace due to their delicious taste, features that increase satiety, low calorie content, and high moisture content. These factors combine to make these foods healthier. In addition, rheological properties and sensory qualities of food hydrogels are essential for their excellent market price [6]. Natural gels made from food biopolymers have made significant strides over the past few decades, and in comparison, to their synthetic analogs, they have a wider range of applications and superior properties in a variety of contexts.

Gels are colloidal systems containing a continuous phase (i.e., a solid matrix) and a dispersed phase (i.e., aqueous solvent), leading to the formation of a semisolid texture [7].

The biopolymers that make up these gels are the primary ingredient that gives them their structure. Polysaccharides and proteins are the two types of biopolymers that are most frequently used. These biopolymers provide a variety of food products with a semisolid consistency. These biopolymer-based gels are composed of networks of various shapes, such as blocks, particles, or fibers, which exhibit a range of behaviors when subjected to mechanical stress [8].

Hydrogels are 3D polymer networks that are elastic because they are bound together by weaker bonding forces in the form of hydrogen or ionic bonds and crosslinked covalent bonds. These factors contribute to the ability of hydrogels to hold their shape. Hydrogels are used in current food design for a variety of purposes, including the creation of complex shapes through 3D printing, the replacement of fats, increased satiety with less food, and the maintenance of metastable structures of products [9]. Gels are also effective for reducing the unpleasant tastes induced by several bioactive compounds and medications because gel formation limits or slows molecular transport [10]. On the other hand, the use of flavor enhancers such as salt and sugar can be reduced by using liquid or brittle gels, which intensify the flavor [11]. If a gel is to be fully integrated into a liquid without altering the flavor of the food, its size must be drastically reduced before it can be used in liquid foods, such as beverages.

In order to satisfy the worldwide demand for food and provide sustenance for the growing human population, several food products that are environmentally friendly, inexpensive, safe, and packed with nutrients are required. In this context, food gels are considered sustainable food matrices containing a high concentration of nutrients. The employment of hydrogels mostly in the industry of food technology is not nearly as widespread as it is in other branches of the scientific community, such as tissue engineering and biomedicine. Therefore, the design of the structure of food hydrogels will require investigation of additional functional applications in order to advance the development of functional foods. Owing to the widespread sensorial acceptance of these medicinal and healthful foods, which were made possible by incorporating food gels, an increase in consumption of such food products has been observed [12]. When utilizing gels in food production systems, it is vital to consider how these substances will react inside the body. As a result, it is essential to take a number of factors into consideration in order to control the discharge or retainment of bioactive substances across the digestive system [13]. Enzymes and the acidic atmosphere of the stomach present the most significant challenge to maintaining the integrity of the shell for subsequent passages through the gastrointestinal tract because the stomach is the first organ that the shell must pass through [14]. Therefore, it is expected that the smart layout of the structure of the hydrogel will lead to an increase in the number of useful applications for new food items. In this review, we focused on the possible implementations and hydrogel interactions pertaining to the microstructure of various hydrogels and their food applications. The novelty of this review lies in the understating of the formation, rheological characteristics, and impact of the microstructure of hydrogels. An understanding of the interlaying mechanism of hydrogels could provide an opportunity for a wider range of applications in the food industry.

2. Gelling Agents

Gelling agents are added to a wide range of foods, such as jellybeans, sweets, and chocolates, to make them thicker and more stable. As a result of gel formation, these ingredients give such foods their distinctive textures. In addition to their other functions, stabilizers and thickeners can also act as gelling agents. The application of plant- and animal-based proteins as gelling agents in the food industry is illustrated in Figure 1. Only a small fraction of gums can be used to make gels, and even among those that can be used for such purposes, there is a wide range of gel characteristics and textures that dictates their use in different food contexts. Natural gums, enzymes, pectins, starches, and agars are just some of the proteins and polysaccharides used as gelling agents. Heteropolysaccharides and hydrocolloids make up the vast majority of gelling polysaccharides (Table 1). Structures can be complex but well-defined repeating units, block copolymers, or either regularly or irregularly branched structures. They come from various sources, such as plant, animal, and microbial sources in nature. Gel formation is the primary focus, although thickening is a possible side effect. Jams, jellies, salad dressings, desserts, marmalade, jujubes, yogurts, etc., are just a few of the uses. However, many proteins are also employed in gel fabrication (Table 2). Corn zein and various animal proteins such as gelatin and whey are among these.

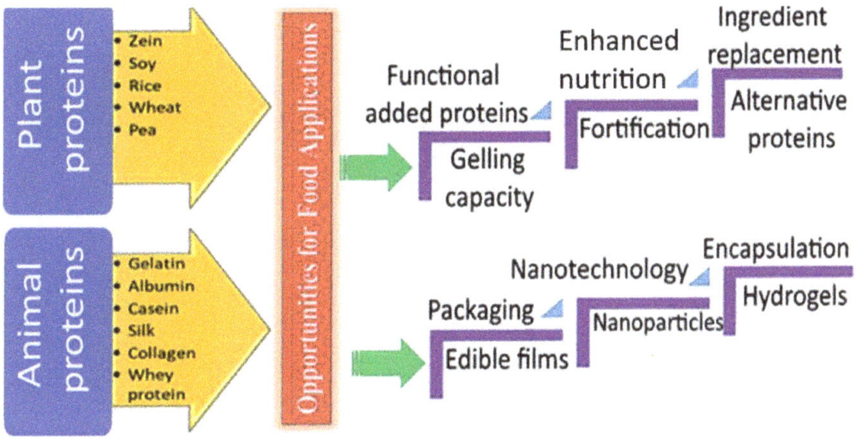

Figure 1. Different types of plant- and animal-based ingredients as gelling agents in the food industry. Figure 1 is reprinted with permission from Munir et al. [15] (Copyright © 2022 by authors).

3. Conditions Necessary for the Formation of Gels

Gel formation occurs as a consequence of the combination of a straightforward polymer dispersion or particle suspension with an extrinsically modifiable temperature or solution composition [16]. As a consequence of this, the process of converting sol-gels typically involves the aggregation of either particles or macromolecules, with the end result being the formation of a network that encompasses the entire volume of the container [15,17]. Gelation reactions can be roughly classified into two categories: those that are driven by physical forces (e.g., heat and pressure) and those that are induced by chemical processes (e.g., acidic, ionic, and enzymatic processes) [18,19]. The gelation of proteins requires a driving force to unfold the native protein structure, followed by an aggregation process, resulting in a three-dimensional organized network of aggregates or strands of molecules crosslinked by non-covalent bonds or, less frequently, by covalent bonds [20]. The numerous physicochemical characteristics covered in the subsequent sections are the primary determinants of the conditions that must exist for gel to form [21]. Figure 2 depicts the primary factors that influence the formation of gels and further information on each of these factors is addressed below in this section.

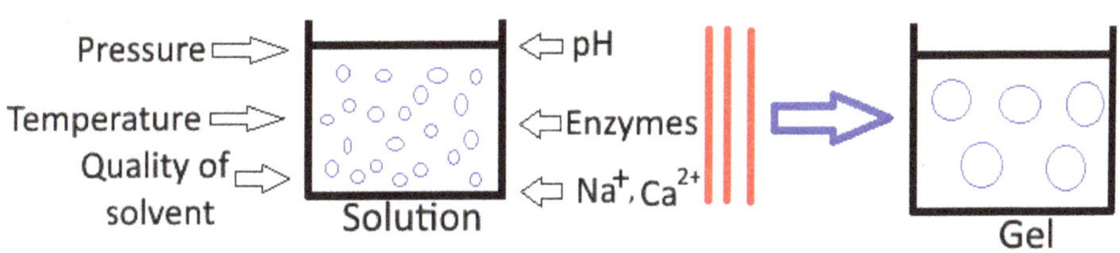

Figure 2. Different factors affecting the formation of food hydrogels.

Table 1. Sources of gelling agents, as well as their textures, rheological properties, and applications.

Gelling Agent and Concentration	Source	Texture and Rheological Properties	Binding Blocks	Applications	Ref
Pectin (0.5–1%, *w*/*w*)	Heteropolysaccharides from higher terrestrial plant cell walls and fruits, such as citrus fruit, guava, and apple	Newtonian behavior	Partially esterified α-(1–4)-linked D-galactouronic and mannuronic acid; rhamnose, galactose, and arabinose can replace galacturonic acid	Jelly, jam, marmalade, fruit chews, and yogurt	[22]
Cellulose (0.5–2%, *w*/*w*)	Chemically modified plant cell walls	Shear thining crossmodelling and viscoelastic behavior	Homopolymer of D-glucose β-(1, 4)	Sweets and salad dressings	[22,23]
Agar (1–2%, *w*/*w*)	Red algae or seaweeds	Thermoreversible gels upon cooling; lowest viscosity at low shear rates; with increasing concentration of agarose, the particle size and particle surface decrease	Utilization of agarose and agaropectin together in a single product	Vegetarian gelatin and laxative	[22,24]
Guar gum (1–5%, *w*/*w*)	Endosperm	Although it is independent of electrolytes and has extremely high low-shear viscosity and high-shear thinning, it degrades and loses viscosity at high temperatures and low pH	Galactomannan molecular sequence	Sweets, yogurt, and various forms of liquid cheese, along with fillings for pastries	[25,26]
Carrageenan (0.5–3%, *w*/*w*)	Red seaweeds	High hardness, compressibility, adhesiveness, and cohesiveness	Sulfated D-galactose and L-anhydrogalactose	Desserts, cell/enzyme immobilization gel	[27,28]
Carob gum or locust bean gum (0.16–1.84%, *w*/*w*)	Carob tree seeds	High-shear thinning; very high low-shear viscosity, despite being degraded and losing viscosity regardless of the presence of electrolytes at high and low pH levels and high temperatures	Galactomannan	Glue	[29]
Xanthan gum (1–3%, *w*/*w*)	Xanthomonas campestris is responsible for the process of fermenting glucose, as well as sucrose	High-shear thinnability; maintains viscosity in the presence of electrolytes, high temperatures, and a wide pH range	Two -D-glucose units are linked at positions 1 and 4 along the polysaccharide chain; the primary structure has two mannose and one glucouronic acid, forming repeating modules of five sugar molecules	As emulsifiers, texture modifiers, fruit salads, and sauces, colloidal oil and solid ingredients are kept from turning into a cream	[30,31]
Alginate (1–2%, *w*/*w*)	Brown seaweeds	Pseudoplastic non-Newtonian and viscoelastic behavior	Different arrangements of (1-4)-linked -D-mannuronate and its C-5 epimer α-L-guluronate residues form a linear copolymer	Appetite-suppressant jellies, divalent ionic gelation, cell immobilization, and encapsulation	[32,33]
Gum Arabic (1–5%, *w*/*w*)	Acacia senegal and Acacia seyal sap	Low-viscosity gum; shear thinning occurs at low shear rates below 10/s, and near-Newtonian behavior occurs above shear rates of 100/s	Combination of saccharides and glycoproteins	Hard gummies, chocolates, and gums	[34,35]

Table 2. Application of proteins as gelling agents.

Gelling Agents	Source	Binding Blocks	Applications	Ref
Whey protein	Acid/sweet dairy whey	β-lactoglobulin and α-lactalbumin	Thickener and gelling agent	[36]
Soya proteins	Soybeans	Glycinin and β-conglycinin interact	Heat-set gel	[37]
Egg proteins	Egg	Albumen is made up of almost 70% globular proteins with ovomucin fibers and 30% egg white	Confectionery gelling and bulking agent	[38]
Zein	Corn	Peptide chain (prolamine)	Encapsulations in candies, nuts, fruit, and pills, along with other foods and baked products	[39]
Gelatin	Animal skin and bones	Glycine–proline protein	Jelly, jam, yogurt, and margarine	[40]

3.1. Pressure

Because high pressure may be used by itself or in conjunction with other processes, such as elevated temperatures, it provides more options for tailoring the functional molecular properties. Reactions that reduce the volume of a system are generally favored under high pressure. Dissociation of water at high pressure lowers its pH value. Gels formed under heat or pressure has distinct visual and rheological properties [41].

3.2. pH

The addition of acids or the fermentation of microorganisms can cause changes in pH, and these changes can alter the net charge of the molecule, which, in turn, can change the attractive and repulsive forces that exist between molecules, as well as the interactions among molecules and the solvent, also known as the hydration properties. In addition, the solubility of salts shifts depending on the pH, which is another factor that might play a role in gel formation. The theory of fractal aggregation may provide an explanation for the technique underlying the formation of acid gel [42].

3.3. Ionic Strength

A gel's band gap can be increased by incorporating monovalent and divalent cations such as Na and Ca into the mixture. Gelation is possible as a result of the reduction or neutralization of the electrostatic repulsive forces that normally exist between molecules. It has been reported that pre-denatured whey proteins can undergo ionically induced gelation, which, in contrast to heat-induced gelation, is referred to as cold gelation [43,44]. In polysaccharide gels, such as xanthan gum, glucans, or guar gum, ionically induced gelation plays a more significant role than in other types of gels.

3.4. Temperature

The majority of gels are obtained by heating their constituents to a critical temperature [45]. The first step in the gelation process is uncovering or molecular fragmentation in response to the energy required, which exposes the reactive sites. Step two involves the joining or sticking together of unfolded compounds to make complexes with a higher molecular weight. Whereas the first process might be reversible; the second is almost always irreversible. Hydrophobic interactions and disulfide (-S-S-) bridges are likely to play essential roles. According to the relative reaction rates of the individual steps over a given temperature range, the unfolding or aggregation reaction can dictate the overall reaction rate.

3.5. Availability of Enzymes

Enzymes control complex procedures and active interaction in natural systems and their potential for building self-regulating soft matter systems is growing [46]. Enzymatically induced gelation works by adding artificial covalent cross links to food proteins. Protein crosslinking works best with reactions set off by transglutaminase (TG), polyphenol oxidase, and lipoxygenase [47].

3.6. Gelling Agent Concentration

Gel formation is only possible once the critical minimum concentration (denoted by the symbol C*), which is unique to each hydrocolloid, has been reached. Agarose can form gels at concentrations as low as 0.2%, whereas acid-thinned starch requires a concentration of at least 15% before gels can be formed [48].

3.7. Solvent Quality

The nature and presence of solvent has a significant effect on gel formation; for instance, concentrated sugar solution is an ineffective solvent for pectin [47]. Only in a concentrated sugar solution can hydrogen bonds form in the junction zones, so gelation can only occur in this type of solvent [49].

4. Gel Types and Mechanism of Gel Formation

Food gels can be categorized in various ways based on their biopolymer networks, gelation mechanisms, morphologies, interactions, etc. Gel structures can be classified either as simple networks (made up of a single biopolymer such as polysaccharides or proteins) binary/mixed networks (made up of two or more biopolymer types), or composite/filled networks (made up of a biopolymer network and various other particles such as fat globules) [18]. Cold-set gels, heat-set gels, ionotropic gels, acid or enzymatically induced gels, etc., are all distinct types of gels that form for different reasons. Gels can be classified as either filamentous or crystalline or as a soft particle suspension, depending on their morphological characteristics.

Gels may contain a variety of interactions, including hydrogen bonding, interactions involving hydrophobic groups, non-covalent interactions, and covalent chemical bonds. Among the many types of chemical bonds, H-bonding is the most common type of interaction. Polymer gels can be divided into three categories: strong, weak, and pseudogels, all of which are fundamentally determined by the three-dimensional layout of the biopolymer network [7]. Hybrid gels of polymers that have been subjected to chemical crosslinking are regarded as particularly robust gels. The crosslinks in these gels are irreversible and cannot be repaired if they become broken. Colloidal gels and certain biopolymer gels are examples of weak gels that contain crosslinks that are capable of being broken and reformed [50,51]. Long-lasting physical interactions within the polymer matrix can function like chemical crosslinks, which gives these materials gel-like properties. Therefore, polymer systems that get tangled up are sometimes called "pseudogels" [51]. If the stress is kept constant, the steady-state response of a pseudogel to an extreme stress is to flow like a fluid. It is possible that different criteria could be used to classify food gels; however, in order to make it simpler to comprehend the methodology of gelation and the microstructure of gels, in this chapter, we describe gels based on the category of the biopolymer system they contain.

As previously stated, gelling agents include any of the polysaccharides and proteins found in plants, animals, and microbes [52]. Polysaccharides, also known as natural gums, are used frequently in the food industry and include ingredients such as agar, guar gum, alginate, xanthan gum, starches, and glucans. The most popular proteins contain pectin, glutenin, skim milk, soy milk, egg albumin, and zein. Many of these proteins and polysaccharides are employed as thickeners in culinary creations. The viscosity and shear modulus can differentiate between thickening and gelling. Gel formers usually increase the solubility until the gel point, when the viscosity becomes (possibly) infinite and the elastic modulus above the gel point is finite [53]. The first steps in the gelling process are the uniform dispersion of the gelling agent, followed by the hydration of the agent. The subsequent formation of the network is responsible for giving the product its texture.

4.1. Polysaccharide Gels

There are just few polysaccharides that can form a gel when exposed to a certain concentration of the gelling agent, which is typically referred to as the critical concentration. Other polysaccharides, on the other hand, are utilized in various foods as thickeners and stabilizers. Polysaccharide critical concentrations are significantly lower than those of protein critical concentrations. After polysaccharides have been completely hydrated, the polymer strands begin interacting (crosslinking) with one another in order to produce junction zones. When the polymer solution (dispersion) reaches a certain critical polymer concentration and a certain degree of crosslinking, the polymer solution eventually transforms into a gel that possesses a stable network structure. The composition of the polysaccharide and the conditions under which the gel is formed can result in a wide variety of junction zone configurations (Figure 3). Some polysaccharides, known as thermos or heat-set gels, crystallize when heated and then cooled again, whereas other polysaccharides, known as cold-set gels, do so at room temperature by utilizing particular types of cations, adjusting the pH of the solution, or incorporating particular cosolutes. Curdlan is a special polysaccharide because it can produce gels that are heat-set as well as cold-set. This ability sets it apart from other polysaccharides [54]. In the following sections, we discuss polysaccharide gels, which are utilized the vast majority of the time [55].

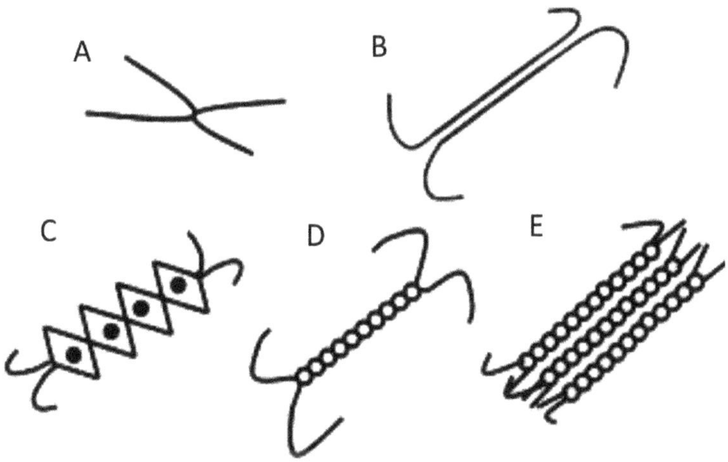

Figure 3. Different junction zones in polysaccharide gels: (**A**) connecting points, (**B**) extended junction zone resembling a block, (**C**) model of crosslinks in alginate and pectin gels based on an egg-box structure, (**D**) the region of the double-helix junction, and (**E**) junction zone for the aggregation of polysaccharide chains. Figure 3 is reprinted with permission from Nazir et al. [8] (Copyright © 2017 Elsevier Ltd., Amsterdam, The Netherlands).

4.1.1. Alginate

In the presence of divalent cations, alginate gels are produced, and the strength of the resulting gels is directly proportional to the type of cation used. Ca^{++}-induced gelation is the most significant among the various cation-induced gelations for functional foods [56]. In general, alginate gels are not affected by heat and are irreversible. Egg-box structures are formed when cations begin a linkage among polyglucuronic acid regions of adjacent polymers. This linkage occurs because buckled conformations in the polymers provide efficient binding sites. Egg-box structures are also known as egg crates. Because alginates can form a gel even when the temperature is just above room temperature, they can be used in a variety of food applications.

4.1.2. Pectin

The degree of esterification of pectin significantly impacts the gelling properties exhibited by pectin [57]. Pectin's with a high methoxyl content only set into a gel when sugars or other cosolutes are present [58], such as carbohydrates, polyols, or monohydric alcohols, and with a sufficiently low acid concentration (3.0 to 4.5) [59]. Low-methoxyl pectins are capable to formgel in the presence of divalent compounds, such as calcium.

4.1.3. Agar

In both bacterial media and food products, agar gels experience a synergistic effect [60]. Agarose, a neutral polysaccharide, and agaropectin, a charged polysaccharide with sulfate groups, combine to form an agar gel. Accordingly, the gel osmotic pressure increases, and the degree of syneresis decreases as the agaropectin fraction increases [61]. In most cases, the total sulfate content has an inverse proportional relationship with the amount of water extracted from the agar gel. The total polymer concentration in the gel follows the same pattern. Intriguingly, the extent of syneresis in agar gel was found to be roughly inversely proportional to the square of the concentration. The uncharged polymer's osmotic pressure scales approximately to this value.

4.1.4. Starch

Amylose, a linear chain, and amylopectin, which is branched, are the two types of polymers that make up starch granules (branched) [62]. Granules are embedded within an amylose matrix in the composite gel that is formed as a result of its formation. A process known as gelatinization is responsible for the gelation of starch. During this process, starch granules are heated, which causes them to absorb a sufficient amount of liquid and swell to several times their original size. After being cooled, the amylose fraction becomes ordered (in the form of single helices) around the swollen granules and is leached out by the granules. The method that causes the starch to gel differs depending on the source of the starch; for example, chitosan-based hydrogel and its mechanism are illustrated in Figure 4A.

4.1.5. Carrageenan

Carrageenan is a type of ionic polymer that, when cooled in the presence of salts (electrolytes), particularly potassium ions, forms helical gels. Both the formation of helices and the setting of gels can be facilitated by particular cations, such as K^+, Rb^+, Cs^+, and NH_4^+. A coil–helix transition is experienced by molecules, followed by the aggregation of helices [63]. The formation of "ordered domains" is a common step in carrageenan gelation [64], which typically involves the association of polymer chains through the formation of intermolecular double helices. Gelation takes place after the subsequent aggregation of these domains, which is mediated by the specific binding of the cations that are responsible for gelation. The transformation of carrageenan from its disordered, random coil state into its ordered, helical state is the first step in the gelation process [65].

4.1.6. Gellan Gum

In terms of their chemical composition, gellan gums can be broken down into two categories: high- and low-acyl gellan gums [66]. Gels with high-acyl gellan tend to be soft and elastic, whereas gels with low-acyl gellan tend to be hard and brittle [67]. The presence of ions facilitates the association of double helices, which, in turn, leads to gelation.

4.1.7. Cellulose Derivative Gels

Carboxymethyl Cellulose (CMC)

CMC is a water-soluble cellulose derivative widely used in the biopolymer industry. It is produced by replacing 2, 3, and 6 hydroxyls on the backbone of cellulose with carboxymethyl groups [68]. Cellulose containing numerous hydroxyl groups is an abundant and inexpensive natural biopolymer, making it a desirable starting material. CMC also exhibits bioactivity, solubility, and biodegradability. CMC is prepared in a nonaqueous monochloroacetic acid/soda solvent medium to achieve the substitution degree via carboxymethylation [69].

CMC-based hydrogel has the potential for use in enzyme immobilization [70], wound healing [71], drug delivery [72], and adsorption [73]. Hydrogels from nanoparticles/CMC can be used for their antimicrobial properties [74], wound healing, drug development, and tissue engineering [75]. The performance of carboxymethyl cellulose hydrogel is enhanced by the addition of nanoparticles [76]. Nanoparticles enhance carboxymethyl cellulose hydrogels by virtue of their superior mechanical, electronic, optical, and physicochemical properties. Carboxymethyl cellulose derived from pineapple plants is an effective vehicle for papain immobilization and forms a strong hydrogen bond between the employed materials [77]. Although CMC can be easily extracted from biomass resources [78], bagasse [79] and empty fruit bunch [80] have also been used to produce carboxymethyl cellulose. Every type of biomass resource imparts unique characteristics to CMC, such as exceptional absorption and adsorption, a high swelling ability, and superior optical properties. In addition to being advantageous for the production of CMC hydrogels, a high level of methylation groups in various types of biomass waste is also beneficial [81].

Hydroxypropyl Methyl Cellulose (HPMC)

Cellulose derivative HPMC is utilized extensively in controlled-release applications, owing to its ability to thicken, gel, and swell. It is also safe to use, easy to compress, has properties that make it swell, and can handle high drug levels.

Owing to its excellent bioactivity, HPMC can be a thermosensitive natural polymer that forms a transparent, highly stable, colorless hydrogel, with positive rheological properties and changes in the texture. Gårdebjer et al. [82] investigated the pore-forming effects of hydroxypropyl methylcellulose, mostly in MFC (micro fibrillated cellulose) film, and made adjustments to the wettability characteristics of the films. The results demonstrate that HPMC can potently react with MFC films, possibly creating H-bonds on the surface of the film.

Hydroxypropyl methylcellulose for use in scaffold engineering was created from crosslinked chitosan by Boyer et al. [83]. After promoting cellular characteristics, they demonstrated that crosslinking hydroxypropyl methylcellulose with chitosan can endow the recovery process with structural strength and shape. The use of HPMC as a composite hydrogel in scaffold engineering was also investigated by Bacakova et al. [84]. Likewise, HPMC is widely used to make composite hydrogels using different polysaccharides (Figure 4B). Overall, HPMC composite hydrogel can facilitate faster healing, a more uniform distribution of cells, and a reduced risk of complications during osteoplasty procedures. Hydrogel scaffolds, films, and membranes are some of the typical applications for HPMC in the sector of medicine.

Hydroxypropyl Cellulose (HPC)

HPC can be dissolved in ice-cold water, and its viscosity changes with pressure. Temperatures higher than 45 °C render HPC insoluble, and unlike MC and HPMC, it does not form a gel. HPC, on the other hand, may be soluble in ethanol and ethanol/water mixtures. A reversible precipitation process begins when the polymer reaches a temperature above 45 °C [85]. However, sucrose or high concentrations of salt can be introduced to lower the deposition temperature [86]. When heated, viscosity drops dramatically—typically by half for every 15 °C [87]. HPC maintains its viscosity over a wide range of pH values, from 2 to 11. However, solutions should be stored at a pH of 6–8 to prevent the loss of viscosity due to decomposition [87].

4.2. Protein Gels

Generally, proteins are the convenient biopolymers for the formation of food-based hydrogels. Moreover, plant and animal-based proteins are widely used in the fabrication of food-based hydrogels. The mecahnisms of protein-based hydrogels are illustarted in Figure 4C.

4.2.1. Gelatin

It is appropriate to begin the discussion with gelatin gel-containing food products, for which the term syneresis was coined. It has been reported that gelatin gels with a pH below or above the isoelectric point swell considerably [88]. These gels are stable at these pH levels and exhibit no syneresis, even at low protein concentrations. The osmotic pressure resulting primarily from the Donnan equilibrium is responsible for the swelling when the gel is close to its isoelectric point of 4.7; however, different results have been observed. Below a 10% concentration, the gel deswells (syneresis), whereas above this threshold, it swells. Osmotic pressure responds more quickly to changes in concentration than the network pressure, comparable to the behavior observed in a chemically crosslinked, uncharged gel. It is abundantly clear that the osmotic pressure of gelatin gel appears to perform the most significant part in the process of triggering syneresis. Increasing the polymer concentration boosts osmotic pressure, which helps to prevent syneresis. It is also possible to increase osmotic pressure by shifting the pH away from the isoelectric point, which causes the polymer charge to increase.

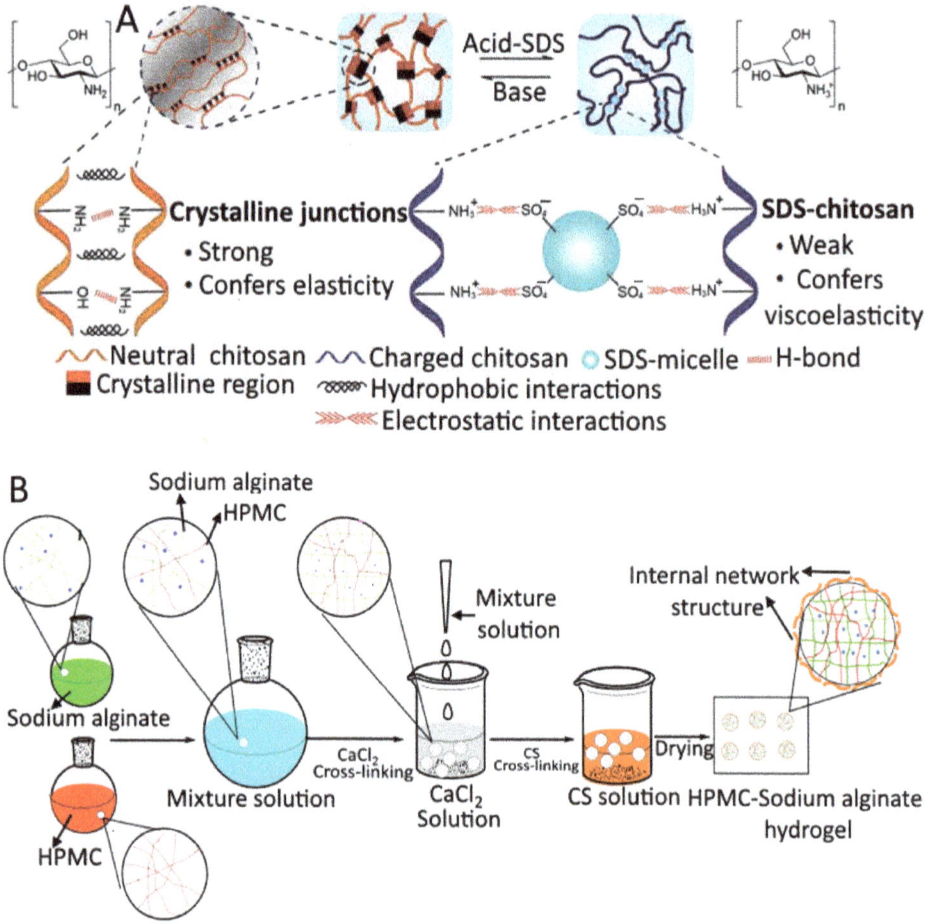

Figure 4. *Cont.*

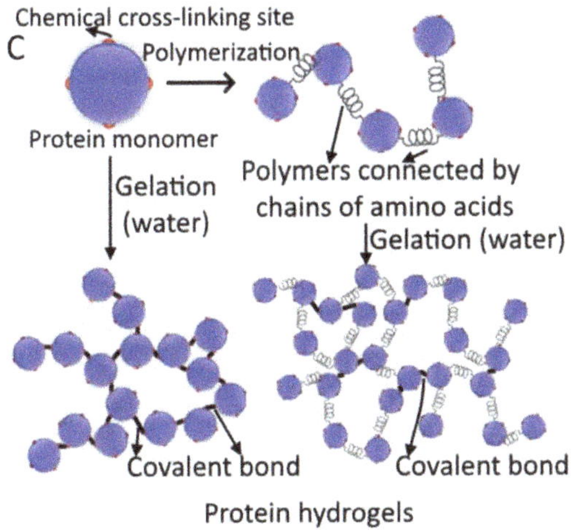

Figure 4. Mechanism of polysaccharide and protein gel formation. Crosslinking of chitosan (**A**); composite hydroxypropyl methyl cellulose-sodium alginate hydrogels (**B**); gelation process of chemically crosslinked protein hydrogels (**C**). (**A**,**B**) are reprinted with permission from He et al. [89] (Copyright © 2017 Wiley-VCH Verlag GmbH & Co. KGaA, Weinheim, Germany) and Hu et al. [90] (Copyright © 2018 Springer Nature B.V., Dordrecht, The Netherlands), respectively, whereas (**C**) is reprinted from Hanson et al. [91] and is an open-access article (Copyright © 2020 by authors) distributed under the terms and conditions of the Creative Commons Attribution (CC BY) license. HPMC, hydroxypropyl methyl cellulose; $CaCl_2$, calcium chloride; CS, chitosan.

4.2.2. Whey Proteins

Whey proteins undergo a series of transitions that are characteristic of globular proteins before and during heat-induced gelation. These transitions include (i) the inactivation of native proteins, also known as their unfolding; (ii) the agglomeration of unfolded compounds; (iii) the formation of strands from aggregates; and (iv) the association of strands into a network. The presence of salts allows for the formation of aggregates to take place [88,92,93].

4.2.3. Egg Albumin

Egg albumin is often thought of as a system made up of many different globular proteins dissolved in water [94]. The formation of egg albumin gel occurs in three steps. First, heating causes molecules to partially unfold; this results in an increase in the number of interactions between molecules. In the second stage, sulfhydryl disulfide exchange ultimately leads to molecular aggregation, and sulfhydryl oxidation occurs both within and between the network-forming aggregates. Multiple hydrogen bonds are generated in the final step of the process, which involves cooling. The process of gel formation can be affected by a number of factors, such as the pH of the solution, the presence of salts or sugars, and the ionic strength.

4.2.4. Soy Proteins

Although acidification (for example, glucono-δ-lactone) can also induce aggregation of denatured protein molecules, gelation can be achieved by heating soybean flour [37] or milk, followed by the addition of salt (for example, Ca^{++} or Mg^{++}), to form a gel or curd.

4.2.5. Milk Proteins

Owing to the high hydrophobicity of casein molecules, submicelles of casein are held together through the formation of hydrophobic bonds and salt bridges. Rennet gelation occurs as a consequence of the enzymatic hydrolysis of k-casein by rennet, which results in the release of CMP (caseinomacropeptide) and the aggregation of micelles [47,95,96].

5. Rheological Characterization of Gels

Rheological characterization is important in determining a gel's composition, structure, and the impact of processing on that structure. In addition, by adjusting the polymer microstructure and surrounding media, a gel with the required characteristics can be created. Modern rheometers can precisely measure the reaction of a complicated material to applied stress or strain. However, understanding the gelling process and manipulating the gel's microstructure with desired sensory properties require a basic knowledge of rheology [97]. The crosslinked polymer networks in gels are what give them their viscoelastic properties. The type and degree of crosslinking in a polymer affect its strength, which is determined by equilibrium modulus, a rheological parameter. The polymer's mean molecular weight increases prior to the process of crosslinking, which ultimately causes the polymer to transform from a liquid (sol) into a solid (gel); this transition point is termed the gel point. There are various rheological methods for determining whether a polymer is a gel. In a creep test, the material is subjected to a constant stress, and the rate of deformation is recorded. A gel is a viscoelastic material; thus, it has two components (elastic and viscous), which produce a characteristic increase in the creep compliance ($J(t) = \gamma(t)/\sigma_0$, where $\gamma(t)$ is the time-dependent shear strain, and σ_0 is the constant stress applied to the sample in the creep step). Creep-recovery compliance is a measurement of the network elasticity of gels [98]. Therefore, it is difficult to achieve a constant deformation in any gel network. Gels differ from liquids and tend to dissolve in solvents, which causes them to swell. A gel's equilibrium modulus affects how much it swells, with harder gels swelling less than softer gels. The most accurate way to determine the critical gel point—whereby a small amount of oscillatory strain (or stress) is applied to a material to see how it reacts—is through linear oscillatory rheology. Owing to the viscoelastic properties of gels, two rheological parameters can be used to describe their viscoelasticity: viscous or loss modulus ($G'' \propto \cos(\omega t)$) and elastic or storage modulus ($G' \propto \sin(\omega t)$). Elastic characters prevail when G' is greater than G'', and viscous characters prevail when G' is smaller than G''. Mechanical spectra can be expressed in terms of the complex modulus (G^*) [99].

$$G^* = \left[\sqrt{G''^2 + G'^2}\right] \tag{1}$$

$$G^* = \left[A_n \omega^{n^*}\right] \tag{2}$$

where A_n is the gel strength of samples, serving another characteristic parameter of the material structural type; n^* is the power law exponent, serving as an index of the viscoelastic nature of the material (a measure of physical crosslinks in the protein network); and ω is angular frequency [99].

The viscoelastic moduli provide a measurement of the shear-deformation resistance in terms of the elastic-deformation resistance (G') and the viscous-deformation resistance (G''). Both moduli are the relation between shear stress (force/unit of area (Pa)) and shear strain. They frequently have a significant amount of influence due to the period of distortion. The relative importance of the elastic modulus (G') and the viscous modulus (G'') is reflected by the phase angle, which is calculated as $\tan\delta = G''/G'$, where δ is the phase angle. The loss factor ($\tan\delta$) provides a measurement of the viscoelastic degree of gels, together with the energy cohesiveness of samples, as it is related to the lifetime of bonds, which conform the polymeric network [100]. The storage modulus (G') surpassing the loss modulus (G'') at a particular frequency, is among the criteria that determine whether a solution will turn into a gel.

5.1. Microrheology of Gels

Despite being macroscopic parameters, the viscoelastic data of gels can provide structural information, as they are fundamental data. Therefore, fundamental tests (oscillatory tests, creep and recovery tests, dynamical thermo mechanical analysis, etc.) are essentially independent of the measurement conditions, reflecting true material properties. These material properties are related to the structural characteristics [101,102]. Recent technological advancements have made available new strategies for comprehending complicated gel structures and dynamics with different characteristic and timeframes. As a result, the concept of microrheology has been introduced, which concerns how materials store and release mechanical energy as a function of length scale. By connecting the microstructure of food systems to their macroscopic characteristics and stability, such knowledge aids food scientists in understanding the fundamental mechanisms that influence the interactions of food components [103].

Particle-tracking rheology, magnetic tweezers, diffusing-wave spectroscopy, atomic force microscopy, laser particle tracking (optical tweezers), quasi-elastic light scattering and piezorheometery are just a few of the methods that can be used to characterize microrheology. Particle-tracking microrheology, also known as video particle-tracking microrheology, is one such method beginning to gain popularity. It has been used to characterize a variety of food gel systems. The local dynamics of soft material are examined in this scenario using the Brownian motion of embedded particles or tracers, without the application of an external driving force. As shown in Figure 5, there are three steps required to calculating linear viscoelasticity using particle tracking. Following tracking of particle trajectories and calculation of the mean-square displacement (MSD) as a function of the lag time of these fluctuations (due to Brownian motion), the data are translated into viscoelastic properties, specifically storage (G') and loss (G'') moduli and creep compliance (J) [104]. With the least possible disruption to the developing gel structure, it can also identify the gel point.

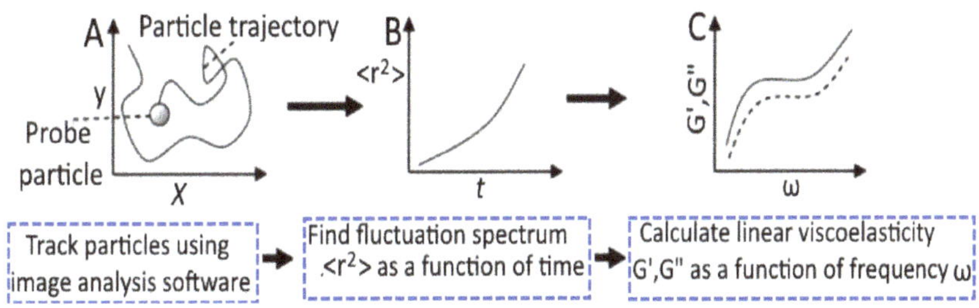

Figure 5. The linear viscoelasticity of low-modulus materials extracted from the fluctuation spectrum using particle-tracking microrheology. (**A**) The probe particle's trajectory is calculated; (**B**) the spectrum of average fluctuations is computed as a function of time (t′); and (**C**) mechanical spectrum in the linear viscoelastic region (LVER). Figure 5 is reprinted with permission from Nazir et al. [8] (Copyright © 2017 Elsevier Ltd., Amsterdam, The Netherlands).

The time-averaged mean-squared displacement in the picture plane (x, y) ($\{\Delta r^2(\tau)\}$) is expressed in Equation (3):

$$\left\{\Delta r^2(\tau) = [x + (t+\tau) - x(t)]^2 + [y + (t+\tau) - y(t)]^2\right\} \tag{3}$$

The calculated $\Delta r^2(\tau)$ corresponds to:

$$\Delta r^2(\tau) = 2dD\tau = 2d\left(\frac{k_B T}{6\pi\eta a}\right)\tau \quad \text{for viscous medium}$$

$$\Delta r^2(\tau) = dk_B T/3\pi a G' \quad \text{for an elastic medium}$$

where D is the diffusion coefficient calculated according to the Stokes–Einstein Equation (4).

$$D = \left(\frac{k_B T}{6\pi\eta a}\right) \quad (4)$$

where k_B is the Boltzmann constant. The calculated $\Delta r^2(\tau)$ is converted into elastic and viscous moduli. Additionally, the estimated $\Delta r^2(\tau)$ can be connected to another significant viscoelastic parameter, i.e., creep compliance ($J(t)$), which is expressed as the time-dependent strain that results from stress values within the linear viscoelastic range (LVER) [98].

5.2. Oral Processing and Texture Perception of Gels

Oral processing refers to the manipulation of food in the mouth, which produces signals transmitted to the brain and result in an impression of texture and mouth feel [11]. Several technical characteristics, such as cutting, elasticity, firmness, effort, extensibility, melting, and the adhesiveness rate in the mouth, provide a more nuanced explanation of a food's texture. Texture is not just how food feels in the mouth, such as smooth or rough, light or heavy, etc. Given that gels are colloidal dispersions, which are typically thick, sticky, and solid-like materials, and thus, $G' > G'' \rightarrow \tan\delta < 1$ (elastic nature), as supported by the continuous phase (solid matrix), consumers may have trouble swallowing them, particularly the elderly and those who have dysphagia. Mastication (the act of chewing food) requires a variety of muscles, including the masseter muscles.

Suprahyoid muscles play a role in the movement of the tongue against the hard palate and the first movement of food during oropharyngeal swallowing. When eating soft foods such as jellies, suprahyoid muscles are more active than masseters; in contrast, eating firm, viscous, and sticky foods or gels causes more masseter activity than suprahyoid activity [105]. Electromyography (EMG), a method for evaluating the participation of masticatory muscles needed for chewing and swallowing various types of food, can be used to measure the activity of these muscles. EMG investigations are therefore helpful in research on food texture [106]. In one study, five different kinds of hydrocolloid gels were chosen, and EMG tests were run utilizing a variety of variables, such as the number of chews, the length of each chew, the quantity of masseter or suprahyoid action, the length of each masseter or suprahyoid action, etc. The findings indicate that chewing requires less effort with gels that melted more readily in the mouth than with hard and slick gels [107]. The fact that all types of gels exhibit strong masseter activity and have high EMG variable values can be explained by the fact that gels are always chewed thoroughly before ingesting.

6. Network Structure and Strength

6.1. Interconnected Polymeric Networks

Generally, the mechanical strength of typical food gels is sufficient for three-dimensional printing and edible film production. However, the networking of gels should be improved for encapsulation application [108]. In this regard, loading capacity is also limited for successful encapsulation of bioactive substances. Owing to the limitations of ordinary hydrogels, interpenetrating network hydrogels with superior mechanical characteristics have been developed. Interpenetrating network hydrogels are polymers composed of at least two networks that are partially intertwined on a molecular scale but not covalently bound to one another. Edible interpenetrating hydrogels are often made with natural polysaccharides and proteins, which are biodegradable and biocompatible. Several methods for producing interpenetrating hydrogels have been devised, including ionic [109], enzymatic [110], and heating and cooling [111] processes. Previously, various researchers developed hydrogels for application in food compositions to improve mechanical characteristics [112,113].

6.2. Strengthening Polymeric Networks

Enhancing the adhesion between distinct gel networks can boost the mechanical strength of hydrogels [114]. In previous research, two primary mechanisms have been implicated in an increase in the strengthening mechanism: polysaccharides incorporated as fillers of the network developed by other biopolymers, such as polysaccharide–protein hydrogels, e.g., k-carrageenan/sodium alginate double networks [115]; a myofibrillar protein embedded with native starch, such as (tapioca or potato starch) [116]; tilapia myofibrillar protein embedded with konjac glucomannan [117]; and myofibrillar protein imbedded in cassava starch [111]. Polysaccharides incorporated in proteins produce a hydrogel with an improved textural profile, possibly as a result of the "filling effect" or "packing effect". Starch particles compactly linked to proteins in these hydrogels exert a "packing effect" by expanding the starch granules and increasing the ensuing internal pressure. As a result, enhanced strengthening of textural and mechanical qualities is observed [118]. Furthermore, other polysaccharides, such carrageenan and konjac glucomannan, primarily operate as fillers to improve the gelling properties of protein gels after hydration, a phenomenon known as the "filling effect". These polysaccharides, when employed as fillers, can improve the mechanical properties of hydrogels by forming a dense and continuous gel network, enhancing the hydrogels' adhesiveness, hardness, and chewiness [119]. Polysaccharides are subsequently interwoven with other biopolymers to improve the density of entangled networks, which are commonly found in polysaccharide–polysaccharide hydrogels [120]. In one investigation, hydrogels made of konjac glucomannan and flaxseed gum were made by heating and cooling the mixture. As the mixture cooled, hydrogen bonds formed between the flaxseed and konjac glucomannan molecules. Hydrogen bonds created strong interactions between molecules that worked together to make the hydrogel hard, springy, and chewy [49].

7. Application of Hydrogel-Based Formulations in the Food Industry

7.1. Fat Replacers

An increase in the consumption of high-fat foods is linked to an increase in the prevalence of chronic diseases, such as heart disease, obesity, and diabetes. As a result, it is essential to develop food products that contain less fat, such as mayonnaise and ice cream. On the other hand, reducing the amount of fat in foods can result in unfavorable changes in the finished product, such as reduced water retention, altered flavor, and a more rigid consistency. For instance, when fat is removed from cakes, the crumb mixture becomes firm and dry. In addition, the functional qualities of salad dressings may suffer as a result of reduced fat content [30]. Owing to their high capacity to hold water, hydrogels made from protein and polysaccharides have proven to useful in maintaining lubricity and texture in a manner comparable to that of full-fat products [119].

While using hydrogels to replace fat, it is preferable to have hydrogels with excellent viscoelasticity and rheological properties. Therefore, various hybrid hydrogels were developed for desirable characteristics, such as pectin and whey protein [119]. Researchers produced mayonnaises with pectin and whey protein hydrogel by replacing 20% of the fat contents. Furthermore, the addition of fat replacers at sixty percent concentration can improve the stability of mayonnaise by reducing the level of flocculation and coalescence of fat droplets [121]. In addition, a number of different biopolymers, such as starch, have been reported for use as fat replacers [122].

7.2. Encapsulation of Bioactive Compounds: Gel-Based Formulations

The addition of odor or bioactive compounds to a food formulation can improve its sensory, nutrient, and antibacterial properties [123]. Hydrogels are becoming increasingly popular as encapsulating agents, owing to their superior encapsulation efficiency, biocompatibility, low price, and ecofriendly characteristics [124]. Curcumin was encapsulated in hydrogels, with a 90.3% success rate [125]. CMC-K-car, like CMC gel, can encapsulate probiotic bacteria, but it does so at a much higher rate (94.7%) than CMC gel (74.7%). Therefore, it can be used to ensure the most effective release of bioactive substances [126]. The amount of curcumin released from LRA-CS (70%) was also reported to be lower than that from water (91%) in an in vitro small intestine model study [125]. To effectively administer lysozyme, a chitosan/sodium alginate hydrogel was developed in a recent study [124]. The relative activity of lysozyme was measured at 87.72%, but the rate at which bacteria were eliminated was as high as 100%. Figure 6 depicts the variety of hydrogel systems and the efficiencies with which they can encapsulate bioactive compounds. In another study, hydrogen was made with food-grade pectin and PEG (polyethylene glycol)-encapsulated Ca, vitamin D, Fe^{2+}, and vitamin C. The gels were used to shield nutrients from artificial stomach acid [127]. Table 3 presents different types of hydrogel-based systems used for the encapsulation of bioactive compounds.

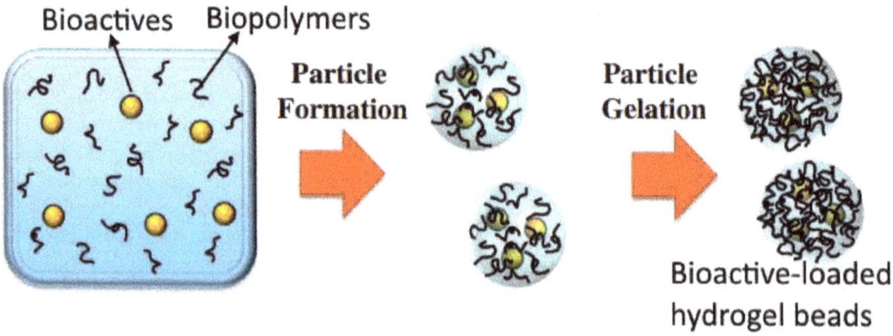

Figure 6. Illustration of a biopolymer-based delivery system for the encapsulation of bioactive compounds. Figure 6 is reprinted with permission from McClements [128] (Copyright © 2016 Elsevier B.V., Amsterdam, The Netherlands).

Table 3. Different types of hydrogel-based systems for the encapsulation of bioactive compounds.

Hydrogels	Form	Bioactive Compound	Encapsulation Efficiency (%)	Load Capacity	Ref.
Pectin-based	Core-shell hydrogel beads	Lactase	72	*	[129]
Xanthan gum	Biopolymer gels	Anthocyanin	*	*	[130]
Gelatin and chitosan	Gelation	Epigallocatechin gallate	95	*	[131]
Alginate-based nanohydrogel	Nanohydrogel	DOX (anticancer drug)	90	35%	[132]
Rice-protein varnish	Nanocomposite	Apigenin	91	$92.5\ mg\cdot g^{-1}$	[133]

*: Not reported.

Bioactive substances provide specific health advantages to humans [134]. However, environmental stressors such as oxygen, temperature, and light can destroy these important molecules. In addition to external stress, the gastrointestinal tract imposes stressful conditions related to enzymes, pH, and other antinutrients in the meal [135,136]. Owing to the presence of dietary fiber, microcapsules have a high potential for target release [137]. As an oral delivery system, corn fiber gum-soy protein isolate hydrogel effectively encapsulates riboflavin. Adding corn fiber gum to pure soy protein isolate hydrogel was reported to enhance the riboflavin release efficiency in the gut [138]. In addition, the drug-carrying capacity alginate can be improved by enhancing its mechanical characteristics and durability [139]. Some researchers studied whether the microcrystalline and microfiber forms of cellulose contained in alginate beads may successfully enhance the mechanical characteristics and drug release of alginate beads in simulated intestinal fluid [140,141].

β-carotene is a bioactive molecule and vitamin A precursor with powerful antioxidant properties [142]. However, owing to its hydrophobic nature, direct application of β-carotene in food products is restricted. In addition, it is chemically unstable under environmental stresses such as oxygen, temperature, and light [143]. However, previous research has demonstrated that emulsion-filled hydrogels are a suitable vehicle for the targeted delivery of β-carotene [144]. In addition to emulsion-filled hydrogels that incorporate useful chemicals (such as beta-carotene) into the lipid phase, hydrogels can release entrapped oil into the intestinal environment by integrating certain biopolymers [144,145].

7.3. Delivery System

Hydrogels are often used in targeted delivery because they can absorb and store a large amount of water or biological fluids in a 3D network structure [146]. Nanoemulsions or Pickering emulsions have also been enclosed in hydrogel beads using Trojan horse nanoparticles [147]. Other hydrogel characteristics including stimulus reactivity to pH, heat, and light are especially helpful for regulated release in food nutrition delivery methods [148].

It should be noted that many studies have been conducted using hydrogels to create smart drug delivery systems, such as microgels. The majority of the time, hydrogels for controlled release medication administration is created based on the swelling or shrinkage caused by pH and temperature signals [149]. There are parallels between the food supply chain and the pharmaceutical supply chain, but there are also some distinctions between them [149]. Therefore, previous studies of the implementation of structured hydrogels with well-designed drug delivery systems serve as instructive examples [149].

7.4. Calorie Control

Reduced-fat or reduced-starch goods must be created immediately. By increasing satiety or decreasing consumption, hydrogels can also support weight loss and calorie restriction. Hydrogel particles made of protein and dietary fiber have a pleasant texture and can serve as a healthier alternative to starch granules [150]. The caloric density of pancakes that were cooked at temperatures significantly higher than the boiling point of water was reported to be reduced using a temperature-insensitive, food-grade, mixed agar–methylcellulose hydrogel [151]. Furthermore, vegetable oils can be used in place of animal oils to create emulsion hydrogels for use in low-fat meals.

7.5. Food Texture Perception

In the food engineering and processing field, one of the most important elements affecting food product quality is the impression of food texture [152]. Hydrogels are soft substances with important textural qualities, such as elasticity, hardness, chewiness, fracture, etc. [153]. Therefore, hydrogels can be utilized to enhance the mouth feel and texture of food, supporting low-calorie consumption [154]. For instance, emulsion hydrogels can affect the texture of food. It is interesting to note a reduction in the amount of calories in food can also be accomplished by replacing meat or grain with hydrogels with excellent textural properties or containing little oil [155].

7.6. Risk Monitoring

Hydrogels have attracted considerable interest in the food industry, as they can be used as a biosensor or signal probe to identify danger or risk elements in food. For instance, a fluorescent DNA hydrogel aptasensor was developed for sensitive ochratoxin A detection. A portable pH meter with an aptamer-responsive, crosslinked hydrogel was also proposed to detect aflatoxin B-1. In a related study, researchers produced a 2D molecularly imprinted photonic crystal hydrogel sensor for detection of oxytetracycline in milk [156]. In recent decades, polymer-based hydrogels and solid-phase extraction have been used as a sampling technique for direct onsite rapid detection of dangerous chemicals such as rhodamine B, mercury (II), and (Hg^{2+}) ions [157].

7.7. Food Packaging Materials

The widespread use and abuse of high-molecular-weight, petroleum-based polymeric materials is a severe problem that must be addressed, and packaging food using hydrogel is a significant step in that direction [158]. Biobased biodegradable materials have been promoted for food packaging by recent research and scientific advancements [159]. The primary function of these components in food packaging is to regulate the humidity inside the container. Hydrogels can be used to reduce water activity (a_w), preventing the growth of mold, microbes, and pathogens that cause spoilage of packaged foods and hygroscopic products. Hydrogel can also protect dry, crunchy items from becoming softer [160,161]. Mechanical characteristics, high absorption, water-absorbing properties, and even breathability and accountability are key properties to be concerned about from the standpoint of the formation of hydrogels. Moreover, traditional considerations such as minimal price, recyclability, and biocompatibility are essential features. Whether the hydrogels function as conventional, active, or smart packaging materials, the hydrogel structure is the most important factor to consider [162].

The headspace or the contained food product can interact directly with active packaging to prevent or inhibit microbial development, retaining the food's nutritional value or enhancing its flavor over time [163]. Therefore, hydrogels have been thoroughly researched in food matrices and in vitro in combination with various antimicrobial chemicals, for example, silver or bioactive compounds [164]. Therefore, by integrating nanoparticles into the polymeric structure of a hydrogel, hybrid hydrogels such as nanocomposite hydrogels can represent a feasible solution to limitations such as the limited resistance to mechanical and hydrostatic stress of standard hydrogels [165]. Hydrogels also demonstrate their potential for use in intelligent packaging systems, whereby they can be used to alert consumers about the state of freshness of the food goods they contain or as a component of a quick test to identify pollutants [166]. The structured design of hydrogels is mostly used in this application area to support their smart response capabilities in the food engineering and processing field.

8. Conclusions and Future Directions

Gels are viscoelastic substances, and the commercial viability of food gels is of extreme significance. To develop gelled products, it is necessary to comprehend sol-gel transformation. One of the most crucial factors for consumers is the gel's texture, which can be affected by a wide range of factors, such as the material's makeup, composition, duration, temperature, pH, etc. The most recent developments in producing food-grade hydrogels for applications involving food were analyzed and discussed in detail in this study. In addition, aspects of hydrogels derived from natural sources were investigated; nevertheless, the gel strength of these substances could be improved by the addition of polysaccharides and proteins. Furthermore, hydrogels are a good source of energy, and when bioactive components are added, they act as a protective carrier for these substances. Owing to these attractive features, they are used in a wide range of food application areas, including for the production of bioactive substances, the protection of different flavors, and the replacement of fat in a wide range of products. In addition, consumers have demonstrated a high level of sensory acceptance for these hydrogel systems as fat replacers, sweetener replacers, etc., offering quality ingredients for a healthier lifestyle. Hydrogels can also be used in other contexts, such as for tissue regeneration, medical technology, and the precise administration of drugs.

Hydrogels have a wide range of uses in food processing, owing to their superior functional characteristics, consistency, mechanical strength, and water permeability. Nonetheless, it is still essential to pay attention to certain aspects of hydrogels. For example, it remains unknown how networks of hydrogels are interconnected. However, it is impossible to exert any control over the interaction of hydrogels. No accessible in-depth research has been conducted in this field to date. Regulation of the interplay between various gels can improve the absorption ability of hydrogels and cause less damage to the surrounding environment. As a result, further research and studies on this aspect of hydrogels are required.

Author Contributions: Conceptualization, P.C.N., P.K.N. and M.S.; methodology, P.C.N. and S.D.; software, K.S. and B.S.I.; validation, P.K.N. and M.S.; formal analysis, P.C.N. and S.D.; investigation, P.C.N. and S.D.; resources, M.S., P.K.N., B.S.I. and K.S.; data curation, P.C.N. and S.D.; writing—original draft preparation, P.C.N., K.S. and S.D.; writing—review and editing, M.S., B.S.I., P.K.N. and K.S.; visualization, M.S., B.S.I., P.K.N. and K.S.; supervision, P.K.N. and M.S.; project administration, P.K.N.; funding acquisition, P.K.N. All authors have read and agreed to the published version of the manuscript.

Funding: This research received no external funding.

Institutional Review Board Statement: Not applicable.

Informed Consent Statement: Not applicable.

Data Availability Statement: Data are available within the article.

Conflicts of Interest: The authors declare no conflict of interest.

References

1. Kaur, R.; Kukkar, D.; Bhardwaj, S.K.; Kim, K.-H.; Deep, A. Potential use of polymers and their complexes as media for storage and delivery of fragrances. *J. Control. Release* **2018**, *285*, 81–95. [CrossRef] [PubMed]
2. Blinov, A.V.; Siddiqui, S.A.; Nagdalian, A.A.; Blinova, A.A.; Gvozdenko, A.A.; Raffa, V.V.; Oboturova, N.P.; Golik, A.B.; Maglakelidze, D.G.; Ibrahim, S.A. Investigation of the influence of Zinc-containing compounds on the components of the colloidal phase of milk. *Arab. J. Chem.* **2021**, *14*, 103229. [CrossRef]
3. Chen, Z.; Wang, Z.; Ren, J.; Qu, X. Enzyme mimicry for combating bacteria and biofilms. *Acc. Chem. Res.* **2018**, *51*, 789–799. [CrossRef] [PubMed]
4. Hussain, I.; Singh, N.; Singh, A.; Singh, H.; Singh, S. Green synthesis of nanoparticles and its potential application. *Biotechnol. Lett.* **2016**, *38*, 545–560. [CrossRef]
5. Zaidi, K.U.; Ali, A.S.; Ali, S.A.; Naaz, I. Microbial tyrosinases: Promising enzymes for pharmaceutical, food bioprocessing, and environmental industry. *Biochem. Res. Int.* **2014**, *2014*, 854687. [CrossRef]
6. Fan, Z.; Cheng, P.; Gao, Y.; Wang, D.; Jia, G.; Zhang, P.; Prakash, S.; Wang, Z.; Han, J. Understanding the rheological properties of a novel composite salecan/gellan hydrogels. *Food Hydrocoll.* **2022**, *123*, 107162. [CrossRef]
7. Ali, A.; Ahmed, S. Recent advances in edible polymer based hydrogels as a sustainable alternative to conventional polymers. *J. Agric. Food Chem.* **2018**, *66*, 6940–6967. [CrossRef]
8. Nazir, A.; Asghar, A.; Maan, A.A. Food gels: Gelling process and new applications. In *Advances in Food Rheology and Its Applications*; Ahmed, J., Ed.; Woodhead Publishing: Sawston, UK, 2017; pp. 335–353.
9. Bashir, S.; Hina, M.; Iqbal, J.; Rajpar, A.; Mujtaba, M.; Alghamdi, N.; Wageh, S.; Ramesh, K.; Ramesh, S. Fundamental concepts of hydrogels: Synthesis, properties, and their applications. *Polymers* **2020**, *12*, 2702. [CrossRef]
10. Sala, G.; Stieger, M.; van de Velde, F. Serum release boosts sweetness intensity in gels. *Food Hydrocoll.* **2010**, *24*, 494–501. [CrossRef]
11. Stieger, M.; van de Velde, F. Microstructure, texture and oral processing: New ways to reduce sugar and salt in foods. *Curr. Opin. Colloid Interface Sci.* **2013**, *18*, 334–348. [CrossRef]
12. Reque, P.M.; Brandelli, A. Encapsulation of probiotics and nutraceuticals: Applications in functional food industry. *Trends Food Sci. Technol.* **2021**, *114*, 1–10. [CrossRef]
13. Boostani, S.; Jafari, S.M. A comprehensive review on the controlled release of encapsulated food ingredients; fundamental concepts to design and applications. *Trends Food Sci. Technol.* **2021**, *109*, 303–321. [CrossRef]
14. Nguyen, M.N.; Tran, P.H.; Tran, T.T. A single-layer film coating for colon-targeted oral delivery. *Int. J. Pharm.* **2019**, *559*, 402–409. [CrossRef]
15. Munir, N.; Hasnain, M.; Waqif, H.; Adetuyi, B.O.; Egbuna, C.; Olisah, M.C.; Chikwendu, C.J.; Uche, C.Z.; C Patrick-Iwuanyanwu, K.; El Sayed, A.M.A. Gelling agents, micro and nanogels in food system applications. In *Application of Nanotechnology in Food Science, Processing and Packaging*; Egbuna, C., Jeevanandam, J., Patrick-Iwuanyanwu, K.C., Onyeike, E.N., Eds.; Springer: Manhattan, NY, USA, 2022; pp. 153–167.
16. Zhu, P.; Huang, W.; Chen, L. Develop and characterize thermally reversible transparent gels from pea protein isolate and study the gel formation mechanisms. *Food Hydrocoll.* **2022**, *125*, 107373. [CrossRef]
17. Catauro, M.; Ciprioti, S.V. Characterization of hybrid materials prepared by sol-gel method for biomedical implementations. A critical review. *Materials* **2021**, *14*, 1788. [CrossRef]
18. Totosaus, A.; Montejano, J.G.; Salazar, J.A.; Guerrero, I. A review of physical and chemical protein-gel induction. *Int. J. Food Sci. Technol.* **2002**, *37*, 589–601. [CrossRef]
19. dos Santos Carvalho, J.D.; Rabelo, R.S.; Hubinger, M.D. Thermo-rheological properties of chitosan hydrogels with hydroxypropyl methylcellulose and methylcellulose. *Int. J. Biol. Macromol.* **2022**, *209*, 367–375. [CrossRef]
20. Kornet, R.; Sridharan, S.; Venema, P.; Sagis, L.M.; Nikiforidis, C.V.; van der Goot, A.J.; Meinders, M.B.; van der Linden, E. Fractionation methods affect the gelling properties of pea proteins in emulsion-filled gels. *Food Hydrocoll.* **2022**, *125*, 107427. [CrossRef]
21. Chen, Y.; Zhang, M.; Sun, Y.; Phuhongsung, P. Improving 3D/4D printing characteristics of natural food gels by novel additives: A review. *Food Hydrocoll.* **2022**, *123*, 107160. [CrossRef]
22. Dickinson, E. Hydrocolloids at interfaces and the influence on the properties of dispersed systems. *Food Hydrocoll.* **2003**, *17*, 25–39. [CrossRef]
23. Gousse, C.; Chanzy, H.; Cerrada, M.; Fleury, E. Surface silylation of cellulose microfibrils: Preparation and rheological properties. *Polymer* **2004**, *45*, 1569–1575. [CrossRef]

24. Saha, D.; Bhattacharya, S. Hydrocolloids as thickening and gelling agents in food: A critical review. *J. Food Sci. Technol.* **2010**, *47*, 587–597. [PubMed]
25. Yang, X.; Li, A.; Li, X.; Sun, L.; Guo, Y. An overview of classifications, properties of food polysaccharides and their links to applications in improving food textures. *Trends Food Sci. Technol.* **2020**, *102*, 587–597.
26. Mudgil, D.; Barak, S.; Khatkar, B.S. Guar gum: Processing, properties and food applications—A review. *J. Food Sci. Technol.* **2014**, *51*, 409–418. [PubMed]
27. Bashir, A.; Sharma, P.K.; Warsi, M.H. Extraction and characterization of xyloglucan (tamarind seed polysaccharide) as pharmaceutical excipient. *WJPPS* **2016**, *6*, 2209–2220.
28. Torres, M.D.; Flórez-Fernández, N.; Dominguez, H. Ultrasound-assisted water extraction of mastocarpus stellatus carrageenan with adequate mechanical and antiproliferative properties. *Mar. Drugs* **2021**, *19*, 280. [CrossRef] [PubMed]
29. Barak, S.; Mudgil, D. Locust bean gum: Processing, properties and food applications—A review. *Int. J. Biol. Macromol.* **2014**, *66*, 74–80. [CrossRef]
30. Liu, R.; Wang, L.; Liu, Y.; Wu, T.; Zhang, M. Fabricating soy protein hydrolysate/xanthan gum as fat replacer in ice cream by combined enzymatic and heat-shearing treatment. *Food Hydrocoll.* **2018**, *81*, 39–47. [CrossRef]
31. Singhvi, G.; Hans, N.; Shiva, N.; Dubey, S.K. Xanthan gum in drug delivery applications. In *Natural Polysaccharides in Drug Delivery and Biomedical Applications*; Hasnain, M.S., Nayak, A.K., Eds.; Academic Press: Cambridge, MA, USA, 2019; pp. 121–144.
32. Jindal, N.; Khattar, J.S. Microbial polysaccharides in food industry. In *Biopolymers for Food Design*; Grumezescu, A.M., Holban, A.M., Eds.; Academic Press: Cambridge, MA, USA, 2018; pp. 95–123.
33. Lee, H.W.; Karim, M.R.; Ji, H.M.; Choi, J.H.; Ghim, H.D.; Park, S.M.; Oh, W.; Yeum, J.H. Electrospinning fabrication and characterization of poly (vinyl alcohol)/montmorillonite nanofiber mats. *J. Appl. Polym. Sci.* **2009**, *113*, 1860–1867. [CrossRef]
34. Roopa, B.; Bhattacharya, S. Alginate gels: I. Characterization of textural attributes. *J. Food Eng.* **2008**, *85*, 123–131. [CrossRef]
35. Yang, Y.; Ye, H.; Zhao, C.; Ren, L.; Wang, C.; Georgiev, M.I.; Xiao, J.; Zhang, T. Value added immunoregulatory polysaccharides of Hericium erinaceus and their effect on the gut microbiota. *Carbohydr. Polym.* **2021**, *262*, 117668. [CrossRef] [PubMed]
36. Gunasekaran, S.; Xiao, L.; Ould Eleya, M. Whey protein concentrate hydrogels as bioactive carriers. *J. Appl. Polym. Sci.* **2006**, *99*, 2470–2476. [CrossRef]
37. Bhattacharya, S.; Jena, R. Gelling behavior of defatted soybean flour dispersions due to microwave treatment: Textural, oscillatory, microstructural and sensory properties. *J. Food Eng.* **2007**, *78*, 1305–1314. [CrossRef]
38. Kleemann, C.; Selmer, I.; Smirnova, I.; Kulozik, U. Tailor made protein based aerogel particles from egg white protein, whey protein isolate and sodium caseinate: Influence of the preceding hydrogel characteristics. *Food Hydrocoll.* **2018**, *83*, 365–374. [CrossRef]
39. Shahidi, F.; Synowiecki, J. Isolation and characterization of nutrients and value-added products from snow crab (*Chionoecetes opilio*) and shrimp (*Pandalus borealis*) processing discards. *J. Agric. Food Chem.* **1991**, *39*, 1527–1532. [CrossRef]
40. Ponsubha, S.; Jaiswal, A.K. Effect of interpolymer complex formation between chondroitin sulfate and chitosan-gelatin hydrogel on physico-chemical and rheological properties. *Carbohydr. Polym.* **2020**, *238*, 116179.
41. Dissanayake, M.; Vasiljevic, T. Functional properties of whey proteins affected by heat treatment and hydrodynamic high-pressure shearing. *J. Dairy Sci.* **2009**, *92*, 1387–1397. [CrossRef]
42. Lucey, J.; Singh, H. Formation and physical properties of acid milk gels: A review. *Food Res. Int.* **1997**, *30*, 529–542.
43. Bryant, C.; McClements, D. Molecular basis of cold-setting whey protein ingredients. *Trends Food Sci. Technol.* **1998**, *9*, 143–151.
44. Alting, A.C.; de Jongh, H.H.; Visschers, R.W.; Simons, J.-W.F. Physical and chemical interactions in cold gelation of food proteins. *J. Agric. Food Chem.* **2002**, *50*, 4682–4689. [CrossRef]
45. Aguilera, J.M.; Kessler, H.G. Properties of mixed and filled-type dairy gels. *J. Food Sci.* **1989**, *54*, 1213–1217. [CrossRef]
46. Heinen, L.; Heuser, T.; Steinschulte, A.; Walther, A. Antagonistic enzymes in a biocatalytic pH feedback system program autonomous DNA hydrogel life cycles. *Nano Lett.* **2017**, *17*, 4989–4995. [CrossRef]
47. Banerjee, S.; Bhattacharya, S. Food gels: Gelling process and new applications. *Crit. Rev. Food Sci. Nutr.* **2012**, *52*, 334–346. [CrossRef] [PubMed]
48. Gharibzahedi, S.M.T.; Chronakis, I.S. Crosslinking of milk proteins by microbial transglutaminase: Utilization in functional yogurt products. *Food Chem.* **2018**, *245*, 620–632. [CrossRef] [PubMed]
49. Jiang, Y.; Reddy, C.K.; Huang, K.; Chen, L.; Xu, B. Hydrocolloidal properties of flaxseed gum/konjac glucomannan compound gel. *Int. J. Biol. Macromol.* **2019**, *133*, 1156–1163. [CrossRef] [PubMed]
50. Burey, P.; Bhandari, B.; Howes, T.; Gidley, M. Hydrocolloid gel particles: Formation, characterization, and application. *Crit. Rev. Food Sci. Nutr.* **2008**, *48*, 361–377. [CrossRef]
51. Richter, S. Recent gelation studies on irreversible and reversible systems with dynamic light scattering and rheology—A concise summary. *Macromol. Chem. Phys.* **2007**, *208*, 1495–1502. [CrossRef]
52. Ahmad, N.H.; Mustafa, S.; Che Man, Y.B. Microbial polysaccharides and their modification approaches: A review. *Int. J. Food Prop.* **2015**, *18*, 332–347. [CrossRef]
53. Vilgis, T.A. Gels: Model systems for soft matter food physics. *Curr. Opin. Food Sci.* **2015**, *3*, 71–84. [CrossRef]
54. Nishinari, K.; Zhang, H. Recent advances in the understanding of heat set gelling polysaccharides. *Trends Food Sci. Technol.* **2004**, *15*, 305–312. [CrossRef]

55. McClements, D.J. Non-covalent interactions between proteins and polysaccharides. *Biotechnol. Adv.* **2006**, *24*, 621–625. [CrossRef] [PubMed]
56. Williams, P.A.; Phillips, G.O. Introduction to food hydrocolloids. In *Handbook of Hydrocolloids*, 3rd ed.; Phillips, G.O., Williams, P.A., Eds.; Woodhead Publishing: Sawston, UK, 2021; pp. 3–26.
57. Lapomarda, A.; Pulidori, E.; Cerqueni, G.; Chiesa, I.; De Blasi, M.; Geven, M.A.; Montemurro, F.; Duce, C.; Mattioli-Belmonte, M.; Tiné, M.R. Pectin as rheology modifier of a gelatin-based biomaterial ink. *Materials* **2021**, *14*, 3109. [CrossRef] [PubMed]
58. Chan, S.Y.; Choo, W.S.; Young, D.J.; Loh, X.J. Pectin as a rheology modifier: Origin, structure, commercial production and rheology. *Carbohydr. Polym.* **2017**, *161*, 118–139.
59. Chan, S.Y.; Owh, C.; Choo, W.S.; Young, D.J.; Loh, X.J.; Li, P.; Huang, W. Pectin as a rheology modifier: Chemistry, production, and rheology. *Mater. Sci. Technol.* **2006**, *24*, 621–625. [CrossRef]
60. Sutherland, I.W. Biotechnology of microbial polysaccharides in food. In *FoodBiotechnology*, 1st ed.; Pometto, A., Shetty, K., Paliyath, G., Levin, R.E., Eds.; CRC Press: Boca Raton, FL, USA, 2005; pp. 220–247.
61. Mizrahi, S. Syneresis in food gels and its implications for food quality. In *Chemical Deterioration and Physical Instability of Food and Beverages*; Skibsted, L.H., Risbo, J., Andersen, M.L., Eds.; Woodhead Publishing: Sawston, UK, 2010; pp. 324–348.
62. Gallant, D.; Bouchet, B.; Buleon, A.; Perez, S. Physical characteristics of starch granules and susceptibility to enzymatic degradation. *Eur. J. Clin. Nutr.* **1992**, *46*, S3–S16. [PubMed]
63. Wang, Q.; Rademacher, B.; Sedlmeyer, F.; Kulozik, U. Gelation behaviour of aqueous solutions of different types of carrageenan investigated by low-intensity-ultrasound measurements and comparison to rheological measurements. *Innov. Food Sci. Emerg. Technol.* **2005**, *6*, 465–472. [CrossRef]
64. Morris, E.R.; Rees, D.A.; Robinson, G. Cation-specific aggregation of carrageenan helices: Domain model of polymer gel structure. *J. Mol. Biol.* **1980**, *138*, 349–362. [CrossRef]
65. Viebke, C.; Piculell, L.; Nilsson, S. On the mechanism of gelation of helix-forming biopolymers. *Macromolecules* **1994**, *27*, 4160–4166. [CrossRef]
66. Babbar, S.; Jain, R.; Walia, N. Guar gum as a gelling agent for plant tissue culture media. *Vitr. Cell. Dev. Biol.-Plant* **2005**, *41*, 258–261. [CrossRef]
67. Jain, R.; Babbar, S.B. Guar gum and isubgol as cost-effective alternative gelling agents for in vitro multiplication of an orchid, *Dendrobium chrysotoxum*. *Curr. Sci.* **2005**, *88*, 292–295.
68. Astrini, N.; Anah, L.; Haryono, A. Crosslinking parameter on the preparation of cellulose based hydrogel with divynilsulfone. *Procedia Chem.* **2012**, *4*, 275–281. [CrossRef]
69. Ibrahim, M.M.; Abd-Eladl, M.; Abou-Baker, N.H. Lignocellulosic biomass for the preparation of cellulose-based hydrogel and its use for optimizing water resources in agriculture. *J. Appl. Polym. Sci.* **2015**, *132*. [CrossRef]
70. Motamedi, E.; Sadeghian Motahar, S.F.; Maleki, M.; Kavousi, K.; Ariaeenejad, S.; Moosavi-Movahedi, A.A.; Hosseini Salekdeh, G. Upgrading the enzymatic hydrolysis of lignocellulosic biomass by immobilization of metagenome-derived novel halotolerant cellulase on the carboxymethyl cellulose-based hydrogel. *Cellulose* **2021**, *28*, 3485–3503.
71. Kanikireddy, V.; Varaprasad, K.; Jayaramudu, T.; Karthikeyan, C.; Sadiku, R. Carboxymethyl cellulose-based materials for infection control and wound healing: A review. *Int. J. Biol. Macromol.* **2020**, *164*, 963–975. [PubMed]
72. Mali, K.; Dhawale, S.; Dias, R.; Dhane, N.; Ghorpade, V. Citric acid crosslinked carboxymethyl cellulose-based composite hydrogel films for drug delivery. *Indian J. Pharm. Sci.* **2018**, *80*, 657–667. [CrossRef]
73. Baiya, C.; Nannuan, L.; Tassanapukdee, Y.; Chailapakul, O.; Songsrirote, K. The synthesis of carboxymethyl cellulose-based hydrogel from sugarcane bagasse using microwave-assisted irradiation for selective adsorption of copper (II) ions. *Environ. Prog. Sustain. Energy* **2019**, *38*, S157–S165. [CrossRef]
74. Turky, G.; Moussa, M.A.; Hasanin, M.; El-Sayed, N.S.; Kamel, S. Carboxymethyl cellulose-based hydrogel: Dielectric study, antimicrobial activity and biocompatibility. *Arab. J. Sci. Eng.* **2021**, *46*, 17–30.
75. Mallakpour, S.; Tukhani, M.; Hussain, C.M. Recent advancements in 3D bioprinting technology of carboxymethyl cellulose-based hydrogels: Utilization in tissue engineering. *Adv. Colloid Interface Sci.* **2021**, *292*, 102415.
76. Sethi, S.; Kaith, B.; Kumar, V. Fabrication and characterization of microwave assisted carboxymethyl cellulose-gelatin silver nanoparticles imbibed hydrogel: Its evaluation as dye degradation. *React. Funct. Polym.* **2019**, *142*, 134–146. [CrossRef]
77. Sharmeen, S.; Rahman, M.S.; Islam, M.M.; Islam, M.S.; Shahruzzaman, M.; Mallik, A.K.; Haque, P.; Rahman, M.M. Application of polysaccharides in enzyme immobilization. In *Functional Polysaccharides for Biomedical Applications*; Maiti, S., Jana, S., Eds.; Woodhead Publishing: Sawston, UK, 2019; pp. 357–395.
78. Kukrety, A.; Singh, R.K.; Singh, P.; Ray, S.S. Comprehension on the synthesis of carboxymethylcellulose (CMC) utilizing various cellulose rich waste biomass resources. *Waste Biomass Valorization* **2018**, *9*, 1587–1595.
79. Koh, M.H. Preparation and Characterization of Carboxymethyl Cellulose from Sugarcane Bagasse. Ph.D. Thesis, Universiti Tunku Abdul Rahman, Petaling Jaya, Malaysia, 2013. Available online: http://eprints.utar.edu.my/895/1/CE-2013-1004861.pdf (accessed on 21 October 2022).
80. Haqiqi, M.T.; Bankeeree, W.; Lotrakul, P.; Pattananuwat, P.; Punnapayak, H.; Ramadhan, R.; Kobayashi, T.; Amirta, R.; Prasongsuk, S. Antioxidant and UV-blocking properties of a carboxymethyl cellulose–lignin composite film produced from oil palm empty fruit bunch. *ACS Omega* **2021**, *6*, 9653–9666. [CrossRef] [PubMed]

81. Nasution, H.; Harahap, H.; Dalimunthe, N.F.; Ginting, M.H.S.; Jaafar, M.; Tan, O.O.; Aruan, H.K.; Herfananda, A.L. Hydrogel and effects of crosslinking agent on cellulose-based hydrogels: A review. *Gels* **2022**, *8*, 568. [CrossRef] [PubMed]
82. Gårdebjer, S.; Larsson, M.; Gebäck, T.; Skepö, M.; Larsson, A. An overview of the transport of liquid molecules through structured polymer films, barriers and composites–Experiments correlated to structure-based simulations. *Adv. Colloid Interface Sci.* **2018**, *256*, 48–64. [CrossRef]
83. Boyer, C.; Figueiredo, L.; Pace, R.; Lesoeur, J.; Rouillon, T.; Le Visage, C.; Tassin, J.-F.; Weiss, P.; Guicheux, J.; Rethore, G. Laponite nanoparticle-associated silated hydroxypropylmethyl cellulose as an injectable reinforced interpenetrating network hydrogel for cartilage tissue engineering. *Acta Biomater.* **2018**, *65*, 112–122. [CrossRef] [PubMed]
84. Bacakova, L.; Pajorova, J.; Bacakova, M.; Skogberg, A.; Kallio, P.; Kolarova, K.; Svorcik, V. Nanocellulose in biotechnology and medicine: Focus on skin tissue engineering and wound healing. *Nanomateirals* **2019**, *9*, 164. [CrossRef]
85. Perinelli, D.R.; Palmieri, G.F.; Cespi, M.; Bonacucina, G. Encapsulation of flavours and fragrances into polymeric capsules and cyclodextrins inclusion complexes: An update. *Molecules* **2020**, *25*, 5878. [CrossRef]
86. Wach, R.A.; Mitomo, H.; Yoshii, F.; Kume, T. Hydrogel of radiation-induced cross-linked hydroxypropylcellulose. *Macromol. Mater. Eng.* **2002**, *287*, 285–295. [CrossRef]
87. Lu, C.; Wang, C.; Zhang, D.; Wang, J.; Yong, Q.; Chu, F. Ultra-strong hydroxypropyl cellulose/polyvinyl alcohol composite hydrogel by combination of triple-network and mechanical training. *Int. J. Biol. Macromol.* **2021**, *184*, 200–208. [CrossRef]
88. Van den Berg, L.; Rosenberg, Y.; Van Boekel, M.A.; Rosenberg, M.; Van de Velde, F. Microstructural features of composite whey protein/polysaccharide gels characterized at different length scales. *Food Hydrocoll.* **2009**, *23*, 1288–1298. [CrossRef]
89. He, H.; Cao, X.; Dong, H.; Ma, T.; Payne, G.F. Reversible programing of soft matter with reconfigurable mechanical properties. *Adv. Funct. Mater.* **2017**, *27*, 1605665. [CrossRef]
90. Hu, Y.; Zhang, S.; Han, D.; Ding, Z.; Zeng, S.; Xiao, X. Construction and evaluation of the hydroxypropyl methyl cellulose-sodium alginate composite hydrogel system for sustained drug release. *J. Polym. Res.* **2018**, *25*, 148. [CrossRef]
91. Hanson, B.S.; Brown, C.P.; Laurent, H.; Hughes, M.D.G.; Dougan, L. Hierarchical biomechanics: Student engagement activities with a focus on biological physics. *Phys. Educ.* **2020**, *55*, 025015. [CrossRef]
92. De Jong, S.; Klok, H.J.; Van de Velde, F. The mechanism behind microstructure formation in mixed whey protein–polysaccharide cold-set gels. *Food Hydrocoll.* **2009**, *23*, 755–764. [CrossRef]
93. Laneuville, S.I.; Turgeon, S.L. Microstructure and stability of skim milk acid gels containing an anionic bacterial exopolysaccharide and commercial polysaccharides. *Int. Dairy J.* **2014**, *37*, 5–15. [CrossRef]
94. Godiya, C.B.; Kumar, S.; Xiao, Y. Amine functionalized egg albumin hydrogel with enhanced adsorption potential for diclofenac sodium in water. *J. Hazard. Mater.* **2020**, *393*, 122417. [CrossRef]
95. Fertsch, B.; Müller, M.; Hinrichs, J. Firmness of pressure-induced casein and whey protein gels modulated by holding time and rate of pressure release. *Innov. Food Sci. Emerg. Technol.* **2003**, *4*, 143–150. [CrossRef]
96. Keogh, M.K.; Lainé, K.I.; O'connor, J.F. Rheology of sodium caseinate-carrageenan mixtures. *J. Texture Stud.* **1996**, *26*, 635–652. [CrossRef]
97. Foegeding, E.A. Rheology and sensory texture of biopolymer gels. *Curr. Opin. Colloid Interface Sci.* **2007**, *12*, 242–250. [CrossRef]
98. Herranz, B.; Tovar, C.A.; Solo-de-Zaldívar, B.; Borderias, A.J. Effect of alkalis on konjac glucomannan gels for use as potential gelling agents in restructured seafood products. *Food Hydrocoll.* **2012**, *27*, 145–153. [CrossRef]
99. Campo-Deaño, L.; Tovar, C. The effect of egg albumen on the viscoelasticity of crab sticks made from Alaska Pollock and Pacific Whiting surimi. *Food Hydrocoll.* **2009**, *23*, 1641–1646. [CrossRef]
100. Moreno, H.M.; Dominguez-Timon, F.; Díaz, M.T.; Pedrosa, M.M.; Borderías, A.J.; Tovar, C.A. Evaluation of gels made with different commercial pea protein isolate: Rheological, structural and functional properties. *Food Hydrocoll.* **2020**, *99*, 105375. [CrossRef]
101. Borderías, A.J.; Tovar, C.A.; Domínguez-Timón, F.; Díaz, M.T.; Pedrosa, M.M.; Moreno, H.M. Characterization of healthier mixed surimi gels obtained through partial substitution of myofibrillar proteins by pea protein isolates. *Food Hydrocoll.* **2020**, *107*, 105976. [CrossRef]
102. Barnes, H.A.; Hutton, J.F.; Walter, K. *An Introduction to Rheology*, 1st ed.; Elsevier Science Publishers B.V.: Amsterdam, The Netherlands, 1989; pp. 1–201.
103. Moschakis, T. Microrheology and particle tracking in food gels and emulsions. *Curr. Opin. Colloid Interface Sci.* **2013**, *18*, 311–323. [CrossRef]
104. Grimm, M.; Jeney, S.; Franosch, T. Brownian motion in a Maxwell fluid. *Soft Matter* **2011**, *7*, 2076–2084. [CrossRef]
105. Ishihara, S.; Nakauma, M.; Funami, T.; Tanaka, T.; Nishinari, K.; Kohyama, K. Electromyography during oral processing in relation to mechanical and sensory properties of soft gels. *J. Texture Stud.* **2011**, *42*, 254–267. [CrossRef]
106. Kohyama, K.; Mioche, L.; Bourdio3, P. Influence of age and dental status on chewing behaviour studied by EMG recordings during consumption of various food samples. *Gerodontology* **2003**, *20*, 15–23. [CrossRef]
107. Pascua, Y.; Koç, H.; Foegeding, E.A. Food structure: Roles of mechanical properties and oral processing in determining sensory texture of soft materials. *Curr. Opin. Colloid Interface Sci.* **2013**, *18*, 324–333. [CrossRef]
108. De Silva, D.A. Characterization of Single Network and Interpenetrating Network Hydrogels of Natural and Synthetic Polymers. Ph.D. Thesis, University of Wollongong, Wollongong, Australia, 2013. Available online: https://ro.uow.edu.au/theses/4042 (accessed on 10 November 2022).

109. Liu, K.; Li, Q.-M.; Zha, X.-Q.; Pan, L.-H.; Bao, L.-J.; Zhang, H.-L.; Luo, J.-P. Effects of calcium or sodium ions on the properties of whey protein isolate-lotus root amylopectin composite gel. *Food Hydrocoll.* **2019**, *87*, 629–636. [CrossRef]
110. Sorde, K.L.; Ananthanarayan, L. Effect of transglutaminase treatment on properties of coconut protein-guar gum composite film. *LWT* **2019**, *115*, 108422. [CrossRef]
111. Fan, M.; Hu, T.; Zhao, S.; Xiong, S.; Xie, J.; Huang, Q. Gel characteristics and microstructure of fish myofibrillar protein/cassava starch composites. *Food Chem.* **2017**, *218*, 221–230. [CrossRef]
112. Yadav, M.; Chiu, F.-C. Cellulose nanocrystals reinforced κ-carrageenan based UV resistant transparent bionanocomposite films for sustainable packaging applications. *Carbohydr. Polym.* **2019**, *211*, 181–194. [CrossRef] [PubMed]
113. Kim, M.H.; Lee, Y.W.; Jung, W.-K.; Oh, J.; Nam, S.Y. Enhanced rheological behaviors of alginate hydrogels with carrageenan for extrusion-based bioprinting. *J. Mech. Behav. Biomed. Mater.* **2019**, *98*, 187–194. [CrossRef]
114. Graham, S.; Marina, P.F.; Blencowe, A. Thermoresponsive polysaccharides and their thermoreversible physical hydrogel networks. *Carbohydr. Polym.* **2019**, *207*, 143–159. [CrossRef]
115. Li, L.; Zhao, J.; Sun, Y.; Yu, F.; Ma, J. Ionically cross-linked sodium alginate/k-carrageenan double-network gel beads with low-swelling, enhanced mechanical properties, and excellent adsorption performance. *Chem. Eng. J.* **2019**, *372*, 1091–1103. [CrossRef]
116. Wu, T.; Huang, J.; Jiang, Y.; Hu, Y.; Ye, X.; Liu, D.; Chen, J. Formation of hydrogels based on chitosan/alginate for the delivery of lysozyme and their antibacterial activity. *Food Chem.* **2018**, *240*, 361–369. [CrossRef] [PubMed]
117. Jian, W.; Wu, H.; Wu, L.; Wu, Y.; Jia, L.; Pang, J.; Sun, Y.-m. Effect of molecular characteristics of Konjac glucomannan on gelling and rheological properties of Tilapia myofibrillar protein. *Carbohydr. Polym.* **2016**, *150*, 21–31. [CrossRef]
118. Mi, H.; Li, Y.; Wang, C.; Yi, S.; Li, X.; Li, J. The interaction of starch-gums and their effect on gel properties and protein conformation of silver carp surimi. *Food Hydrocoll.* **2021**, *112*, 106290. [CrossRef]
119. Yang, X.; Gong, T.; Lu, Y.-h.; Li, A.; Sun, L.; Guo, Y. Compatibility of sodium alginate and konjac glucomannan and their applications in fabricating low-fat mayonnaise-like emulsion gels. *Carbohydr. Polym.* **2020**, *229*, 115468. [CrossRef] [PubMed]
120. Zheng, J.; Zeng, R.; Zhang, F.; Kan, J. Effects of sodium carboxymethyl cellulose on rheological properties and gelation behaviors of sodium alginate induced by calcium ions. *LWT* **2019**, *103*, 131–138. [CrossRef]
121. Sun, C.; Liu, R.; Liang, B.; Wu, T.; Sui, W.; Zhang, M. Microparticulated whey protein-pectin complex: A texture-controllable gel for low-fat mayonnaise. *Food Res. Int.* **2018**, *108*, 151–160. [CrossRef]
122. Diamantino, V.R.; Costa, M.S.; Taboga, S.R.; Vilamaior, P.S.; Franco, C.M.; Penna, A.L.B. Starch as a potential fat replacer for application in cheese: Behaviour of different starches in casein/starch mixtures and in the casein matrix. *Int. Dairy J.* **2019**, *89*, 129–138. [CrossRef]
123. Abaee, A.; Mohammadian, M.; Jafari, S.M. Whey and soy protein-based hydrogels and nano-hydrogels as bioactive delivery systems. *Trends Food Sci. Technol.* **2017**, *70*, 69–81. [CrossRef]
124. Wu, M.; Wang, J.; Ge, Q.; Yu, H.; Xiong, Y.L. Rheology and microstructure of myofibrillar protein–starch composite gels: Comparison of native and modified starches. *Int. J. Biol. Macromol.* **2018**, *118*, 988–996. [CrossRef] [PubMed]
125. Liu, K.; Zha, X.-Q.; Shen, W.-D.; Li, Q.-M.; Pan, L.-H.; Luo, J.-P. The hydrogel of whey protein isolate coated by lotus root amylopectin enhance the stability and bioavailability of quercetin. *Carbohydr. Polym.* **2020**, *236*, 116009. [CrossRef]
126. Dafe, A.; Etemadi, H.; Zarredar, H.; Mahdavinia, G.R. Development of novel carboxymethyl cellulose/k-carrageenan blends as an enteric delivery vehicle for probiotic bacteria. *Int. J. Biol. Macromol.* **2017**, *97*, 299–307. [CrossRef]
127. Gautam, M.; Santhiya, D. Pectin/PEG food grade hydrogel blend for the targeted oral co-delivery of nutrients. *Colloids Surf. A Physicochem. Eng. Asp.* **2019**, *577*, 637–644. [CrossRef]
128. McClements, D.J. Designing biopolymer microgels to encapsulate, protect and deliver bioactive components: Physicochemical aspects. *Adv. Colloid Interface Sci.* **2017**, *240*, 31–59. [CrossRef]
129. Cargnin, M.A.; Gasparin, B.C.; dos Santos Rosa, D.; Paulino, A.T. Performance of lactase encapsulated in pectin-based hydrogels during lactose hydrolysis reactions. *LWT* **2021**, *150*, 111863. [CrossRef]
130. Jin, W.; Xiang, L.; Peng, D.; Liu, G.; He, J.; Cheng, S.; Li, B.; Huang, Q. Study on the coupling progress of thermo-induced anthocyanins degradation and polysaccharides gelation. *Food Hydrocoll.* **2020**, *105*, 105822. [CrossRef]
131. Gómez-Mascaraque, L.G.; Soler, C.; Lopez-Rubio, A. Stability and bioaccessibility of EGCG within edible micro-hydrogels. Chitosan vs. gelatin, a comparative study. *Food Hydrocoll.* **2016**, *61*, 128–138. [CrossRef]
132. Zhou, T.; Li, J.; Liu, P. Ionically crosslinked alginate-based nanohydrogels for tumor-specific intracellular triggered release: Effect of chemical modification. *Colloids Surf. A Physicochem. Eng. Asp.* **2018**, *553*, 180–186. [CrossRef]
133. Wang, A.; Lin, J.; Zhong, Q. Enteric rice protein-shellac composite coating to enhance the viability of probiotic *Lactobacillus salivarius* NRRL B-30514. *Food Hydrocoll.* **2021**, *113*, 106469. [CrossRef]
134. Khan, M.K.I.; Ghauri, Y.M.; Alvi, T.; Amin, U.; Khan, M.I.; Nazir, A.; Saeed, F.; Aadil, R.M.; Nadeem, M.T.; Babu, I. Microwave assisted drying and extraction technique; kinetic modelling, energy consumption and influence on antioxidant compounds of fenugreek leaves. *Food Sci. Technol.* **2021**, *42*. [CrossRef]
135. Yan, W.; Zhang, B.; Yadav, M.P.; Feng, L.; Yan, J.; Jia, X.; Yin, L. Corn fiber gum-soybean protein isolate double network hydrogel as oral delivery vehicles for thermosensitive bioactive compounds. *Food Hydrocoll.* **2020**, *107*, 105865. [CrossRef]

136. Alvi, T.; Asif, Z.; Khan, M.K.I. Clean label extraction of bioactive compounds from food waste through microwave-assisted extraction technique—A review. *Food Biosci.* **2022**, *46*, 101580. [CrossRef]
137. Chen, K.; Zhang, M.; Adhikari, B.; Wang, M. Microencapsulation of Sichuan pepper essential oil in soybean protein isolate-Sichuan pepper seed soluble dietary fiber complex coacervates. *Food Hydrocoll.* **2022**, *125*, 107421. [CrossRef]
138. Kandekar, U.; Ramdasi, S.; Thakurdesai, P. Fenugreek: Novel delivery technologies and versatile formulation excipients. In *Fenugreek*, 1st ed.; Ghosh, D., Thakurdesai, P., Eds.; CRC Press: Boca Raton, FL, USA, 2022; pp. 363–402.
139. Yadav, H.; Agrawal, R.; Panday, A.; Patel, J.; Maiti, S. Polysaccharide-silicate composite hydrogels: Review on synthesis and drug delivery credentials. *J. Drug Deliv. Sci. Technol.* **2022**, *74*, 103573. [CrossRef]
140. Khalesi, H.; Lu, W.; Nishinari, K.; Fang, Y. Fundamentals of composites containing fibrous materials and hydrogels: A review on design and development for food applications. *Food Chem.* **2021**, *364*, 130329. [CrossRef]
141. Liu, S.; Qamar, S.A.; Qamar, M.; Basharat, K.; Bilal, M. Engineered nanocellulose-based hydrogels for smart drug delivery applications. *Int. J. Biol. Macromol.* **2021**, *181*, 275–290. [CrossRef]
142. Almagro, L.; Correa-Sabater, J.M.; Sabater-Jara, A.B.; Pedreño, M.Á. Biotechnological production of β-carotene using plant in vitro cultures. *Planta* **2022**, *256*, 41. [CrossRef]
143. Park, S.; Mun, S.; Kim, Y.-R. Effect of xanthan gum on lipid digestion and bioaccessibility of β-carotene-loaded rice starch-based filled hydrogels. *Food Res. Int.* **2018**, *105*, 440–445. [CrossRef] [PubMed]
144. Mun, S.; Kim, Y.-R.; McClements, D.J. Control of β-carotene bioaccessibility using starch-based filled hydrogels. *Food Chem.* **2015**, *173*, 454–461. [CrossRef] [PubMed]
145. Zhang, Z.; Zhang, R.; Chen, L.; Tong, Q.; McClements, D.J. Designing hydrogel particles for controlled or targeted release of lipophilic bioactive agents in the gastrointestinal tract. *Eur. Polym. J.* **2015**, *72*, 698–716. [CrossRef]
146. Gyles, D.A.; Castro, L.D.; Silva, J.O.C., Jr.; Ribeiro-Costa, R.M. A review of the designs and prominent biomedical advances of natural and synthetic hydrogel formulations. *Eur. Polym. J.* **2017**, *88*, 373–392. [CrossRef]
147. Montes, C.; Villaseñor, M.J.; Rios, A. Analytical control of nanodelivery lipid-based systems for encapsulation of nutraceuticals: Achievements and challenges. *Trends Food Sci. Technol.* **2019**, *90*, 47–62. [CrossRef]
148. Alvarez-Lorenzo, C.; Blanco-Fernandez, B.; Puga, A.M.; Concheiro, A. Crosslinked ionic polysaccharides for stimuli-sensitive drug delivery. *Adv. Drug Deliv. Rev.* **2013**, *65*, 1148–1171. [CrossRef]
149. Qiu, Y.; Park, K. Environment-sensitive hydrogels for drug delivery. *Adv. Drug Deliv. Rev.* **2001**, *53*, 321–339. [CrossRef]
150. Rayner, M.; Östbring, K.; Purhagen, J. Application of natural polymers in food. In *Natural Polymers*; Springer: Cham, Switzerland, 2016; pp. 115–161.
151. Thompson, B.R.; Horozov, T.S.; Stoyanov, S.D.; Paunov, V.N. Structuring and calorie control of bakery products by templating batter with ultra melt-resistant food-grade hydrogel beads. *Food Funct.* **2017**, *8*, 2967–2973. [CrossRef]
152. Khalesi, H.; Lu, W.; Nishinari, K.; Fang, Y. New insights into food hydrogels with reinforced mechanical properties: A review on innovative strategies. *Adv. Colloid Interface Sci.* **2020**, *285*, 102278. [CrossRef]
153. Stribiţcaia, E.; Krop, E.M.; Lewin, R.; Holmes, M.; Sarkar, A. Tribology and rheology of bead-layered hydrogels: Influence of bead size on sensory perception. *Food Hydrocoll.* **2020**, *104*, 105692. [CrossRef]
154. Gheorghita Puscaselu, R.; Lobiuc, A.; Dimian, M.; Covasa, M. Alginate: From food industry to biomedical applications and management of metabolic disorders. *Polymers* **2020**, *12*, 2417. [CrossRef] [PubMed]
155. Zhang, H.; Zhang, F.; Yuan, R. Applications of natural polymer-based hydrogels in the food industry. In *Hydrogels Based on Natural Polymers*; Chen, Y., Ed.; Elsevier: Amsterdam, The Netherlands, 2020; pp. 357–410.
156. Qi, F.; Meng, Z.; Xue, M.; Qiu, L. Recent advances in self-assemblies and sensing applications of colloidal photonic crystals. *Anal. Chim. Acta* **2020**, *1123*, 91–112. [CrossRef] [PubMed]
157. Wang, Z.; Meng, Z.; Xue, M.; Zhang, H.; Shea, K.J.; Kang, L. Detection of lysozyme in body fluid based on two-dimensional colloidal crystal sensor. *Microchem. J.* **2020**, *157*, 105073. [CrossRef]
158. Nesic, A.R.; Seslija, S.I. The influence of nanofillers on physical–chemical properties of polysaccharide-based film intended for food packaging. In *Food Packaging: Nanotechnology in the Agri-Food Industry*; Grumezescu, A.M., Ed.; Elsevier Academic Press: Cambridge, MA, USA, 2017; Volume 7, pp. 637–697.
159. Haghighi, H.; Licciardello, F.; Fava, P.; Siesler, H.W.; Pulvirenti, A. Recent advances on chitosan-based films for sustainable food packaging applications. *Food Packag. Shelf Life* **2020**, *26*, 100551. [CrossRef]
160. Scarfato, P.; Di Maio, L.; Incarnato, L. Recent advances and migration issues in biodegradable polymers from renewable sources for food packaging. *J. Appl. Polym. Sci.* **2015**, *132*. [CrossRef]
161. Attaran, S.A.; Hassan, A.; Wahit, M.U. Materials for food packaging applications based on bio-based polymer nanocomposites: A review. *J. Thermoplast. Compos. Mater.* **2017**, *30*, 143–173. [CrossRef]
162. Petkoska, A.T.; Daniloski, D.; D'Cunha, N.M.; Naumovski, N.; Broach, A.T. Edible packaging: Sustainable solutions and novel trends in food packaging. *Food Res. Int.* **2021**, *140*, 109981. [CrossRef]
163. Véronique, C. Bioactive packaging technologies for extended shelf life of meat-based products. *Meat Sci.* **2008**, *78*, 90–103.
164. Kerry, J.; O'grady, M.; Hogan, S. Past, current and potential utilisation of active and intelligent packaging systems for meat and muscle-based products: A review. *Meat Sci.* **2006**, *74*, 113–130. [CrossRef]

165. Vedadghavami, A.; Minooei, F.; Mohammadi, M.H.; Khetani, S.; Kolahchi, A.R.; Mashayekhan, S.; Sanati-Nezhad, A. Manufacturing of hydrogel biomaterials with controlled mechanical properties for tissue engineering applications. *Acta Biomater.* **2017**, *62*, 42–63. [CrossRef]
166. Kraśniewska, K.; Galus, S.; Gniewosz, M. Biopolymers-based materials containing silver nanoparticles as active packaging for food applications–A review. *Int. J. Mol. Sci.* **2020**, *21*, 698. [CrossRef] [PubMed]

gels MDPI

Review

Milk Protein-Based Nanohydrogels: Current Status and Applications

Manpreet Kaur [1], Aarti Bains [2], Prince Chawla [1,*], Rahul Yadav [3], Anil Kumar [3], Baskaran Stephen Inbaraj [4], Kandi Sridhar [5,*] and Minaxi Sharma [6,*]

1 Department of Food Technology and Nutrition, Lovely Professional University, Phagwara 144411, Punjab, India; manpreetk.oberoi997@gmail.com
2 Department of Biotechnology, CT Institute of Pharmaceutical Sciences, South Campus, Jalandhar 144020, Punjab, India; aarti05888@gmail.com
3 Shoolini Life Sciences Pvt. Ltd., Shoolini University, Solan 173229, Himachal Pradesh, India; yrahul1857@gmail.com (R.Y.); kumaranil_16@rediffmail.com (A.K.)
4 Department of Food Science, Fu Jen Catholic University, New Taipei City 24205, Taiwan; sinbaraj@yahoo.com or 138547@mail.fju.edu.tw
5 UMR1253, Science et Technologie du Lait et de L'œuf, INRAE, L'Institut Agro Rennes-Angers, 65 Rue de Saint Brieuc, F-35042 Rennes, France
6 Laboratoire de Chimie Verte et Produits Biobasés, Département Agro Bioscience et Chimie, Haute Ecole Provinciale du Hainaut-Condorcet, 11, Rue de la Sucrerie, 7800 Ath, Belgium
* Correspondence: princefoodtech@gmail.com (P.C.); sridhar4647@gmail.com or sridhar.kandi@agrocampus-ouest.fr (K.S.); minaxi86sharma@gmail.com or m.sharma@carah.be (M.S.)

Abstract: Milk proteins are excellent biomaterials for the modification and formulation of food structures as they have good nutritional value; are biodegradable and biocompatible; are regarded as safe for human consumption; possess valuable physical, chemical, and biological functionalities. Hydrogels are three-dimensional, cross-linked networks of polymers capable of absorbing large amounts of water and biological fluids without dissolving and have attained great attraction from researchers due to their small size and high efficiency. Gelation is the primary technique used to synthesize milk protein nanohydrogels, whereas the denaturation, aggregation, and gelation of proteins are of specific significance toward assembling novel nanostructures such as nanohydrogels with various possible applications. These are synthesized by either chemical cross-linking achieved through covalent bonds or physical cross-linking via noncovalent bonds. Milk-protein-based gelling systems can play a variety of functions such as in food nutrition and health, food engineering and processing, and food safety. Therefore, this review highlights the method to prepare milk protein nanohydrogel and its diverse applications in the food industry.

Keywords: nanohydrogel; milk proteins; characteristics of milk-protein-based nanohydrogels; food applications

Citation: Kaur, M.; Bains, A.; Chawla, P.; Yadav, R.; Kumar, A.; Inbaraj, B.S.; Sridhar, K.; Sharma, M. Milk Protein-Based Nanohydrogels: Current Status and Applications. *Gels* **2022**, *8*, 432. https://doi.org/10.3390/gels8070432

Academic Editor: Pietro Matricardi

Received: 7 June 2022
Accepted: 7 July 2022
Published: 10 July 2022

1. Introduction

Hydrophilic gels, or more commonly designated as hydrogels, are three-dimensional networks of polymers that can swell in water and have the potential to retain a high amount within their structure without dissolving [1]. The history of the development of hydrogels is grouped into three generations: first-generation or conventional super-porous hydrogels, and second-generation and third-generation hydrogels [2]. First-generation hydrogels were introduced in 1900 and are termed as a colloidal gel of inorganic salt prepared via the polymerization of water-soluble monomers with multifunctional cross-linkers or with hydrophilic polymers through cross-linking, and they possess a great swelling capacity and good mechanical properties [3]. Second-generation hydrogels came in 1970 and were prepared using environmental factors such as light, pH, heat, and ionic strength and resulted in better control over properties such as drug release, gel formation, biodegradability, and dissolution, e.g., poly(N-iso-propyl acrylamide) [4]. Third-generation hydrogels came in the 1990s and mainly focused on improving mechanical strength and elasticity, and the research is still underway for improving the techno-functional properties of hydrogels [5]. However, recently, hydrogels were used in environmental engineering [6], soft robotics [7], and wastewater treatment [8]. Moreover, hydrogels possess several potential properties such as biodegradability, metabolizability, surface modification, stimulus response, water holding and retaining ability, nonantigenicity, greater stability during storage, easy preparation, as well as controllable size [9]. Due to the unique functional properties, hydrogels discover likely applications in biomedical engineering (drug conveyance, tissue designing, and drug discharge) [10], medical science, agriculture (soil moisturizing, nutrient carrier, and erosion control) [11], textiles, construction [12], diagnostics, regenerative medicines [13], electrical [14], flocculation [15], wastewater treatment, sensors and actuators [16], personal healthcare and hygiene products [17], as well as the food industry (food safety, food nutrition, and food engineering). The covalent and noncovalent interactions such as electrostatic interactions, hydrogen bonds, van der Waals interactions, and intermolecular hydrophobic interactions constitute the chemistry of the hydrogels, and the presence of hydroxyl, amines, carboxyl, ethers, and sulfate groups is responsible for soft and pliable structure [1]. Hydrogels can be synthesized from natural origin (proteins and polysaccharides) as well as synthetic sources (polyvinyl alcohol and polyethylene oxide) [18].

Nanohydrogels, also referred to as aqua gels, are three-dimensional networks of hydrophilic or amphiphilic polymers with diameters less than 100 nm [19]. The word "Nanohydrogel" (NanoGel) was coined to describe the networking and crosslinking of the polyanions and nonionic polymers in the preparation of polynucleotide transport frameworks [20]. Henceforth, these are emerging as creative materials for the development of active packaging in food processing for designing the delivery systems intended for the precise release of nutraceuticals at specific sites of action in the body [21]. Milk is an emulsion produced by the lacteal secretion of mammals. The major components of milk are water, lactose, protein, fat, and minerals. The minor constituents are phospholipids, vitamins, enzymes, sterols, and pigments. The real constituents are milk fat, casein protein, and lactose sugar [22]. The proteins represent the most 'value-added' component owing to the budding recognition that bovine milk proteins possess superior dietary and functional characteristics paralleled with other protein sources. Milk comprises around 4–5% of the protein. Milk-based protein nanohydrogels are natural and, hence, categorized as GRAS (Generally Recognized as Safe), inexpensive, nontoxic, and possessing high dietary benefit and exceptional functional, biological, and sensory properties [23]. Milk proteins are made of two significant proteins: caseins and whey proteins. Nanohydrogels can be prepared from caseins, whey proteins (primarily β-lactoglobulin), and their derivatives such as whey protein isolate (WPI) having a protein content of more than 90% and whey protein concentrate (WPC) having a protein content between 50 and 85% on a dry basis [24]. The most popular and successful method to produce whey protein nanohydrogels is the gelation technique that includes both heat and cold-gelation techniques. Gelation of food proteins can be achieved by physical (heat, high pressure), chemical (ionic, pH), and biochemical (enzymes) methods [25]. These are generally utilized as drug delivery frameworks and in the food industry; applications include food nutrition, food safety, food engineering, and food packaging. In this perspective, Li et al. [26] encapsulated epigallocatechin-3-gallate (EGCG), the key catechin in green tea and a powerful antioxidant in nanohydrogels of β-lactoglobulin, and were able to achieve a stable and clear nanosystem at pH 6.4–7.0 and the maximum protection of EGCG antioxidant activity at 85 °C and a molar ratio of 1:2 (β-lactoglobulin: EGCG). However, several reports have been undertaken over the last few decades to create hydrogels with stimuli-responsive attributes that correlate to typical hydrogel qualities including porosity, swelling, and physical organization. Therefore, this review article emphasizes milk-protein-based nanohydrogels, the current status of milk-protein-based nanohydrogels in the food industry, along with the methods to formulate milk-protein-based nanohydrogels. In addition, potential applications of nanohydrogel are discussed with schematic diagrams.

2. Classification of Hydrogels

Hydrogels are classified based on several characteristics such as cross-linking, stimuli response, preparation source, degradability, electrical charge, physical properties, polymeric composition, configuration, anisotropy, physical appearance, and physical structure [27]. They can be natural or synthetic, where natural hydrogels include polysaccharides (cellulose, alginate, and pullulan) and proteins (milk proteins and soy proteins) and are biocompatible and biodegradable but possess weak stability and mechanical strength [28,29]. However, synthetic hydrogels are prepared through the polymerization of monomers such as polyvinyl alcohol (PVA) and polyacrylamide, and these are stable and have good mechanical strength [30]. The polymer composition can be a homopolymer, copolymer, or multi-polymer interpenetrating polymer. Herein, the homopolymer comprises a single species of monomer, whereas the copolymer comprises two or more diverse polymer species with at least one hydrophilic component, while the multi-polymer interpenetrating polymeric hydrogel comprises two independent cross-linked synthetic and/or natural polymers [31,32]. Furthermore, hydrogels have unique properties such as the ability to be crystalline, semi-crystalline, or amorphous. However, crystalline hydrogels are highly branched polymeric network structures with a definite order of crystallization. In addition, semicrystalline hydrogels contain both crystalline and amorphous regions [33]. Amorphous hydrogels are random network structures at the molecular level. Hydrogels can be prepared in the form of a matrix, film, or microsphere. Similarly, based on their interaction type, a hydrogel can be classified into four categories: ionic (contains anionic and/or cationic groups), nonionic (neutral), amphoteric (contains both acidic and basic groups), and zwitterionic (contains anionic and cationic groups in each structural repeating unit). In this context, several studies have revealed that chemical crosslinked hydrogels are synthesized through covalent bonding, e.g., disulfide formation, and hence have permanent junctions, while physical hydrogels are molecular arrangements prepared through hydrogen bonds, hydrophobic interactions, and van der Waals forces [34]. The response of the hydrogels depends on various physical or chemical conditions. Thus, physical stimuli include temperature, magnetic field, light, pressure, sound, and electric field [35], and chemical stimulations include pH, ionic strength, solvent composition, and molecular species [36]. Milk-protein-based nanohydrogel is natural and considered as stimuli-responsive hydrogels as these are triggered by environmental variables. Milk-protein-based nanohydrogels can be prepared by physical cross-linking as well as chemical cross-linking. In Figure 1, detailed classification of hydrogels are given.

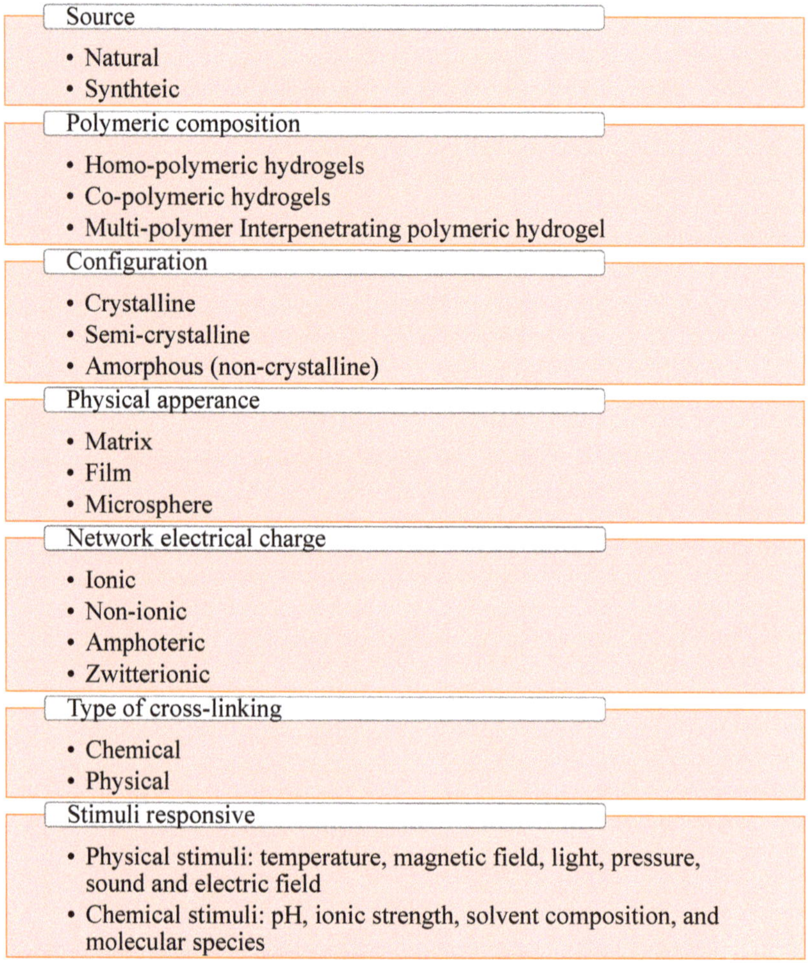

Figure 1. Detailed representation of classification of hydrogels.

3. Synthesis of Milk-Protein-Based Nanohydrogels

Milk-protein-based nanohydrogels possess high nutritional value as their role in the delivery of bioactive compounds and essential amino acids, specifically sulfur-containing amino acids such as methionine and cysteine [37]. Moreover, the high swelling capability of nanohydrogel provides a proper arrangement for the development of various nanocarriers including nanohydrogel, nanocapsules, nanoemulsion, and nanocomposites. The gels are entirely biodegradable and offer a biological functionality that comprises digestibility conformation and immune system regulation [38]. Therefore, they are completely regarded as safe for human consumption and nontoxic. Milk protein gels especially prepared from β-lactoglobulin, whey protein isolate, and whey protein concentrate can entrap functional components owing to the microscopic size with an enormous internal network for multivalent bioconjugation.

3.1. Factors Influencing Gel Fomation

The gel phase transition occurs as a result of a competitive balance between a repulsive force that seeks to expand the polymer network and an attractive force that acts to shrink the network. The electrostatic interaction between polymer charges of the same sort is the most powerful repulsive force, and it may be induced on a gel by injecting ionization into the network. The osmotic pressure created by counter ions contributes to the increasing pressure. Van der Waals contacts, hydrophobic interactions, ion–ion interactions of opposing types, and hydrogen bonding are examples of attractive interactions. The phase change was identified in gels generated by all of the basic forces [39].

The covalent and noncovalent bonds constitute the chemistry of milk protein hydrogels. The main types of noncovalent interactions are electrostatic interactions, hydrogen bonds, van der Waals forces, hydration interactions, and intermolecular hydrophobic interactions [40,41]. Whey protein nanohydrogels are most commonly prepared by the gelation technique [42]. The gel characteristics depend upon various factors such as the protein content, salt concentration, type of ion, pH, ionic strength, the extent of chemical bonding agents utilized, and temperature [43]. Gelation occurs after the unfolding of the original protein structure to prepare a three-dimensional network. Heating is the primary factor behind the formation of food gels containing globular proteins, while other physical (pressure), chemical (acid, ions), and biological (enzymes) forces can be utilized in tandem [44].

Thermal gelation is a phenomenon comprising three stages: (i) primary aggregation via covalent (disulfide bridges) and noncovalent (hydrogen bonds, hydrophobic interactions, and van der Waals interactions) bonds; (ii) secondary aggregation through the joining of primary protein aggregates; (iii) lastly, when the quantity of protein secondary aggregates reaches a threshold concentration, a 3-D network with the potential to entrap water forms [21]. The extent to which proteins denature when heated principally depends upon two factors, intrinsic factors and extrinsic factors. Intrinsic factors comprise solution properties such as pH, protein concentration, and ionic strength, whereas extrinsic factors include heating circumstances such as temperature, heating rate, heating time, and heating technique [45]. New technologies such as dielectric heating (microwave), electric field, and high pressure can also be used in place of heat denaturation [21].

The microwave (MW) is a type of dielectric heating in which an alternating electromagnetic field interacts with polar molecules such as water and ionic species, compelling them to continually realign themselves by reversing an electric field surrounding the food product, causing heat generation [46]. This molecule movement is exceedingly quick due to the high frequency of the field, which can range from 300 to 3000 MHz. According to research by de Pomerai et al. [47], exposure to MW radiation has also been shown to increase protein aggregation, change protein shape without bulk heating, and stimulate the production of particular structures such as amyloid fibrils. Due to its weak penetration capacity, MW heating is frequently claimed to provide uneven heating, which might result in uneven processing. The principal downsides of MW prospecting include nonuniform heating, complexity, expensive equipment costs, difficulties in assuring homogeneity, and a lack of appropriate packing materials. Consequently, this may lead to a number of concerns relating to poor final quality, overheating, and a variety of safety-related issues [48].

The High-Voltage Electric Field (HVEF) is a food processing method that may guarantee product safety while maintaining its characteristics owing to the electric current's minimally negative effects [49]. The PEF technique has evident advantages because of its low-energy needs and the ability to induce protein structure and functionality alteration without heat side-effects such as the thermal destruction of relevant chemicals. However, heating avoidance is not always practicable, and the nature of the electric pulses (i.e., high voltage) makes full control and automation of the process problematic. Furthermore, the high initial investment cost, as well as the expense of the rigorous maintenance and repair of PEF equipment, prevents widespread industrial application of this technology [50].

Isostatic high pressure (HP) could be employed for food texture engineering due to its influence on the properties of food proteins. At pressures more than 400 MPa (>100 MPa for β-Lg), α-La denaturation is noticed as a decrease in solubility at pH 4.6. However, pressure-induced aggregation resulted in porous gels prone to exudation, as opposed to heat-induced gels with a finely stranded network and good water retention. Pressure denaturation of proteins is a complicated event that is affected by a variety of parameters, including protein structure, pressure range, temperature, pH, and solvent composition [51]. Despite its demonstrated utility in protein functionalization, the extreme complexity of the process, as well as the lack of knowledge of the underlying principles involved, necessitates additional effort to validate HP as a tool in bioscience. The use of technology is also constrained by problems including high equipment prices, high maintenance needs, and scale-up restrictions [50].

3.2. Method of Cross-Linking

Gels are formed either by physical cross-linking or chemical cross-linking [52]. Physical hydrogels are also entitled reversible or pseudo-gels, which are networks of heterogenous clusters prepared by molecular entanglements via weak hydrophobic links, ionic interactions, or hydrogen bonding. For instance, a simple thermal process was used to make nanohydrogels from bovine lactoferrin that were resistant to pH (from 3 to 11) and salt (from 0 to 200 mM NaCl) concentration, thus regarded as transporters or functional components in food [21]. Likewise, Relkin et al. [53] used high pressures of 1200 bar to encapsulate α-tocopherol in nanostructures of whey protein dispersals (4% and pH 6.5) and observed a decline in particle charges (to 47 mV) and particle sizes (to 212 nm) and a substantial destabilization of protein structure, though only a 30% vitamin degradation during processing and no degradation during 8 weeks of storage were observed. In another study, the development of lactoferrin nanoparticles by thermal gelation for iron delivery was investigated, where the nanoparticles exhibited an iron-binding efficiency value of ≈20%, were stable to temperature (4–60 °C) and pH (pH 2–11), and achieved a shelf-life of 76 days at 4 °C and, hence, presented a pH-dependent behavior [54]. Caffeine (hydrophilic) and curcumin (lipophilic) were proficiently encapsulated in lactoferrin-glycomacropeptide (LF-GMP) nanohydrogels with great encapsulation efficiencies (>90%) by the thermal gelation technique [55]. The release mechanism of these bioactive compounds at different pH was performed and a pH-dependent release profile was observed. Similarly, thermally induced whey protein isolated hydrogels have also been used to encapsulate anthocyanin-rich bilberry extracts, demonstrating a Fickian diffusion-based release mechanism of bioactive phenolic compounds in simulated gastric juices [56].

On the other hand, chemical hydrogels also termed as irreversible or permanent gels are polymeric networks prepared by covalent bonding at definite sites and, thus, can absorb water, swell, and hold until equilibrium is attained [57]. Chemically cross-linked hydrogels can form a superior hydrogel in terms of swelling and gelling capacity. In a study conducted by Donato et al. [58], the influence of mono- and divalent salts (NaCl and $CaCl_2$, respectively) concentration on heat-precipitated bovine serum albumin gels at pH 7.0 was determined, and it was concluded that repulsive forces were decreased at excessive salt ranges and $CaCl_2$ changed into a greater green color than NaCl in identical ionic energy upon decreasing those interactions.

3.3. Type of Gel Formed

Globular proteins may form both hot and cold-set gels [59], and the production of hot-set gels is initiated by the aggregation of the unfolded protein moieties of an intermolecular β-sheet network [60]. As these gels have a high strength, consistent structure, and ability to form a gel at ambient temperature [61], cold-set gels are exceptional for formulating functional meals with better control over shape, structure, and texture along with targeted delivery [62]. The formation of cold-set gels occurs in two stages. The first stage is to heat the protein at neutral pH (above isoelectric point), below the protein gelling ability, and at low ionic strength, which causes protein denaturation and poor folding. The second stage is acidification to approach the isoelectric pH of protein (acid-induced cold gelation) or salt supplementation (salt-induced cold gelation) to generate cross-linkages among protein clusters and reduce the inter-protein repulsion [63]. For the cold gelation of globular proteins, calcium chloride and sodium chloride are utilized [64]. Alternatively, cations such as Fe^{2+} [61] and Mg^{2+} [65] were employed to create whey protein cold-set gels having better functional and nutritional value. In another study, the characteristics of ion-encased whey protein cold-set gels can be influenced by the heat treatment of protein solution. Martin et al. [61] investigated the impact of different heat treatments (mild, intermediate, and severe) on the iron encapsulation effectiveness of a whey protein cold-set gel. In comparison to mild (85 °C, 30 min, and pH 7.0) and severe (pH 2.0 and 80 °C for extended period), cold-set gel synthesized from whey protein bearing intermediate heating conditions (85 °C, 3 h, and pH 3.35) exhibited a higher iron release at natural pH owing to the formation of different structural entities as a function of heating conditions. At acidic pH, however, there was no discernible difference in iron release among the samples.

Casein protein is present as a calcium caseinate-phosphate complex [66]. It is in a colloidal state and forms more than 80% of the complete protein in milk [67]. Transglutaminase can covalently join glutamine deposits with the remaining lysine due to the low degree of secondary and tertiary arrangements, converting casein micelles to nanogel particles [68]. However, casein hydrogels can also be made using genipin, a naturally derived chemical that can join amino groups of lysine, hydroxylysine, or arginine residues in various polypeptide chains utilizing monomeric or oligomeric cross-linkers. The encapsulation of bioactive compounds such as probiotic organisms [69], fats and oils [59], and vitamin B12 is the most promising application of casein hydrogels [70].

Consequently, proteins can be blended with polysaccharides effectively to form hydrogels to be used for encapsulation and delivery of nutrients. In this context, Ozel et al. [71] blended polysaccharides (xanthan gum, pectin, and gum tragacanth) with heat-set whey protein and utilized it to encapsulate the dark carrot extricate. Likewise, for encapsulation and delivery of hydrophobic nutraceuticals such as w-3 fatty acids, Salimen and Weiss [72] created a stable nanohydrogel produced from a compound of protein-polysaccharide (i.e., β-Lg-Pectin) and achieved stability against oxidation during storage (Figure 2).

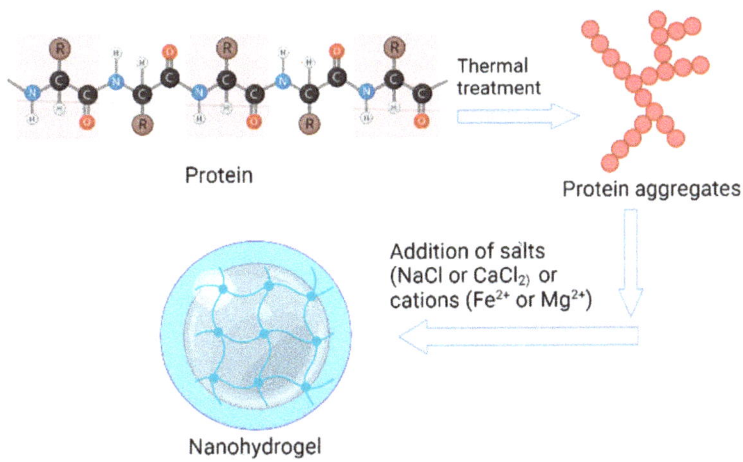

Figure 2. Schematic representation of synthesis of milk-protein-based nanohydrogel.

4. Characterization of Milk Protein Hydrogels

4.1. Fourier Transform Infrared Spectroscopy (FTIR)

Characterization of formulated gels based on various techniques is an important feature and in Figure 3 all the instrumental techniques are revealed. FTIR is an analytical method performed to investigate the presence of chemical bonds or functional groups in the hydrogel by the use of an infrared absorption spectrum [73]. The presence of the amide bond in the protein or peptide-based hydrogel can be identified using FTIR. The results of FTIR as concluded by several authors demonstrated that hydrogel prepared by the protein-polysaccharide complex consists of C = O, C–O, O–H, C–C, C–H, C–N, and N–H. For instance, curcumin- and caffeine-encapsulated Lactoferrin-Glycomacropeptide (Lf-GMP) nanohydrogel showed characteristic bands of caffeine in Lf-GMP as peaks at 1707 cm, 974, and 1359 cm^{-1} corresponding to C = C, C–C, and C–H stretching [74]. Additionally, it is possible to assure the encapsulation of caffeine in Lf-GMP nanohydrogels owing to changes in the amide I band (1700–1600 cm^{-1}), mainly C = O stretching, and amide II band (1600–1500 cm^{-1}), C = N stretching coupled with the N–H bending mode, signifying a hydrophilic interaction between the caffeine and Lf-GMP nano hydrogel; meanwhile, the major groups involved are C = O, C–N, and N–H [75,76]. Consequently, several gums have shown great complex binding with milk protein. In this context, Pilevaran et al. [77] synthesized the hydrogel using whey protein and xanthan gum. In their study, different amounts of gums were used for the synthesis of the hydrogel. However, the FTIR test clearly showed that the peak observed at 1649 cm^{-1} was related to the C–N stretching of the amide II band and the C–N–H in-plane bending and correlated to C–H bending. The two strong peaks perceived at 1073 and 1155 cm^{-1} relate to saccharides.

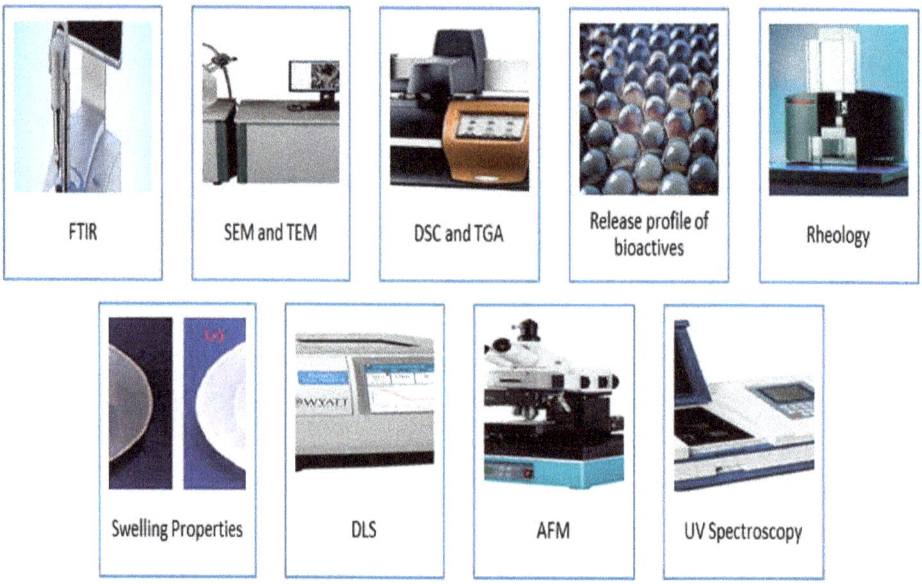

Figure 3. Schematic representation of characterization of milk protein nanohydrogels.

4.2. *Morphological Characterization of Nanohydrogels*

Scanning electron microscopy (SEM) is used to describe the microstructure of the surface morphology of the polymer. The porosity of the hydrogel, three-dimensional network, and surface morphology are analyzed using scanning electron microscopy (SEM). SEM is the most commonly used technique for morphological studies of hydrogels irrespective of their composition [78,79]. Hamcerencu et al. [80] conducted a study on whey protein concentrate-Xanthan gum (WPC-XG) hydrogel and showed spherical and large aggregates of protein with high porosity and various sizes, whereas WPC hydrogel showed a more compact structure with lower porosity in comparison with WPC–XG hydrogels.

Similarly, Transmission Electron Microscopy (TEM) was used to understand the morphology of Lf-GMP nanohydrogels. TEM images show that Lf-GMP nanohydrogels with bioactive compounds encapsulated maintain the spherical shape with a solid dense structure of sizes around 120 nm and 125 nm for caffeine and curcumin, respectively.

4.3. Differential Scanning Calorimetry

The thermal stability change in the structural strength of protein-based nanohydrogels with temperature is measured via a differential scanning calorimeter (DSC) or thermogravimetric analysis (TGA). The relative change in the structure of the prepared hydrogel with temperature is compared with the standard protein hydrogel. Parameters such as glass transition temperature and melting point are important characteristics to reveal the thermoplastic behavior of protein nanohydrogels determined by a DSC or TGA [81,82]. In the DSC technique, the difference in the quantity of heat required to raise the temperature of a sample compared to a reference is measured as a function of temperature. A single endothermic peak was detected for all the hydrogels. Barsett et al. [83] performed DSC on Lf-GMP hydrogel and observed the highest peak temperature at 104 °C, the onset temperature at 61.5 °C, and the end set temperature at 130 °C related to XG 0.01%; the lowest peak temperature at 98 °C, the onset temperature at 64.5 °C, and the end set temperature at 128 °C related to WPC hydrogel. On the other hand, the WPC–XG composite hydrogel at 0.3 and 0.6% of XG had more thermal stability. In addition, DSC revealed that XG and WPC interacted and promoted the formation of a 3D network structure.

4.4. Release Profile of Bioactives

In a study performed by Vesely et al. [84], the experiments of bioactive compounds released from Lf-GMP nanohydrogels were conducted at 37 °C at two different pH values: 2 and 7. These conditions were used to simulate the release mechanisms of these bioactive compounds when subjected to digestion in the human gastrointestinal system. The release profile of caffeine from Lf-GMP nanohydrogels revealed that at pH 2, a higher amount of caffeine was released from Lf-GMP nanohydrogels and, thus, it was concluded that bioactive molecules are on the surface of the carrier and are released faster than if they were entrapped. Unlike caffeine, curcumin release from Lf-GMP nanohydrogels is pH-dependent, as, at pH 2, there is a clear release profile of curcumin from nanohydrogel, whereas, at pH 7, no curcumin was released and, hence, it was concluded that curcumin is more triggered under acidic conditions than under a basic environment [85].

4.5. Rheological Characterization of Nanohydrogels

Rheological evaluation of the protein nanohydrogels demonstrates the impact of environmental variables such as pH, temperature, enzyme concentration, and ion concentration on rheological characteristic parameters such as viscoelastic characteristics, gelation time, gel strength, yield-strain, and various properties such as stiffness, gelation kinetics, viscoelastic regions, relative liquid–solid properties, and relaxation time scale [86]. Oscillation rheology is used to study the rheological behavior. In oscillation rheology, sinusoidal shear is applied and the resulting stress is measured as a function of time [87].

4.6. Determination of Swelling Properties

Hydrogels are cross-linked polymer networks swollen in a liquid state. The absorbed liquid behaves as a selective filter and allows free diffusion of few solute molecules, whereas the polymer network acts as a matrix to hold liquid together. Hydrogels can soak up to a thousand times their dry weight. It can be estimated by measuring the dry weight and the swollen state weight and calculating either the water uptake or a volume of adsorbed solvent [88]. The assessment of swelling also acts a measure for other hydrogel properties such as cross-linking degree, mechanical properties, and degradation rate:

$$\text{Water uptake (in \%)} = \text{swollen weight-dry weight/dry weight} * 100$$

$$\text{Volume of adsorbed solvent (in \%)} = \text{swollen weight-dry weight/water density} * 100.$$

Hydrogels immersed in the aqueous medium absorb water due to the osmotic pressure difference between the gel network and surrounding environment [89]. In a study conducted by Taskooh [90] on freeze-dried hydrogel samples placed in 25 mL of buffer solution (pH 7, 0.1 M) at 21 °C, the swelling ratio increased from 15.5 ± 6.3 to 20.34 ± 5.75 ($p < 0.01$) with varying iron concentration from 10 mM to 70 mM. By increasing the cross-linker (Fe) concentration, the density of cross-links was enhanced so that the water absorption capacity and swelling ratio improved.

4.7. Dynamic Light Scattering (DLS)

DLS is important to measure size changes for micro- as well as nanosized polymer gels. Nanoscale aggregation phenomena during the initial stages' steps of milk protein aggregation as affected by the thermal treatment are assessed by DLS. In a study conducted by Silva et al. [91], DLS was performed on the casein hydrogels prepared using 5 mM, 10 mM, and 20 mM in a casein dispersion of 2.5%, and it showed a decrease in the hydrodynamic diameter (Dh) of the casein micelles as a function of genipin concentration. In the control sample (without GP), it was 179 nm, whereas it was 160 nm with the 20 mM sample. The dispersion viscosities also decreased when higher genipin concentrations were tested assuming values from 1.655 mPa.s^{-1} (control) to 1.441 mPas.s^{-1} (20 mM genipin).

4.8. Atomic Force Microscopy (AFM)

AFM has been used to determine biomolecule morphological features such as distinct molecular shapes and surface coverage. It is also a nondestructive imaging approach that is commonly used to investigate the architecture of biological systems, such as proteins. This was another microscopic approach employed in the investigation of casein hydrogels carrying insulin performed by Alibolandi et al. [92] and is an appropriate way for evaluating the surface of hydrogel samples. The linkages between molecules and the loops were the prominent and concave components, respectively. The tapping-mode AFM pictures of hydrogel and its height profile were illustrated and validated the creation of a three-dimensional network structure in the casein hydrogel.

4.9. UV Spectroscopy

Protein hydrogels absorb small wavelengths in the UV region of the spectrum presenting an absorption peak at 280 nm [93] and low absorption in the visible and near-infrared. Whey protein hydrogel was prepared using copper as a cross-linking agent ($CuSo_4$) and the absorbance was measured. The use of whey proteins led to an increase in the absorbance of the film due to changes in the structure of the hydrogel as a higher absorption (smaller transmittance) was obtained by increasing the Cu^{2+} films.

5. Applications of Milk Protein-Based Nanohydrogels

Nanotechnology refers to the study of particles/objects/materials/matter with sizes ranging from 1 to 100 nm [94]. Herein, Figure 4 is revealing all the applications which are directly or indirectly related to the milk protein-based nanohydrogels. In the food industry, nanotechnology is applied to all the sectors such as food processing (encapsulation and defined discharge of nutraceuticals or food additives with improved bioavailability), food packaging (active and smart packaging systems for better food safety and biosecurity), food safety assurance (detection of contaminants or foreign objects or undesirable microorganisms using nanosensors, risk monitoring, adsorption, and removal operations), and food quality control (calorie count and food texture perception) [95].

Micronutrients such as minerals, vitamins, and nutraceuticals are more explicitly targeted by nanodelivery in the food industry [96]. However, significant challenges arise in the absorption and distribution of micronutrients through diet such as poor chemical stability, less oral bioavailability, poor solubility, and undesirable sensory attributes. These constraints must be overcome by consolidated logic and designing a methodology to be more specific; the job of nanodelivery for micronutrients includes improved gastrointestinal stability, security against oxidation, controlled and designated discharge, enhanced bioavailability as well as bioactivity, and assurance of the active compound during food handling stockpiling and circulation [97].

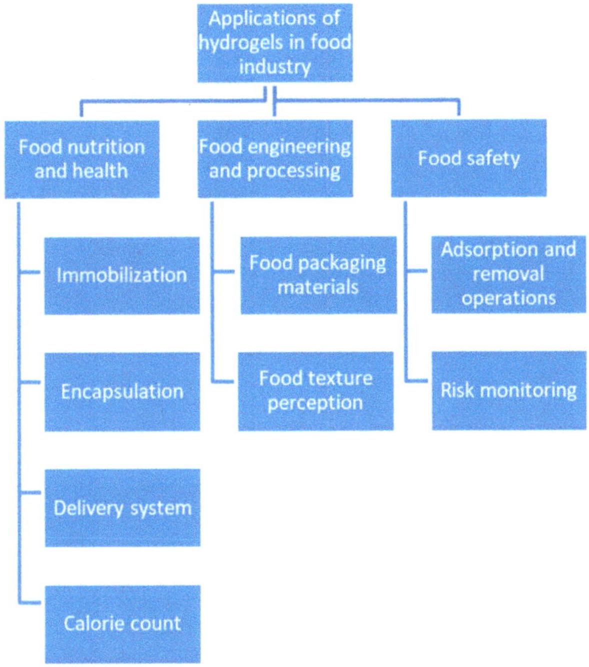

Figure 4. Applications of nanohydrogels in the food industry.

The ideal characteristics of a delivery system include properties such as (i) the ability to deliver active compounds exactly at the target place, (ii) efficiency of maintaining the active compound at desired levels during storage, and (iii) ensuring availability at a specific rate and targeted time [98]. All these features are fulfilled by the protein-based nanohydrogels, particularly the whey-protein-based nanohydrogels with greater efficiency, and hence, these are considered as an ultimate drug delivery arrangement due to their high stability, optimal drug loading ability, a multi-response approach aiming at a specific position, biological consistency, and reaction to a variety of external environmental stimuli [99].

Whey-protein-based hydrogels can entrap both the hydrophilic (e.g., caffeine) and lipophilic (e.g., curcumin) compounds without affecting their activity much, thus contributing to the development of novel functional foods [100]. Owing to their biocompatibility feature, hydrogels are regarded as a blessing in tissue engineering and medicine delivery [101]. Moreover, when nanoparticles are successfully integrated into hydrogels, they can be targeted specifically to tumor locations. Nanohydrogels play an essential role in the delivery of nanomedicine through the epitome of biomolecules [102]. Table 1 lists the applications of whey-protein-based nanohydrogels in the food industry.

Whey proteins are extensively employed in the food sector due to their exceptional functional characteristics such as emulsifying, foaming, gelling, as well as solubility [103]. The functional properties of proteins directly play a key role in organoleptic (color and flavor), kinesthetic (mouthfeel, smoothness, and texture), as well as textural (chewiness, elasticity, adhesiveness, and cohesiveness) features [104]. In other words, owing to their inherent viscoelastic properties, nanohydrogels can act as emulsifying and foaming agents in food stabilization and form firm nanocomplexes with polysaccharides, and hence, they are considered imperative aspects in the preparation of nanohydrogels for applications in the food industry [105].

The capacity of globular proteins to form gels under cold circumstances has recently received attention due to its applicability in innovative food and nonfood areas [106].

Apart from calcium and sodium ions, iron may be employed as an effective delivery vehicle via protein gelation as the amino acids improve iron ion bioavailability; hence, the concentration of iron influences the structure of the gels as, at low iron concentrations, gels are prepared as filamentous microstructures due to hydrophobic interactions, whereas the gels set at high iron ion concentrations have random spherical aggregates with significant van der Waals forces [107]. Filamentous nanohydrogels make up an outstanding matrix for the transportation of iron, thus stimulating its absorption and permitting the development of novel functional foods with targeted mineral deficiency [108].

As hydrogels possess a three-dimensional porous structure, hydrogels may absorb up to many thousands of times their dry weight, making them an ideal encapsulation technology for water-soluble components [21]. Immobilization (enzyme, cell, and microorganism) also takes place within the hydrogel network and on the hydrogel's surface, where the surface and internal structure characteristics of hydrogels have a significant impact on their immobilization capacity [109]. Hydrogels are excellent for use in delivery systems due to their unique ability to absorb and store significant volumes of water or biological fluids inside their three-dimensional network [110].

The main role of the hydrogel in food packaging is to control humidity inside the package as it lowers the water activity and retards the growth of spoilage microorganisms on foods [111]. Hydrogels possess great antimicrobial activity when incorporated with different antimicrobial compounds such as nanoparticles (silver) or bioactive compounds [112]. In addition to this, hydrogels show potential applications in smart food packaging as they either act as an indicator regarding the freshness of the food or detect the presence of contaminants [113]. Hydrogels have become important in food safety as they act as sensor or signal probes to detect hazards in food. Natural hydrogels are being investigated for the detection of biological hazards in foods. Hydrogels can be used to improve the texture or mouthfeel (elasticity, hardness, and chewiness) of the food owing to their soft texture [114]. Hydrogels play a vital role in lowering the calorie content of the food by either enhancing satiety or by reducing intake [115]. Wu et al. [116] fabricated a hydrogel using protein-dietary fiber, which acted as healthier replacements for starch granules. Hydrogels act as an efficient platform for adsorption and removal operations owing to their high-water retention capacity, high porosity, and reusability [117].

Table 1. Applications of whey-protein-based nanohydrogels in the food industry.

Gel Components	Carrier/Cargo	Gelation Technique	Application	References
Whey protein isolate/lauric acid	Echium oil	Physical self-assembly	Encapsulation/delivery	[118]
Whey protein concentrate/Pectin	D-Limonene	Heating	Encapsulation/delivery	[119]
β-Lactoglobulin nanoparticles	Caffeine	Thermal gelation	Delivery	[120]
Whey protein isolate	Iron	Salt-induced gelation	Fortification of food systems and site-specific delivery of iron	[121]
β-lactoglobulin/alginate	Quercetin	Heating	Encapsulation/ Delivery	[122]
β-Lactoglobulin/Chlorogenic acid	Epigallocatechin-3-gallate	Gelation	Encapsulation/delivery	[123]
Whey protein isolate and polysaccharides	Black carrot extract	Heating	Organized delivery conditions for bioactive agents	[71]
Whey protein isolate	Caffeine	Heating	Delivery of nutraceuticals	[124]
Whey protein isolate and niosomes	α-tocopherol	Acid-induced gelation	Intestinal delivery and improved bioavailability of α-tocopherol	[125]

Table 1. *Cont.*

Gel Components	Carrier/Cargo	Gelation Technique	Application	References
α-Lactalbumin	Curcumin	Temperature-induced gelation	Delivery of bioactive therapeutic agent helping treat various human diseases	[126]
Whey protein concentrate	Phytosterols	Gelation	Encapsulation/delivery	[127]
Lactoferrin and Glycomacropeptide	Curcumin and caffeine	Thermal gelation	Bioactive compound carrier	[55]
β-Lactoglobulin	Vitamin B2	Gelation	Encapsulation/delivery	[128]
Whey protein concentrate	Folic acid	Electrospray particles	Encapsulation of bioactive compounds	[129]
Sodium caseinate, whey protein isolate, and soy protein isolate	β-Lactoglobulin	Heating	Encapsulation/delivery	[130]
Whey Protein Isolate/Pectin	Anthocyanin	Heating	Encapsulation/delivery	[131]
Bovine serum albumin and Polyethylene glycol	5-Fluorouracil	Heating	Injectable drug transport medium	[132]
β-Lactoglobulin/Zein	Tangerine	Gelation	Encapsulation/delivery	[133]
β-Lactoglobulin/Dextran	β-carotene	Temperature-induced gelation	Encapsulation/delivery	[134]
Whey protein isolate	Fe^{2+} and ascorbate	Salt-induced gelation	Increase in Fe^{2+} bioavailability, formulation development for fortification of food with iron	[61]
Whey protein concentrate	α-Tocopherol	Heating and high pressure	Encapsulation/delivery	[53]
Whey protein isolate	Zinc	Heating and ethanol desolvation	Encapsulation/delivery	[135]
β-Lactoglobulin	Catechin	Heating	Encapsulation/delivery	[136]
β-Lactoglobulin	Epigallocatechin-3-gallate	Thermal gelation	Encapsulation/delivery	[26]
Whey protein isolate	Bilberry extract	Heating	Whey-protein-based acidic gels are used for the encapsulation and stabilization of anthocyanin-rich bilberry extract	[137]
β-Lactoglobulin	Fe^{2+}	Salt-induced gelation	Development of filamentous gel matrix for intestinal delivery of iron	[138]
Whey protein concentrate, alginate	Caffeine	Heating	Hydrogels resistant to proteolytic enzymes in the stomach	[139]
β-Lactoglobulin	Fe^{2+}	Salt-induced gelation	Increase in the bioavailability of iron ion	[140]
β-Lactoglobulin and alginate	α-tocopherol	Salt-induced gelation	Intestinal delivery and bioavailability improvement of α-tocopherol	[141]
Whey protein isolate	Ethyl hexanoate	Heating and ethanol desolvation	Encapsulation/delivery	[142]
Whey protein isolate and tara gum	Magnesium	Salt-induced gelation	Preparation of gels with a wide range of textural qualities for use in the food industry	[65]

Table 1. *Cont.*

Gel Components	Carrier/Cargo	Gelation Technique	Application	References
Whey protein concentrate and honey	-	Heating	In the formulation of desserts such as flans, cakes, and tart fillings	[143]
Bovine serum albumin and acrylamide	Salicylic acid or sodium benzoate	Copolymerization of vinylated bovine serum albumin and acrylamide	Constant drug discharge agent for substances binding with albumin	[144]
Whey protein isolate		Thermal gelation	Structuring	[145]
Methacrylate-derivatized bovine serum albumin and methacrylic acid sodium salt	Diflunisal and β-propranolol	Free radical polymerization	Oral drug carriers with a high degree of swelling and increased water affinity	[146]
β-lactoglobulin and low methoxy pectin	ω-3 fatty acids	Physical self-assembly	Encapsulation/delivery	[147]
β-Lactoglobulin	α-tocopherol	Temperature-induced gelation	Encapsulation/delivery	[141]
β-Lactoglobulin	Curcumin	Temperature-induced gelation	Encapsulation/delivery	[148]

6. Conclusions

Protein nanohydrogels are being investigated for use in a variety of fields, including the delivery of chemicals. The nature of the component protein, other conjugated polymeric systems, pH, temperature, types of ionic structure, oxidative-redox conditions, and ionic strength influences the characteristics of nanohydrogels. The release profile of bioactive chemicals from the hydrogel may be altered by fine-tuning the aforementioned properties. Nowadays, protein nanohydrogels are being investigated as adaptable vaccine delivery vehicles to elicit a vigorous immune cell reaction. Protein hydrogels can shield cells, peptides, biomolecules, and pharmaceuticals from a hostile environment. Nanoscale delivery devices that contain active compounds are effective in protecting them from deterioration throughout processing, transport, and storage. Furthermore, nanoencapsulation can conceal off-odors and flavors of bioactives and ease food integration. Aside from these advantages, nanoencapsulation assists in the regulated, prolonged release of the nutrient at the active site, hence increasing its bioavailability. Whey proteins are incredibly adaptable, nutritious, and cost-effective dietary items that may be employed as rich matrices to build a variety of nanostructures in diverse ways owing to their reactivity to diverse environmental variables (e.g., pH, temperature, electric field, and ionic strength). Although whey protein nanostructures have the potential to be useful in a wide range of consumer food products, significant challenges remain to be overcome, including their large-scale production, their stability in challenging processing and storage environments, physical and chemical interactions between foods encapsulating sensitive molecules, their sturdiness and flexibility in the gastrointestinal tract, and the sense of awareness and satisfaction among consumers. At present, a better understanding of the action mechanism of the milk-protein-based nanohydrogels in the gastrointestinal tract will help in the optimization and expansion of nanotechnology in the drug delivery sector. In order to attract consumers, an in-depth study of the bioavailability, permeability, and toxicity needs to be addressed. Hence, it is crucial to develop predictive and validated toxicological tests to identify potential risks to humans, which could include oxidative damage, lesions of the kidney and liver, inflammation of the gastrointestinal tract, and cancer. Overall, the expansion of milk protein nanosystems for oral delivery of bioactive compounds has sparked research, positioning itself not only as a budding solution to food industry problems but also as an inventive tool for pharmaceutical uses. As a result, the potential of nanotechnology for solving various challenges is recognized by the US-FDA and it is providing support to overcome these nanotechnology-based challenges.

Author Contributions: Conceptualization, A.B. and P.C.; investigation, K.S. and P.C.; writing—original draft preparation, M.K. and K.S.; writing—review and editing, K.S., R.Y., M.S. and B.S.I.; visualization, A.B., M.S., K.S., A.K. and B.S.I.; supervision, A.B. and P.C. All authors have read and agreed to the published version of the manuscript.

Funding: This study received no external funding.

Institutional Review Board Statement: Not applicable.

Informed Consent Statement: Not applicable.

Data Availability Statement: Data sharing is not applicable to this article.

Acknowledgments: Support provided by Lovely Professional University, Pubjab, India is gratefully acknowledged.

Conflicts of Interest: The authors declare no conflict of interest.

References

1. Ahmad, Z.; Salman, S.; Khan, S.A.; Amin, A.; Rahman, Z.U.; Al-Ghamdi, Y.O.; Akhtar, K.; Bakhsh, E.M.; Khan, S.B. Versatility of Hydrogels: From Synthetic Strategies, Classification, and Properties to Biomedical Applications. *Gels* **2022**, *8*, 167. [CrossRef]
2. Laftah, W.A.; Hashim, S.; Ibrahim, A.N. Polymer hydrogels: A review. *Polym.-Plast. Technol. Eng.* **2011**, *50*, 1475–1486. [CrossRef]
3. Buwalda, S.J.; Boere, K.W.; Dijkstra, P.J.; Feijen, J.; Vermonden, T.; Hennink, W.E. Hydrogels in a historical perspective: From simple networks to smart materials. *J. Control. Release* **2014**, *190*, 254–273. [CrossRef] [PubMed]
4. Raman, R.; Langer, R. Biohybrid Design Gets Personal: New Materials for Patient-Specific Therapy. *Adv. Mater.* **2020**, *32*, 1901969. [CrossRef] [PubMed]
5. Omidian, H.; Rocca, J.G.; Park, K. Advances in superporous hydrogels. *J. Contrl. Rel.* **2005**, *102*, 3–12. [CrossRef] [PubMed]
6. Anthony, C.Y.; Hernandez, H.L.; Kim, A.H.; Stapleton, L.M.; Brand, R.J.; Mellor, E.T.; Bauer, C.P.; McCurdy, G.D.; Wolff, A.J.; Chan, D.; et al. Wildfire prevention through prophylactic treatment of high-risk landscapes using viscoelastic retardant fluids. *Proc. Natl. Acad. Sci. USA* **2019**, *116*, 20820–20827. [CrossRef]
7. Migliorini, L.; Santaniello, T.; Yan, Y.; Lenardi, C.; Milani, P. Low-voltage electrically driven homeostatic hydrogel-based actuators for underwater soft robotics. *Sens. Actuators B Chem.* **2016**, *228*, 758–766. [CrossRef]
8. Mittal, H.; Ray, S.S.; Okamoto, M. Recent progress on the design and applications of polysaccharide-based graft copolymer hydrogels as adsorbents for wastewater purification. *Macromol. Mater. Eng.* **2016**, *301*, 496–522. [CrossRef]
9. McClements, D.J. Designing biopolymer microgels to encapsulate, protect and deliver bioactive components: Physicochemical aspects. *Adv. Colloid Interface Sci.* **2017**, *240*, 31–59. [CrossRef]
10. Haag, S.L.; Bernards, M.T. Polyampholyte hydrogels in biomedical applications. *Gels* **2017**, *3*, 41. [CrossRef]
11. Guilherme, M.R.; Aouada, F.A.; Fajardo, A.R.; Martins, A.F.; Paulino, A.T.; Davi, M.F.; Rubira, A.F.; Muniz, E.C. Superabsorbent hydrogels based on polysaccharides for application in agriculture as soil conditioner and nutrient carrier: A review. *Eur. Polym. J.* **2015**, *72*, 365–385. [CrossRef]
12. Snoeck, D.; Pel, L.; De Belie, N. The water kinetics of superabsorbent polymers during cement hydration and internal curing visualized and studied by NMR. *Sci. Rep.* **2017**, *7*, 1–14. [CrossRef] [PubMed]
13. Zhang, L.; Li, K.; Xiao, W.; Zheng, L.; Xiao, Y.; Fan, H.; Zhang, X. Preparation of collagen–chondroitin sulfate–hyaluronic acid hybrid hydrogel scaffolds and cell compatibility in vitro. *Carbohydr. Polym.* **2011**, *84*, 118–125. [CrossRef]
14. Cunha, I.; Barras, R.; Grey, P.; Gaspar, D.; Fortunato, E.; Martins, R.; Pereira, L. Reusable cellulose-based hydrogel sticker film applied as gate dielectric in paper electrolyte-gated transistors. *Adv. Funct. Mater.* **2017**, *27*, 1606755. [CrossRef]
15. Das, R.; Ghorai, S.; Pal, S. Flocculation characteristics of polyacrylamide grafted hydroxypropyl methyl cellulose: An efficient biodegradable flocculant. *Chem. Eng. J.* **2013**, *229*, 144–152. [CrossRef]
16. Deng, K.; Bellmann, C.; Fu, Y.; Rohn, M.; Guenther, M.; Gerlach, G. Miniaturized force-compensated hydrogel-based pH sensors. *Sens. Actuators B Chem.* **2018**, *255*, 3495–3504. [CrossRef]
17. Ma, J.; Li, X.; Bao, Y. Advances in cellulose-based superabsorbent hydrogels. *RSC Adv.* **2015**, *5*, 59745–59757. [CrossRef]
18. Zhao, W.; Jin, X.; Cong, Y.; Liu, Y.; Fu, J. Degradable natural polymer hydrogels for articular cartilage tissue engineering. *J. Chem. Technol. Biotechnol.* **2013**, *88*, 327–339. [CrossRef]
19. Ayoubi-Joshaghani, M.H.; Seidi, K.; Azizi, M.; Jaymand, M.; Javaheri, T.; Jahanban-Esfahlan, R.; Hamblin, M.R. Potential Applications of Advanced Nano/Hydrogels in Biomedicine: Static, Dynamic, Multi-Stage, and Bioinspired. *Adv. Funct. Mater.* **2020**, *30*, 2004098. [CrossRef]
20. Kabanov, A.V.; Vinogradov, S.V. Nanogels as pharmaceutical carriers: Finite networks of infinite capabilities. *Angew. Chem. Int. Ed.* **2009**, *48*, 5418–5429. [CrossRef]
21. Ramos, O.L.; Pereira, R.N.; Martins, A.; Rodrigues, R.; Fucinos, C.; Teixeira, J.A.; Pastrana, L.; Malcata, F.X.; Vicente, A.A. Design of whey protein nanostructures for incorporation and release of nutraceutical compounds in food. *Crit. Rev. Food Sci. Nutr.* **2017**, *57*, 1377–1393. [CrossRef]

22. Guetouache, M.; Guessas, B.; Medjekal, S. Composition and nutritional value of raw milk. *J. Issues Biol. Sci. Pharm. Res.* **2014**, *2350*, 1588. [CrossRef]
23. Poonia, A. Potential of milk proteins as nanoencapsulation materials in food industry. In *Nanoscience in Food and Agriculture*; Springer: Cham, Switzerland, 2017; Volume 5, pp. 139–168. [CrossRef]
24. Ramos, Ó.L.; Reinas, I.; Silva, S.I.; Fernandes, J.C.; Cerqueira, M.A.; Pereira, R.N.; Vicente, A.A.; Poças, M.F.; Pintado, M.E.; Malcata, F.X. Effect of whey protein purity and glycerol content upon physical properties of edible films manufactured therefrom. *Food Hydrocoll.* **2013**, *30*, 110–122. [CrossRef]
25. Norton, J.E.; Espinosa, Y.G.; Watson, R.L.; Spyropoulos, F.; Norton, I.T. Functional food microstructures for macronutrient release and delivery. *Food Funct.* **2015**, *6*, 663–678. [CrossRef]
26. Li, B.; Du, W.; Jin, J.; Du, Q. Preservation of (−)-epigallocatechin-3-gallate antioxidant properties loaded in heat treated β-lactoglobulin nanoparticles. *J. Agric. Food Chem.* **2012**, *60*, 3477–3484. [CrossRef]
27. Ullah, F.; Othman, M.B.H.; Javed, F.; Ahmad, Z.; Akil, H.M. Classification, processing and application of hydrogels: A review. *Mater. Sci. Eng. C* **2015**, *57*, 414–433. [CrossRef]
28. Zhang, Y.; Huang, Y. Rational design of smart hydrogels for biomedical applications. *Front. Chem.* **2021**, *1288*, 615665. [CrossRef]
29. Benwood, C.; Chrenek, J.; Kirsch, R.L.; Masri, N.Z.; Richards, H.; Teetzen, K.; Willerth, S.M. Natural biomaterials and their use as bioinks for printing tissues. *Bioengineering* **2021**, *8*, 27. [CrossRef]
30. Bashir, S.; Hina, M.; Iqbal, J.; Rajpar, A.H.; Mujtaba, M.A.; Alghamdi, N.A.; Wageh, S.; Ramesh, K.; Ramesh, S. Fundamental concepts of hydrogels: Synthesis, properties, and their applications. *Polymers* **2020**, *12*, 2702. [CrossRef]
31. Akram, M.; Hussain, R. Nano hydrogels: History, development, and applications in drug delivery. *Nanocellul. Nanohydrogel Matrices Biotechnol. Biomed. Appl.* **2017**, *31*, 297–330. [CrossRef]
32. Dragan, E.S. Design and applications of interpenetrating polymer network hydrogels. A review. *Chem. Eng. J.* **2014**, *243*, 572–590. [CrossRef]
33. Shi, W.; Chen, X.; Li, B.; Weitz, D.A. Spontaneous Creation of Anisotropic Polymer Crystals with Orientation-Sensitive Birefringence in Liquid Drops. *ACS Appl. Mater. Interfaces* **2020**, *12*, 3912–3918. [CrossRef] [PubMed]
34. Hacker, M.C.; Krieghoff, J.; Mikos, A.G. Synthetic polymers. In *Principles of Regenerative Medicine*; Academic Press: Cambridge, MA, USA, 2019; pp. 559–590. [CrossRef]
35. Hiratani, T.; Kose, O.; Hamad, W.Y.; MacLachlan, M.J. Stable and sensitive stimuli-responsive anisotropic hydrogels for sensing ionic strength and pressure. *Mater. Horiz.* **2018**, *5*, 1076–1081. [CrossRef]
36. Meleties, M.; Katyal, P.; Lin, B.; Britton, D.; Montclare, J.K. Self-assembly of stimuli-responsive coiled-coil fibrous hydrogels. *Soft Matter* **2021**, *17*, 6470–6476. [CrossRef]
37. Ramos, O.L.; Fernandes, J.C.; Silva, S.I.; Pintado, M.E.; Malcata, F.X. Edible films and coatings from whey proteins: A review on formulation, and on mechanical and bioactive properties. *Crit. Rev. Food Sci. Nutr.* **2012**, *52*, 533–552. [CrossRef] [PubMed]
38. Jafari, S.M.; McClements, D.J. Nanotechnology approaches for increasing nutrient bioavailability. *Adv. Food Nutr. Res.* **2017**, *81*, 1–30. [CrossRef]
39. Tanaka, T.; Annaka, M.; Ilmain, F.; Ishii, K.; Kokufuta, E.; Suzuki, A.; Tokita, M. Phase transitions of gels. In *Mechanics of Swelling*; Springer: Berlin/Heidelberg, Germany, 1992; pp. 683–703. [CrossRef]
40. Davis, H.E.; Leach, J.K. Designing bioactive delivery systems for tissue regeneration. *Ann. Biomed. Eng.* **2011**, *39*, 1–13. [CrossRef] [PubMed]
41. Nicolai, T.; Britten, M.; Schmitt, C. β-Lactoglobulin and WPI aggregates: Formation, structure and applications. *Food Hydrocoll.* **2011**, *25*, 1945–1962. [CrossRef]
42. Akhtar, A.; Aslam, S.; Khan, S.; McClements, D.J.; Khalid, N.; Maqsood, S. Utilization of diverse protein sources for the development of protein-based nanostructures as bioactive carrier systems: A review of recent research findings (2010–2021). *Crit. Rev. Food Sci. Nutr.* **2021**, 1–19. [CrossRef]
43. Kharlamova, A.; Nicolai, T.; Chassenieux, C. Calcium-induced gelation of whey protein aggregates: Kinetics, structure and rheological properties. *Food Hydrocoll.* **2018**, *79*, 145–157. [CrossRef]
44. Yang, Z.; Chen, L.; McClements, D.J.; Qiu, C.; Li, C.; Zhang, Z.; Miao, M.; Tian, Y.; Zhu, K.; Jin, Z. Stimulus-responsive hydrogels in food science: A review. *Food Hydrocoll.* **2022**, *124*, 107218. [CrossRef]
45. Akharume, F.U.; Aluko, R.E.; Adedeji, A.A. Modification of plant proteins for improved functionality: A review. *Compr. Rev. Food Sci. Food Saf.* **2021**, *20*, 198–224. [CrossRef] [PubMed]
46. Jadhav, A.A.; Devale, R.P. Review on Microwave, The General purpose in Microwave Assisted Synthesis for Green Chemistry. *Asian J. Res. Chem.* **2022**, *15*, 182–185. [CrossRef]
47. de Pomerai, D.I.; Smith, B.; Dawe, A.; North, K.; Smith, T.; Archer, D.B.; Duce, I.R.; Jones, D.; Candido, E.P.M. Microwave radiation can alter protein conformation without bulk heating. *FEBS Lett.* **2003**, *543*, 93–97. [CrossRef]
48. Vadivambal, R.; Jayas, D.S. Non-uniform temperature distribution during microwave heating of food materials—A review. *Food Bioprocess Technol.* **2010**, *3*, 161–171. [CrossRef]
49. Mohamed, M.E.; Eissa, A.H.A. Pulsed electric fields for food processing technology. *Struct. Funct. Food Eng.* **2012**, *11*, 275–306. [CrossRef]
50. Pataro, G.; Ferrari, G. Limitations of pulsed electric field utilization in food industry. In *Pulsed Electric Fields to Obtain Healthier and Sustainable Food for Tomorrow*; Academic Press: Cambridge, MA, USA, 2020; pp. 283–310. [CrossRef]

51. Roche, J.; Royer, C.A. Lessons from pressure denaturation of proteins. *J. R. Soc. Interface* **2018**, *15*, 20180244. [CrossRef]
52. Maitra, J.; Shukla, V.K. Cross-linking in hydrogels-a review. *Am. J. Polym. Sci.* **2014**, *4*, 25–31. [CrossRef]
53. Relkin, P.; Shukat, R. Food protein aggregates as vitamin-matrix carriers: Impact of processing conditions. *Food Chem.* **2012**, *134*, 2141–2148. [CrossRef]
54. Martins, J.T.; Santos, S.F.; Bourbon, A.I.; Pinheiro, A.C.; González-Fernández, Á.; Pastrana, L.M.; Cerqueira, M.A.; Vicente, A.A. Lactoferrin-based nanoparticles as a vehicle for iron in food applications–Development and release profile. *Food Res. Int.* **2016**, *90*, 16–24. [CrossRef]
55. Bourbon, A.I.; Cerqueira, M.A.; Vicente, A.A. Encapsulation and controlled release of bioactive compounds in lactoferrin-glycomacropeptide nano hydrogels: Curcumin and caffeine as model compounds. *J. Food Eng.* **2016**, *180*, 110–119. [CrossRef]
56. Abaee, A.; Madadlou, A.; Saboury, A.A. The formation of non-heat-treated whey protein cold-set hydrogels via non-toxic chemical cross-linking. *Food Hydrocoll.* **2017**, *63*, 43–49. [CrossRef]
57. Cerqueira, M.A.; Pinheiro, A.C.; Silva, H.D.; Ramos, P.E.; Azevedo, M.A.; Flores-López, M.L.; Rivera, M.C.; Bourbon, A.I.; Ramos, O.L.; Vicente, A.A. Design of bio-nanosystems for oral delivery of functional compounds. *Food Eng. Rev.* **2014**, *6*, 1–19. [CrossRef]
58. Donato, L.; Garnier, C.; Doublier, J.L.; Nicolai, T. Influence of the NaCl or $CaCl_2$ concentration on the structure of heat-set bovine serum albumin gels at pH 7. *Biomacromolecules* **2005**, *6*, 2157–2163. [CrossRef]
59. Sun, H.; Yang, H.; Huang, W.; Zhang, S. Immobilization of laccase in a sponge-like hydrogel for enhanced durability in enzymatic degradation of dye pollutants. *J. Colloid Interface Sci.* **2015**, *450*, 353–360. [CrossRef] [PubMed]
60. Pucci, C.; Tardani, F.; La Mesa, C. Formation and Properties of Gels Based on Lipo-plexes. *J. Phys. Chem. B* **2014**, *118*, 6107–6116. [CrossRef]
61. Martin, A.H.; De Jong, G.A.H. Enhancing the in vitro Fe2+ bio-accessibility using ascorbate and cold-set whey protein gel particles. *Dairy Sci. Technol.* **2012**, *92*, 133–149. [CrossRef]
62. O'Neill, G.J.; Egan, T.; Jacquier, J.C.; O'Sullivan, M.; O'Riordan, E.D. Whey microbeads as a matrix for the encapsulation and immobilisation of riboflavin and peptides. *Food Chem.* **2014**, *160*, 46–52. [CrossRef]
63. Kuhn, K.R.; Cavallieri, Â.L.F.; Da Cunha, R.L. Cold-set whey protein gels induced by calcium or sodium salt addition. *Int. J. Food Sci. Technol.* **2010**, *45*, 348–357. [CrossRef]
64. Kuhn, K.R.; Cavallieri, Â.L.F.; Da Cunha, R.L. Cold-set whey protein–flaxseed gum gels induced by mono or divalent salt addition. *Food Hydrocoll.* **2011**, *25*, 1302–1310. [CrossRef]
65. da Silva, M.V.; Delgado, J.M.P.Q.; Goncalves, M.P. Impact of Mg2+ and tara gum concentrations on flow and textural properties of WPI solutions and cold-set gels. *Int. J. Food Prop.* **2010**, *13*, 972–982. [CrossRef]
66. Holt, C.; Carver, J.A.; Ecroyd, H.; d Thorn, D.C. Invited review: Caseins and the casein micelle: Their biological functions, structures, and behavior in foods. *J. Dairy Sci.* **2013**, *96*, 6127–6146. [CrossRef] [PubMed]
67. Ramos, O.L.; Pereira, R.N.; Rodrigues, R.; Teixeira, J.A.; Vicente, A.A.; Malcata, F.X. Physical effects upon whey protein aggregation for nano-coating production. *Food Res. Int.* **2014**, *66*, 344–355. [CrossRef]
68. ABD EL-SALAM, M.H.; El-Shibiny, S. Formation and potential uses of milk proteins as nano delivery vehicles for nutraceuticals: A review. *Int. J. Dairy Technol.* **2012**, *65*, 13–21. [CrossRef]
69. Heidebach, T.; Först, P.; Kulozik, U. Influence of casein-based microencapsulation on freeze-drying and storage of probiotic cells. *J. Food Eng.* **2010**, *98*, 309–316. [CrossRef]
70. Song, F.; Zhang, L.M.; Shi, J.F.; Li, N.N. Novel casein hydrogels: Formation, structure and controlled drug release. *Colloids Surf. B Biointerfaces* **2010**, *79*, 142–148. [CrossRef]
71. Ozel, B.; Cikrikci, S.; Aydin, O.; Oztop, M.H. Polysaccharide blended whey protein isolate-(WPI) hydrogels: A physicochemical and controlled release study. *Food Hydrocoll.* **2017**, *71*, 35–46. [CrossRef]
72. Weiss, J.; Gibis, M.; Schuh, V.; Salminen, H. Advances in ingredient and processing systems for meat and meat products. *Meat Sci.* **2010**, *86*, 196–213. [CrossRef]
73. Ellerbrock, R.H.; Ahmed, M.A.; Gerke, H.H. Spectroscopic characterization of mucilage (Chia seed) and polygalacturonic acid. *J. Plant Nutr. Soil Sci.* **2019**, *182*, 888–895. [CrossRef]
74. Ruchi, V.; Lalit, K. Characterization of caffeine isolated from Camellia sinensis leaves of Sikkim Himalayan Region. *J. Chem. Pharm. Res.* **2010**, *2*, 194–198.
75. Bagheri, L.; Madadlou, A.; Yarmand, M.; Mousavi, M.E. Spray-dried alginate microparticles carrying caffeine-loaded and potentially bioactive nanoparticles. *Food Res. Int.* **2014**, *62*, 1113–1119. [CrossRef]
76. Li, M.; Ma, Y.; Ngadi, M.O. Binding of curcumin to β-lactoglobulin and its effect on antioxidant characteristics of curcumin. *Food Chem.* **2013**, *141*, 1504–1511. [CrossRef] [PubMed]
77. Pilevaran, M.; Tavakolipour, H.; Naji-Tabasi, S.; Elhamirad, A.H. Investigation of functional, textural, and thermal properties of soluble complex of whey protein–xanthan gum hydrogel. *J. Food Process Eng.* **2021**, *44*, e13751. [CrossRef]
78. McMahon, R.E.; Hahn, M.S.; Pendleton, M.W.; Ellis, E.A. A Simple Preparation Method for Mesh Fibrin Hydrogel Composites for Conventional SEM. *Microsc. Microanal.* **2010**, *16* (Suppl. S2), 1030–1031. [CrossRef]
79. Gun'ko, V.M.; Savina, I.N.; Mikhalovsky, S.V. Properties of water bound in hydrogels. *Gels* **2017**, *3*, 37. [CrossRef] [PubMed]
80. Hamcerencu, M.; Desbrieres, J.; Popa, M.; Khoukh, A.; Riess, G. New unsaturated derivatives of Xanthan gum: Synthesis and characterization. *Polymer* **2007**, *48*, 1921–1929. [CrossRef]

81. Wattie, B.; Dumont, M.J.; Lefsrud, M. Synthesis and properties of feather keratin-based superabsorbent hydrogels. *Waste Biomass Valorization* **2018**, *9*, 391–400. [CrossRef]
82. Mondal, M.I.H. *Cellulose-Based Superabsorbent Hydrogels*; Springer: Berlin/Heidelberg, Germany, 2019.
83. Barsett, H.; Ebringerová, A.; Harding, S.E.; Heinze, T.; Hromádková, Z.; Muzzarelli, C.; Muzzraelli, R.A.A.; Paulsen, B.S.; ElSEOUD, O.A. *Polysaccharides I: Structure, Characterisation and Use*; Springer Science & Business Media: Berlin, Germany, 2005; Volume 186. [CrossRef]
84. Vesely, D. Diffusion of liquids in polymers. *Int. Mater. Rev.* **2008**, *53*, 299–315. [CrossRef]
85. Patra, D.; Sleem, F. A new method for pH triggered curcumin release by applying poly (l-lysine) mediated nanoparticle-congregation. *Anal. Chim. Acta* **2013**, *795*, 60–68. [CrossRef]
86. Panahi, R.; Baghban-Salehi, M. Protein-based hydrogels. In *Cellulose-Based Superabsorbent Hydrogels*; Springer: Cham, Switzerland, 2019; pp. 1561–1600. [CrossRef]
87. Rahman, M.S.; Islam, M.M.; Islam, M.S.; Zaman, A.; Ahmed, T.; Biswas, S.; Sharmeen, S.; Rashid, T.U.; Rahman, M.M. Morphological characterization of hydrogels. *Cellul.-Based Superabsorbent Hydrogels* **2019**, 819–863. [CrossRef]
88. Puppi, D.; Migone, C.; Grassi, L.; Pirosa, A.; Maisetta, G.; Batoni, G.; Chiellini, F. Integrated three-dimensional fiber/hydrogel biphasic scaffolds for periodontal bone tissue engineering. *Polym. Int.* **2016**, *65*, 631–640. [CrossRef]
89. Shi, Y.; Xiong, D.; Liu, Y.; Wang, N.; Zhao, X. Swelling, mechanical and friction properties of PVA/PVP hydrogels after swelling in osmotic pressure solution. *Mater. Sci. Eng. C* **2016**, *65*, 172–180. [CrossRef]
90. Hoshino, K.I.; Nakajima, T.; Matsuda, T.; Sakai, T.; Gong, J.P. Network elasticity of a model hydrogel as a function of swelling ratio: From shrinking to extreme swelling states. *Soft Matter* **2018**, *14*, 9693–9701. [CrossRef] [PubMed]
91. Silva, N.F.N.; Saint-Jalmes, A.; de Carvalho, A.F.; Gaucheron, F. Development of casein microgels from cross-linking of casein micelles by genipin. *Langmuir* **2014**, *30*, 10167–10175. [CrossRef] [PubMed]
92. Alibolandi, M.; Alabdollah, F.; Sadeghi, F.; Mohammadi, M.; Abnous, K.; Ramezani, M.; Hadizadeh, F. Dextran-b-poly (lactide-co-glycolide) polymersome for oral delivery of insulin: In vitro and in vivo evaluation. *J. Control. Release* **2016**, *227*, 58–70. [CrossRef] [PubMed]
93. Demchenko, A.P. *Ultraviolet Spectroscopy of Proteins*; Springer Science & Business Media: Berlin, Germany, 2013.
94. Madkour, L.H. Introduction to nanotechnology (NT) and nanomaterials (NMs). In *Nanoelectronic Materials*; Springer: Cham, Switzerland, 2019; pp. 1–47. [CrossRef]
95. Li, J.; Jia, X.; Yin, L. Hydrogel: Diversity of structures and applications in food science. *Food Rev. Int.* **2021**, *37*, 313–372. [CrossRef]
96. Daniloski, D.; Petkoska, A.T.; Lee, N.A.; Bekhit, A.E.D.; Carne, A.; Vaskoska, R.; Vasiljevic, T. Active edible packaging based on milk proteins: A route to carry and deliver nutraceuticals. *Trends Food Sci. Technol.* **2021**, *111*, 688–705. [CrossRef]
97. Rashidi, L. Different nano-delivery systems for delivery of nutraceuticals. *Food Biosci.* **2021**, *43*, 101258. [CrossRef]
98. Rai, S.; Singh, N.; Bhattacharya, S. Concepts on smart nano-based drug delivery system. *Recent Pat. Nanotechnol.* **2022**, *16*, 67–89. [CrossRef]
99. Grumezescu, A.M. (Ed.) *Organic Materials as Smart Nanocarriers for Drug Delivery*; William Andrew: Norwich, NY, USA, 2018.
100. Loveday, S.M.; Rao, M.A.; Singh, H. Food protein nanoparticles: Formation, properties and applications. *Food Mater. Sci. Eng.* **2012**, 263–294. [CrossRef]
101. Zhou, Z.F.; Sun, T.W.; Chen, F.; Zuo, D.Q.; Wang, H.S.; Hua, Y.Q.; Cai, Z.D.; Tan, J. Calcium phosphate-phosphorylated adenosine hybrid microspheres for anti-osteosarcoma drug delivery and osteogenic differentiation. *Biomaterials* **2017**, *121*, 1–14. [CrossRef] [PubMed]
102. Guo, H.C.; Ye, E.; Li, Z.; Han, M.Y.; Loh, X.J. Recent progress of atomic layer deposition on polymeric materials. *Mater. Sci. Eng. C* **2017**, *70*, 1182–1191. [CrossRef] [PubMed]
103. Pires, A.F.; Marnotes, N.G.; Rubio, O.D.; Garcia, A.C.; Pereira, C.D. Dairy by-products: A review on the valorization of whey and second cheese whey. *Foods* **2021**, *10*, 1067. [CrossRef] [PubMed]
104. Lesme, H.; Rannou, C.; Famelart, M.H.; Bouhallab, S.; Prost, C. Yogurts enriched with milk proteins: Texture properties, aroma release and sensory perception. *Trends Food Sci. Technol.* **2020**, *98*, 140–149. [CrossRef]
105. Manzoor, M.; Singh, J.; Bandral, J.D.; Gani, A.; Shams, R. Food hydrocolloids: Functional, nutraceutical and novel applications for delivery of bioactive compounds. *Int. J. Biol. Macromol.* **2020**, *165*, 554–567. [CrossRef]
106. Guo, X.; Xu, D.; Yuan, H.; Luo, Q.; Tang, S.; Liu, L.; Wu, Y. A novel fluorescent nanocellulosic hydrogel based on carbon dots for efficient adsorption and sensitive sensing in heavy metals. *J. Mater. Chem. A* **2019**, *7*, 27081–27088. [CrossRef]
107. Pereira, R.N.; Rodrigues, R.M.; Altinok, E.; Ramos, Ó.L.; Malcata, F.X.; Maresca, P.; Ferrari, G.; Teixeira, J.A.; Vicente, A.A. Development of iron-rich whey protein hydrogels following application of ohmic heating–Effects of moderate electric fields. *Food Res. Int.* **2017**, *99*, 435–443. [CrossRef]
108. Kazemi-Taskooh, Z.; Varidi, M. Food-based iron delivery systems: A review. *Trends Food Sci. Technol.* **2021**, *116*, 75–89. [CrossRef]
109. Song, J.; He, W.; Shen, H.; Zhou, Z.; Li, M.; Su, P.; Yang, Y. Self-assembly of a magnetic DNA hydrogel as a new biomaterial for enzyme encapsulation with enhanced activity and stability. *Chem. Commun.* **2019**, *55*, 2449–2452. [CrossRef]
110. Wei, Z.; Volkova, E.; Blatchley, M.R.; Gerecht, S. Hydrogel vehicles for sequential delivery of protein drugs to promote vascular regeneration. *Adv. Drug Deliv. Rev.* **2019**, *149*, 95–106. [CrossRef]
111. Batista, R.A.; Espitia, P.J.P.; Quintans, J.D.S.S.; Freitas, M.M.; Cerqueira, M.Â.; Teixeira, J.A.; Cardoso, J.C. Hydrogel as an alternative structure for food packaging systems. *Carbohydr. Polym.* **2019**, *205*, 106–116. [CrossRef] [PubMed]

112. Benito-Peña, E.; González-Vallejo, V.; Rico-Yuste, A.; Barbosa-Pereira, L.; Cruz, J.M.; Bilbao, A.; Alvarez-Lorenzo, C.; Moreno-Bondi, M.C. Molecularly imprinted hydrogels as functional active packaging materials. *Food Chem.* **2016**, *190*, 487–494. [CrossRef]
113. Baek, S.; Kim, D.; Jeon, S.L.; Seo, J. Preparation and characterization of pH-responsive poly (N, N-dimethyl acrylamide-co-methacryloyl sulfadimethoxine) hydrogels for application as food freshness indicators. *React. Funct. Polym.* **2017**, *120*, 57–65. [CrossRef]
114. Dickinson, E. Emulsion gels: The structuring of soft solids with protein-stabilized oil droplets. *Food Hydrocoll.* **2012**, *28*, 224–241. [CrossRef]
115. Cao, Y.; Mezzenga, R. Design principles of food gels. *Nat. Food* **2020**, *1*, 106–118. [CrossRef]
116. Wu, B.C.; Degner, B.; McClements, D.J. Soft matter strategies for controlling food texture: Formation of hydrogel particles by biopolymer complex coacervation. *J. Phys. Condens. Matter* **2014**, *26*, 464104. [CrossRef] [PubMed]
117. Gonçalves, J.O.; Santos, J.P.; Rios, E.C.; Crispim, M.M.; Dotto, G.L.; Pinto, L.A.A. Development of chitosan based hybrid hydrogels for dyes removal from aqueous binary system. *J. Mol. Liq.* **2017**, *225*, 265–270. [CrossRef]
118. Azizi, M.; Kierulf, A.; Lee, M.C.; Abbaspourrad, A. Improvement of physicochemical properties of encapsulated echium oil using nanostructured lipid carriers. *Food Chem.* **2018**, *246*, 448–456. [CrossRef]
119. Ghasemi, S.; Jafari, S.M.; Assadpour, E.; Khomeiri, M. Nanoencapsulation of d-limonene within nanocarriers produced by pectin-whey protein complexes. *Food Hydrocoll.* **2018**, *77*, 152–162. [CrossRef]
120. Guo, Y.; Harris, P.; Kaur, A.; Pastrana, L.; Jauregi, P. Characterisation of β-lactoglobulin nanoparticles and their binding to caffeine. *Food Hydrocoll.* **2017**, *71*, 85–93. [CrossRef]
121. Onsekizoglu Bagci, P.; Gunasekaran, S. Iron-encapsulated cold-set whey protein isolate gel powder–Part 1: Optimisation of preparation conditions and in vitro evaluation. *Int. J. Dairy Technol.* **2017**, *70*, 127–136. [CrossRef]
122. Mirpoor, S.F.; Hosseini, S.M.H.; Nekoei, A.R. Efficient delivery of quercetin after binding to beta-lactoglobulin followed by formation soft-condensed core-shell nanostructures. *Food Chem.* **2017**, *233*, 282–289. [CrossRef] [PubMed]
123. Fan, Y.; Zhang, Y.; Yokoyama, W.; Yi, J. β-Lactoglobulin–chlorogenic acid conjugate-based nanoparticles for delivery of (−)-epigallocatechin-3-gallate. *RSC Adv.* **2017**, *7*, 21366–21374. [CrossRef]
124. Zand-Rajabi, H.; Madadlou, A. Citric acid cross-linking of heat-set whey protein hydrogel influences its textural attributes and caffeine uptake and release behaviour. *Int. Dairy J.* **2016**, *61*, 142–147. [CrossRef]
125. Abaee, A.; Madadlou, A. Niosome-loaded cold-set whey protein hydrogels. *Food Chem.* **2016**, *196*, 106–113. [CrossRef]
126. Lou, J.; Hu, W.; Tian, R.; Zhang, H.; Jia, Y.; Zhang, J.; Zhang, L. Optimization and evaluation of a thermoresponsive ophthalmic in situ gel containing curcumin-loaded albumin nanoparticles. *Int. J. Nanomed.* **2014**, *9*, 2517. [CrossRef]
127. Cao, W.J.; Ou, S.Y.; Lin, W.F.; Tang, C.H. Food protein-based phytosterol nanoparticles: Fabrication and characterization. *Food Funct.* **2016**, *7*, 3973–3980. [CrossRef]
128. Madalena, D.A.; Ramos, Ó.L.; Pereira, R.N.; Bourbon, A.I.; Pinheiro, A.C.; Malcata, F.X.; Teixeira, J.A.; Vicente, A.A. *In vitro* digestion and stability assessment of β-lactoglobulin/riboflavin nanostructures. *Food Hydrocoll.* **2016**, *58*, 89–97. [CrossRef]
129. Pérez-Masiá, R.; López-Nicolás, R.; Periago, M.J.; Ros, G.; Lagaron, J.M.; López-Rubio, A. Encapsulation of folic acid in food hydrocolloids through nanospray drying and electrospraying for nutraceutical applications. *Food Chem.* **2015**, *168*, 124–133. [CrossRef]
130. Yi, J.; Lam, T.I.; Yokoyama, W.; Cheng, L.W.; Zhong, F. Beta-carotene encapsulated in food protein nanoparticles reduces peroxyl radical oxidation in Caco-2 cells. *Food Hydrocoll.* **2015**, *43*, 31–40. [CrossRef]
131. Arroyo-Maya, I.J.; McClements, D.J. Biopolymer nanoparticles as potential delivery systems for anthocyanins: Fabrication and properties. *Food Res. Int.* **2015**, *69*, 1–8. [CrossRef]
132. Wang, X.; Hu, H.; Yang, Z.; He, L.; Kong, Y.; Fei, B.; Xin, J.H. Smart hydrogel-functionalized textile system with moisture management property for skin application. *Smart Mater. Struct.* **2014**, *23*, 125027. [CrossRef]
133. Lohcharoenkal, W.; Wang, L.; Chen, Y.C.; Rojanasakul, Y. Protein nanoparticles as drug delivery carriers for cancer therapy. *BioMed Res. Int.* **2014**, *2014*, 180549. [CrossRef] [PubMed]
134. Yi, J.; Lam, T.I.; Yokoyama, W.; Cheng, L.W.; Zhong, F. Controlled release of β-carotene in β-lactoglobulin–dextran-conjugated nanoparticles' *in vitro* digestion and transport with Caco-2 monolayers. *J. Agric. Food Chem.* **2014**, *62*, 8900–8907. [CrossRef] [PubMed]
135. Gülseren, İ.; Fang, Y.; Corredig, M. Zinc incorporation capacity of whey protein nanoparticles prepared with desolvation with ethanol. *Food Chem.* **2012**, *135*, 770–774. [CrossRef]
136. Shpigelman, A.; Cohen, Y.; Livney, Y.D. Thermally-induced β-lactoglobulin–EGCG nanovehicles: Loading, stability, sensory and digestive-release study. *Food Hydrocoll.* **2012**, *29*, 57–67. [CrossRef]
137. Betz, M.; Kulozik, U. Whey protein gels for the entrapment of bioactive anthocyanins from bilberry extract. *Int. Dairy J.* **2021**, *21*, 703–710. [CrossRef]
138. Remondetto, G.E.; Beyssac, E.; Subirade, M. Iron availability from whey protein hydrogels: An *in vitro* study. *J. Agric. Food Chem.* **2004**, *52*, 8137–8143. [CrossRef]
139. Gunasekaran, S.; Ko, S.; Xiao, L. Use of whey proteins for encapsulation and controlled delivery applications. *J. Food Eng.* **2007**, *83*, 31–40. [CrossRef]
140. Remondetto, G.E.; Paquin, P.; Subirade, M. Cold Gelation of β-lactoglobulin in the Presence of Iron. *J. Food Sci.* **2002**, *67*, 586–595. [CrossRef]

141. Somchue, W.; Sermsri, W.; Shiowatana, J.; Siripinyanond, A. Encapsulation of α-tocopherol in protein-based delivery particles. *Food Res. Int.* **2009**, *42*, 909–914. [CrossRef]
142. Giroux, H.J.; Britten, M. Encapsulation of hydrophobic aroma in whey protein nanoparticles. *J. Microencapsul.* **2011**, *28*, 337–343. [CrossRef] [PubMed]
143. Yamul, D.K.; Lupano, C.E. Properties of gels from whey protein concentrate and honey at different pHs. *Food Res. Int.* **2003**, *36*, 25–33. [CrossRef]
144. Tada, D.; Tanabe, T.; Tachibana, A.; Yamauchi, K. Drug release from hydrogel containing albumin as crosslinker. *J. Biosci. Bioeng.* **2005**, *100*, 551–555. [CrossRef]
145. Puyol, P.; Perez, M.D.; Horne, D.S. Heat-induced gelation of whey protein isolates (WPI): Effect of NaCl and protein concentration. *Food Hydrocoll.* **2001**, *15*, 233–237. [CrossRef]
146. Iemma, F.; Spizzirri, U.G.; Puoci, F.; Muzzalupo, R.; Trombino, S.; Cassano, R.; Leta, S.; Picci, N. pH-Sensitive hydrogels based on bovine serum albumin for oral drug delivery. *Int. J. Pharm.* **2006**, *312*, 151–157. [CrossRef]
147. Zimet, P.; Livney, Y.D. Beta-lactoglobulin and its nanocomplexes with pectin as vehicles for ω-3 polyunsaturated fatty acids. *Food Hydrocoll.* **2009**, *23*, 1120–1126. [CrossRef]
148. Sneharani, A.H.; Karakkat, J.V.; Singh, S.A.; Rao, A.A. Interaction of curcumin with β-lactoglobulin stability, spectroscopic analysis, and molecular modeling of the complex. *J. Agric. Food Chem.* **2010**, *58*, 11130–11139. [CrossRef]

gels MDPI

Review

Food Emulsion Gels from Plant-Based Ingredients: Formulation, Processing, and Potential Applications

Canice Chun-Yin Yiu [1], Sophie Wenfei Liang [2], Kinza Mukhtar [3], Woojeong Kim [1], Yong Wang [1,*] and Cordelia Selomulya [1]

[1] School of Chemical Engineering, UNSW Sydney, Kensington, NSW 2052, Australia; chun_yin.yiu@student.unsw.edu.au (C.C.-Y.Y.); woojeong.kim@unsw.edu.au (W.K.); cordelia.selomulya@unsw.edu.au (C.S.)
[2] Agrotechnology and Food Sciences Group, Wageningen University & Research, Droevendaalsesteeg 4, 6708 PB Wageningen, The Netherlands; sophie.liang@wur.nl
[3] National Institute of Food Science and Technology, University of Agriculture, Faisalabad 38000, Pakistan; kinzaulemaan786@gmail.com
* Correspondence: yong.wang2@unsw.edu.au

Abstract: Recent advances in the understanding of formulations and processing techniques have allowed for greater freedom in plant-based emulsion gel design to better recreate conventional animal-based foods. The roles of plant-based proteins, polysaccharides, and lipids in the formulation of emulsion gels and relevant processing techniques such as high-pressure homogenization (HPH), ultrasound (UH), and microfluidization (MF), were discussed in correlation with the effects of varying HPH, UH, and MF processing parameters on emulsion gel properties. The characterization methods for plant-based emulsion gels to quantify their rheological, thermal, and textural properties, as well as gel microstructure, were presented with a focus on how they can be applied for food purposes. Finally, the potential applications of plant-based emulsion gels, such as dairy and meat alternatives, condiments, baked goods, and functional foods, were discussed with a focus on sensory properties and consumer acceptance. This study found that the implementation of plant-based emulsion gel in food is promising to date despite persisting challenges. This review will provide valuable insights for researchers and industry professionals looking to understand and utilize plant-based food emulsion gels.

Keywords: plant-based; emulsion gel; food application

Citation: Yiu, C.C.-Y.; Liang, S.W.; Mukhtar, K.; Kim, W.; Wang, Y.; Selomulya, C. Food Emulsion Gels from Plant-Based Ingredients: Formulation, Processing, and Potential Applications. *Gels* **2023**, *9*, 366. https://doi.org/10.3390/gels9050366

Academic Editors: Baskaran Stephen Inbaraj, Kandi Sridhar and Minaxi Sharma

Received: 23 March 2023
Revised: 12 April 2023
Accepted: 18 April 2023
Published: 26 April 2023

1. Introduction

Food emulsion gels are ubiquitous in the food industry to create texture or a sensory experience for low-fat products or as a vehicle to deliver functional food ingredients via encapsulation [1,2]. Emulsion gels are semi-solid systems with a gel network structure, often embedded with oil droplets [3]. They integrate the dual characteristics of emulsion and gel, improving both the stability of the emulsion system and the rheological and nutritional properties of a hydrogel, making them a unique and versatile format for developing new foods [4]. The difference between an emulsion gel and a simple emulsion with oil droplets as the inner phase lies in the presence of the gel network. In an ordinary emulsion, oil droplets are dispersed in a continuous phase with emulsifiers to lower the interfacial tension between the two immiscible liquids. However, in an emulsion gel, the dispersed oil droplets are not only stabilized by emulsifiers but are also trapped within a continuous gel network, which provides additional stability and unique characteristics to the system [5].

Emulsion gel systems were originally applied to reduce the fat content of food products by incorporating the gelled water phase while still retaining sensory properties [6]. Emulsion gels can be used to replace fats in a variety of food products, including baked goods, processed meat, dairy products, functional foods, and edible 3D printing inks [7]. They are composed of an oil phase and gel matrix, and the gel matrix comprises components with gel properties, mainly edible proteins and polysaccharides. The generation of emulsion gels is based on the interaction forces between the filler and the matrix, while those without interaction forces require external processing or the addition of emulsifiers [8].

In addition to their potential as a fat replacer, emulsion gels show promising prospects as a delivery medium for functional ingredients through encapsulation [9]. They can effectively inhibit the release of the ingredients and improve the efficiency of bioactive substances, while also controlling the release rate of the encapsulated ingredients. According to research, the use of a protein–polysaccharide for encapsulation may yield better results than systems that utilize only protein or polysaccharides [9]. This hybrid matrix has demonstrated improved loading rates and enhanced stability of encapsulated bioactive molecules. These findings have significant implications for the development of functional foods [10].

Traditionally, emulsion gels are stabilized by animal-sourced ingredients, particularly dairy and meat proteins [10]. However, increasing the global consumption of plant proteins has been widely recognized as a critical approach to ensuring food security and sustainability through 2050 [11]. Despite their potential as a healthy alternative to traditional fats, plant-based emulsion gels have yet to be utilized to their full potential. One reason for this is the poor functionality of plant proteins, particularly in terms of solubility and emulsification capability [12], which is especially true for pea protein [13]. Many recent attempts to produce emulsion gels utilizing plant-based ingredients have proven the possibility to partially or totally replace animal ingredients [14]. Nonetheless, further studies are needed to expand the food applications of plant ingredients [2,15,16]. The properties of emulsion gels are closely related to the dispersed phase and continuous phase, and the content, ratio, and type of both directly impact the final properties of the gel. Therefore, systematic understandings of emulsion formation are still needed to promote the application of plant-based ingredients in emulsion gels.

This paper aims to review the current research on emulsion gels using plant-based ingredients, including the formulation, processing, and potential applications. Specifically, it will focus on the composition and properties of emulsion gels, the interaction forces between the oil phase and the gel matrix, the different processing methods for protein and polysaccharide matrices, and the characterization techniques to analyze food emulsion gels. This review will provide a comprehensive understanding of the current state of knowledge on emulsion gels as well as insights into their potential applications in the food industry.

2. Formulation of Plant-Based Food Emulsion Gels

The composition of an emulsion gel is fundamental to the stability and physical and mechanical characteristics of the product. Emulsion gels are typically differentiated by the material(s) from which they derive their structure. Emulsion gels may be formed by both proteins and polysaccharides. These hydrocolloids may be used in isolation or in conjunction to stabilize oil-in-water emulsions and provide structure. For emulsion gels that utilized both protein and polysaccharides in their formulation, they may be referred to as "mixed gels" [5,10]. The following sections discuss the functionality of these hydrocolloids and examine their roles in the formation of an emulsion gel and the characteristics they provide to these gels both alone and in a mixed gel formulation. A summary of major techniques for emulsion gel formation and mechanism is presented in Figure 1.

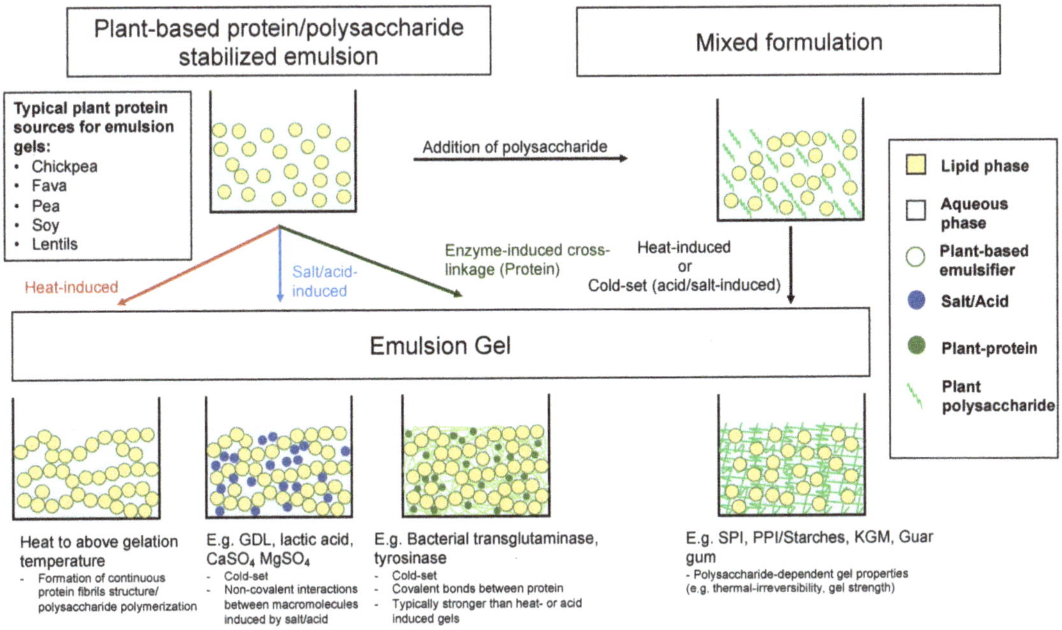

Figure 1. Summary of various methods outlined in Section 2, in which plant-based proteins, polysaccharides, and lipids are used in the formation of plant-based emulsion gel. GDL: Glucono-δ-lactone. SPI: soy protein isolate. PPI: pea protein isolate. KGM: Konjac glucomannan.

2.1. Protein

Protein in a plant-based emulsion gel may be used alone as a structural component or with other components as an emulsifier to stabilize the lipophilic phase [5]. Proteins from legumes and cereals were frequently chosen to make emulsion gels due to their desirable characteristics (emulsion stabilizing and gelation properties) and economy from high-protein plant sources [17]. As such, proteins from soy, peas, chickpeas, fava beans, potatoes, and wheat were among the most studied as structural components and emulsifiers. Additional considerations may be applied when selecting plant proteins, as protein extracts from various sources contain varying protein profiles and allergenicity that may be critical to human health [18]. A prime example of which would be potato protein, where essential amino acid content is among the best in plant-based protein sources [18].

2.1.1. Plant Protein-Based Emulsion Gels

Plant protein-only emulsion gels may be categorized into two major types, characterized by gelation methods. This is namely heat-induced gelation and cold-set gelation with the aid of added coagulants. The formation of plant-protein-based emulsion gels typically follows a two-step process [15]. The first step involves the homogenization of oil and a suspended protein phase to form an emulsion stock. Subsequently, the emulsion is gelled by heat or the addition of coagulants [15,19]. Heat-induced gelation, where globular proteins from plants denature and aggregate to form strong fibrils at elevated temperatures, has been a widely studied method to form a strong, solid-like gel [20]. In heat-set gels, temperatures above 80 °C are typically used for plant protein gelation. Nevertheless, temperatures as high as 95 °C were frequently reported [21,22]. In heat-induced gels, gelation onset temperature may be determined by the oil volume fraction of a gel matrix. In a study on soy protein emulsion gel, a higher oil volume fraction resulted in a lower gelation onset temperature [23]. Structures of heat-induced gels were also found to be more homogenous than those of acid or enzyme-induced gels due to fewer aggregates forming [24]. Moreover, characteristics of heat-induced gels may be further tuned by the total protein content, pH, ionic strength, and ratios between structural proteins within the gel matrix (e.g., glycinin and conglycinin in soy protein gel) [24,25]. However, with the high temperature during heat-induced gelation, heat-set gels may not be suitable if heat-sensitive material is involved.

In response, cold-set gels were investigated to reduce the degradation of heat-sensitive components during gelation. In these gels, gelation was induced by adding coagulants (e.g., acids, salts, and enzymes) to negate heating after the loading of lipids. However, denaturation of plant proteins was still required before cold-setting to expose the sites for aggregation and agglomeration [26]. Acids (e.g., glucono-δ-lactone (GDL) and lactic acid) and salts (magnesium, calcium, and sodium salts) were frequently used to coagulate protein gels. Klost and Drusch [26] reported that lactic acid fermentation of pea protein was able to create a gel suitable for plant-based yogurt [26]. In acid-induced gels, GDL has been a common ingredient in the traditional manufacturing of soft tofu. In emulsion gels, GDL-coagulated soy protein gels were found to exhibit weak rheological behaviors due to non-covalent intermolecular forces that exist within the gel [27]. In salt-induced gels, salt type and salt content are equally significant in gel stability and rheology. Various magnesium and calcium salts ($MgCl_2$, $MgSO_4$, and $CaSO_4$) were also tested on soy protein isolate by Wang et al. [28]. In their study, magnesium salts ($MgCl_2$ and $MgSO_4$) showed higher aggregation power than $CaSO_4$ as a stronger gel in terms of firmness and rigidity was formed [28]. Optimization of salt content would also provide freeze-thaw stability and rigidity to salt-induced protein gels [14]. Excessive salt was found to promote protein aggregation, leading to a coarse and inhomogeneous gel in NaCl-induced soy protein gels [14].

Enzymes from bacterial sources such as transglutaminase and tyrosinase were used to cross-link proteins from soy, pea, zein, and potato origins. Glusac et al. [29] investigated the use of bacterial tyrosinase to cross-link zein and potato protein in encapsulating olive oil. Additionally, a protease inhibitor in potato protein was added to provide a cross-linkage site with $\alpha-$zeins resulting in a gel that exhibits superior storage stability to zein-only emulsions after a month of storage at room temperature [29]. In experiments where only the coagulant was altered, enzyme-induced gelation was reported to have a significantly stronger gel structure and elasticity than acid- or salt-induced soy protein emulsion gels. This is indicated by a 31–33% increase in yield stress when comparing the GDL and $CaCO_3$ gels to the transglutaminase gel [27]. Enzymes promote the formation of permanent covalent bonds between macromolecules, which results in the creation of a "classical polymer gel" [5]. This is distinct from physical gels (e.g., heat-induced gel), where gel-like properties arise from intermolecular interactions. This fundamental difference explains the typical higher resistance observed for cross-linked gels [5].

2.1.2. Plant Protein as Emulsifier

Dispersion of the oil phase in the form of an emulsion is essential for producing both protein-only and mixed emulsion gels. Proteins could act as a surface-active material that stabilizes the oil/water interface and thereby adsorbs at the interface of the dispersed oil phase, owing to their amphiphilicity from containing both polar and non-polar amino acids [18]. This phenomenon is seen in native isolates of plant proteins [18,30]. Moreover, thermal denaturation of protein in emulsion systems was linked to increasing droplet flocculation due to increased exposure to hydrophobic sites on the protein surface [31]. However, some studies have demonstrated that the pre-denaturation of plant proteins may increase the creaming stability of plant protein-stabilized emulsions [32,33]. This was attributed to the more viscous heat-treated emulsion and the formation of supramolecular structure at the droplet interface, which impeded creaming [32,33]. The emulsification of oil using plant proteins may create an active-filler gel where the dispersed droplets are mechanically linked to the surrounding structural matrix. This is particularly true for heat-set protein gels, as a higher gel strength is typically observed as a result [5].

Since an emulsion is first produced before gelation, it is also important to maximize emulsion stability to give it flexibility for gelation conditions and further processing. The use of dairy proteins to stabilize a hydrophobic phase has been widely established [34]. Likewise, the use of plant protein to stabilize an emulsion has been extensively studied in recent years, with soy, pea, and other lentil-based proteins being adopted by many studies to replace animal-sourced protein in emulsification [31,35]. Gumus et al. investigated the emulsification capability and stability of pea, lentil, and fava beans-stabilized emulsion under different pH, temperature, and salt conditions. In their study, lentil protein displayed superior stability across all environmental stressors [36]. Although the reason for lentils' superior performance was not specified, it was suggested that extreme pH and salt content may have a more profound effect on the electrostatic repulsion of pea and fava proteins. Hydrophobic and steric interactions of lentil protein may also be responsible for the better temperature stability observed [36]. Recently, attention was also given to the use of globular protein fibrils over native plant protein isolates to improve emulsion stability and rheology. Micron-length protein fibrils are produced from native protein isolate via acid heat treatment at a pH lower than the isoelectric point of a protein [37]. Pang et al. investigated the use of rice bran protein fibrils to stabilize fish oil. It was shown that heating rice bran protein for 420 min achieved the highest emulsification capability with fish oil. This is attributed to an increase in the hydrophobicity of the fibrils and molecule flexibility at the interface [37].

2.2. *Polysaccharide*

2.2.1. Structural Roles of Plant Polysaccharide in Emulsion Gel

Polysaccharides had been typically used as the structural component of an emulsion gel, both alone and in conjunction with protein in a mixed gel regimen. In a mixed gel, an emulsion is stabilized by a surfactant before the addition of polysaccharide and subsequent gelation. Similarly, the dispersed phase may act as both an active and inactive filler depending on the specific interactions between the selected protein/surfactant and the continuous polysaccharide phase [1,15]. A wide range of plant-based polysaccharides, such as methylcellulose, inulin, pectin, and native and modified starches, have been reported for use in emulsion gels to provide stability and alter the mechanical properties of the gels [38–42].

The gelation of polysaccharides may be induced by heat or by a coagulant (e.g., salt and acid) to form polymeric structures within the aqueous phase of an emulsion [15]. Particularly, interactions between the polysaccharide and the protein may provide enhanced performance over gels that use either component in a mixed gel. For example, the synergistic relationship between pea protein isolate (PPI) and inulin in an emulsion gel was studied by Xu et al. [39]. In their studies, when PPI and inulin were used together, a lower inulin content was required for gelation in contrast to pure inulin gel [39]. This was because inulin introduced more hydrophobic and hydrogen bond sites without adversely affecting protein electrostatic interactions, which greatly contributed to the firmness and improved properties of the gel [39]. Thus, the presence of synergetic interactions between proteins and polysaccharides was typically referenced for adopting a mixed gel regime [43–45].

The rheological properties and stability of emulsion gels may also be modified through the addition of polysaccharides. Polysaccharides such as xanthan gum, guar gum, and potato starch were shown to increase the viscosity of emulsion gels in 3D printing, a structurally critical application [41,46]. In contrast to salt-induced soy protein emulsion gel, where the gel is too weak to hold a 3D structure, emulsion gels with xanthan and guar gums were able to hold structure with minimal loss in feature [46]. Additionally, increasing the viscosity of a gel during mixing may also result in smaller and more uniform droplets. Figure 2 shows confocal laser scanning microscopy (CLSM) images of PPI and curdlan gum heat-induced emulsion gel fabricated in our laboratory with or without konjac glucomannan (KGM), a plant-based polysaccharide. The emulsion gels in both images are identical in terms of total oil (20% *w*/*w*), PPI (5% *w*/*w*), and polysaccharide concentrations

(6% w/w). Due to the high water absorption ability of KGM [47], viscosity was greatly increased in pre-gelled emulsion gel after its addition (data not shown). The two images only differ in polysaccharide composition, with Figure 2a showing gel with 6% curdlan and Figure 2b showing gel with 4.8% curdlan and 1.2% KGM. Droplets were larger and less uniform in curdlan-only emulsion gels compared to those with substituted KGM, as shown in the droplet size distribution in Figure 2c. The addition of polysaccharides that induce an increase in viscosity was seen to reduce the coalescence of the dispersed oil droplets during processing. Smaller droplet size is often associated with higher stability and gel strength in emulsion gels [48,49]. In addition, smaller droplet sizes may reduce oil release during gastric digestion and hence speed up gastric emptying [10].

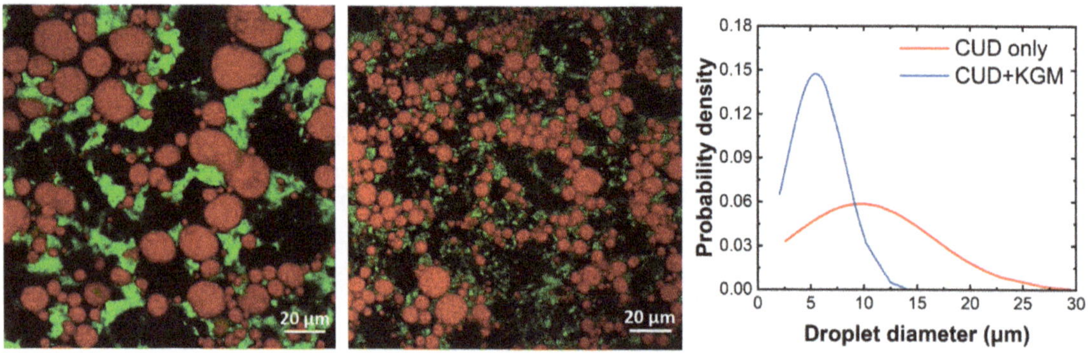

(**a**) (**b**) (**c**)

Figure 2. CLSM images of PPI and curdlan (CUD) emulsion gels with and without Konjac glucomannan (KGM). Both emulsion gels have identical total protein (5% w/w), oil (20% w/w), and total polysaccharide content (6% w/w): (**a**) Emulsion gel with 6% CUD; (**b**) Emulsion gel with 4.8% curdlan and 1.2% KGM; (**c**) Normal distribution of droplet diameters from (**a**,**b**). 50 droplets were randomly sampled from the respective images and processed using ImageJ®. Red: lipid fraction; green: protein and polysaccharide.

Thermal stability of the gel structure may also be achieved using plant-based polysaccharides. This may be critical for certain applications as phase change may not be desirable at elevated temperatures. As such, KGM was used to create thermally irreversible emulsion gels for cooking stability [38,50]. Huang et al. [50] created a soy protein isolate (SPI) and KGM crosslinked composite gel that was able to maintain structure and texture after 20 min of cooking at 95 °C. Similarly, a combination of deacetylated KGM (DKG) and methylcellulose (MC) was reported by Jeong et al. [38]. Across all ratios between DKG and MC tested, all gels showed thermal irreversibility at 80 °C [38].

2.2.2. Plant Polysaccharide as Emulsifier

Recently, interests have also pivoted to develop polysaccharide-only emulsion gels where polysaccharide is responsible for both emulsion stabilization and structure formation. Jiang et al. [51] explored the use of regenerated cellulose (RC) to stabilize a sunflower oil in water emulsion with curdlan, a bacterial gum, as the major structural component. The RC stabilized emulsion was shown to have high stability as no noticeable flocculation or separation was observed after 4 h of storage at room temperature and 1 h at 80 °C [51]. A plant-based emulsion gel was demonstrated by Zhou et al. [52] using polysaccharides extracted from psyllium husk. High stability was observed from the psyllium polysaccharide stabilized emulsion, as no visible phase separation was observed after prolonged storage at room temperature [52]. The reduction in polysaccharide content led to a softer gel as the polysaccharides were adsorbed at the oil/water interface rather than forming a continuous gel structure [52]. The psyllium gel was also found to have self-healing properties after the destruction of the gel structure due to the rapid re-formation of hydrogen bonds [52]. Moreover, psyllium husk is a prebiotic that promotes the health of gut microbiota and has been used to treat gastrointestinal diseases (e.g., constipation and diarrhea) [53]. The advantageous physical and bioactive properties displayed by polysaccharide-based emulsion gel in these two studies underline possible effective application of these gels in food and pharmaceuticals.

2.3. Lipid

The lipid phase in an emulsion gel is bound by a surfactant, and its interaction with the continuous phase is often guided through the surfactant. These dispersed droplets may be active or inactive, depending on whether the dispersed droplets directly interact with the continuous structural component. For instance, a non-ionic surfactant stabilized lipid phase is often deemed an inactive filler with little contribution to the gel structure. As a result, the presence of these droplets in a gel matrix impedes gel strength as it is analogous to porous structures [1,5]. Conversely, in protein-stabilized emulsion gels, a layer of protein is adsorbed to the surface of oil droplets. Interactions between the protein and the gel matrix in the continuous phase effectively make the lipid phase an integral part of the structure [1,5]. In practice, a variety of lipids have been exploited for their respective physical, chemical, and nutritional properties to deliver desirable product characteristics.

2.3.1. Health Consideration on Lipid Selection

Most plant-extracted lipids exist as oils, with a higher proportion of mono- or polyunsaturated fatty acids (MUFA and PUFA) than animal fat [54]. Since consumption of high PUFA or MUFA oils is linked to better cardiovascular health than saturated fat, oils extracted from canola, soybean, sunflower, and olive are the usual candidates for emulsion gels [55,56]. Naturally occurring plant fats such as coconut oil, cocoa butter, and palm oil were also commonly used in food due to their solid structure at room temperature and higher saturated fat content [54]. Moreover, the use of fully hydrogenated oil had emerged as an alternative to using natural plant fats, owing to its lower cost compared to natural plant fats while still being solid at room temperature [57]. This is in contrast to partially hydrogenated oils that contain trans-fat and have been proven to have worse cardiovascular outcomes [57].

2.3.2. Enhancement of Textural Properties by Lipid

Aside from health benefits, the choice of lipid was also typically linked to the texturization of the final gel [15]. Oil types were found to have profound effects on the rheology and mechanical properties of a gel. Gels containing solid plant fat are typically stiffer and harder than those that use plant oil. Gu et al. [58] analyzed the effect of different oil types on the SPI emulsion gels. In their study, sunflower oil, soy oil, and palm stearin were emulsified and gelled in an SPI matrix. Palm stearin gel was a stiffer and harder gel than both of its oil counterparts, independent of the gelation method (GDL and heat) [58]. It was believed that palm stearin's crystalline structure allowed for the creation of an effectively absorbed protein layer around the dispersed droplets during emulsification, which increased the rigidity of the gel [58]. In more specific applications such as animal fat analog, Jeong et al. [38] showed that the use of coconut oil often resulted in superior hardness, springiness, and chewiness in textural analysis over canola oil at identical gel formulation at room temperature. Samples prepared using coconut oil showed a difference of up to 3.6 times in hardness over canola oil at 25 °C, with differences diminishing as the two samples were heated up to 80 °C [38].

2.3.3. Lipid as a Carrier of Nutrients

The lipid phase has also been used as a carrier of lipophilic additives to create functional foods using emulsion gel to increase the bioaccessibility and stability under heat and light of bioactive compounds [10]. Common plant oils, including corn, flaxseed, and soybean oil, are seen as carriers of bioactive ingredients ranging from vitamins to phenolic compounds such as cholecalciferol, β-carotene, and curcumin in plant-based emulsion gel [59–61]. In these studies, no indication was observed that a specific oil type was selected based on its functional attributes or suitability as a carrier.

Regardless, recent studies indicated that solid fat content (SFC) and oils high in antioxidants may increase the stability of the loaded material in addition to solubilization. In a whey protein isolate (WPI) and coconut oil emulsion gel created by Lu et al. [62], β-carotene was found to be more stable in an emulsion gel that has a higher SFC in both UV treatment and elevated temperature storage (55 °C). This was attributed to the inability of free radicals to travel through solid fat and the higher stability of coconut oil compared to other oils (e.g., corn), as well as the antioxidative ability of the medium-chain triglyceride (MCT) component in coconut oil [62]. The additional antioxidative effects of MCT in coconut oil were also demonstrated in an emulsion designed for lycopene delivery, where lycopene retention was approximately 40% better at 37 °C storage after 2 weeks than that of other oil types (e.g., sesame, linseed, and walnut oil) [63].

3. Processing of Plant-Based Food Emulsion Gels

As the applications of emulsion gels expand to encompass a growing variety of functions, diverse processing techniques and configurations have been proposed to achieve the desired food-like properties. These promising approaches are employed for the processing, transportation, and targeted release of food additives, functional ingredients, and bioactive substances, offering flexibility in tailoring food disintegration and sensory properties, as well as structural and functional parameters. The processing of emulsion gels is important due to their potential as fat substitutes in various animal-based products. For example, these emulsions can enhance the nutritional properties of meat products, as they are more adept at transporting and safeguarding oxidized lipids in food while also preserving flavor compounds and bioactive compounds. Soft or hard-textured emulsion gels are preferred for fat replacement over conventional emulsions without gel formation. It can better imitate the physical attributes of animal fats (lard), such as texture and water-holding capacity. Emulsion gel is formed using various technologies such as high internal phase, ultrasonication, high-pressure homogenization, and microfludization [64].

In a study on the formation of protein-based emulsion gel (yogurt and tofu analogs) using fava beans, whole nutrients were utilized [64]. Fava bean was processed through thermal pre-treatment, dehulling, milling, plant oil addition, homogenization, starch gelation prevention, and finally, the inducement of protein gelation. Starch removal and hydrolysis were used to prevent starch from gelation [64]. Starch hydrolysis showed better results in yogurt analog production, as this process increased the gel's strength and viscosity. In addition, hydrolysis utilized the whole floor with no waste production. Moreover, the tofu analog was better prepared when formed with starch removal, as opposed to starch hydrolysis, because it decreased the gel strength and water-holding capacity of both products. These two methods are preferred to generate protein-based emulsion gel from whole fava bean flour.

In many studies, the fabrication of an emulsion gel generally begins with high-shear mixing between the oil phase and aqueous phase, where the surfactant and or biopolymer are solubilized or suspended [65]. Although high-shear mixing is generally straightforward, other methods have been proposed to further improve stability and modify the final gel properties [66,67]. This includes high-pressure homogenization (HPH), ultrasonication (UF), and microfluidization (MF). Various emulsion gels can be prepared using these methods and exhibit differing gel structures and interactions among droplets with an impact on mechanical and release characteristics. Figure 3 illustrates the preparation of plant-based emulsion gel through high-pressure homogenization, microfluidization, and ultrasonication.

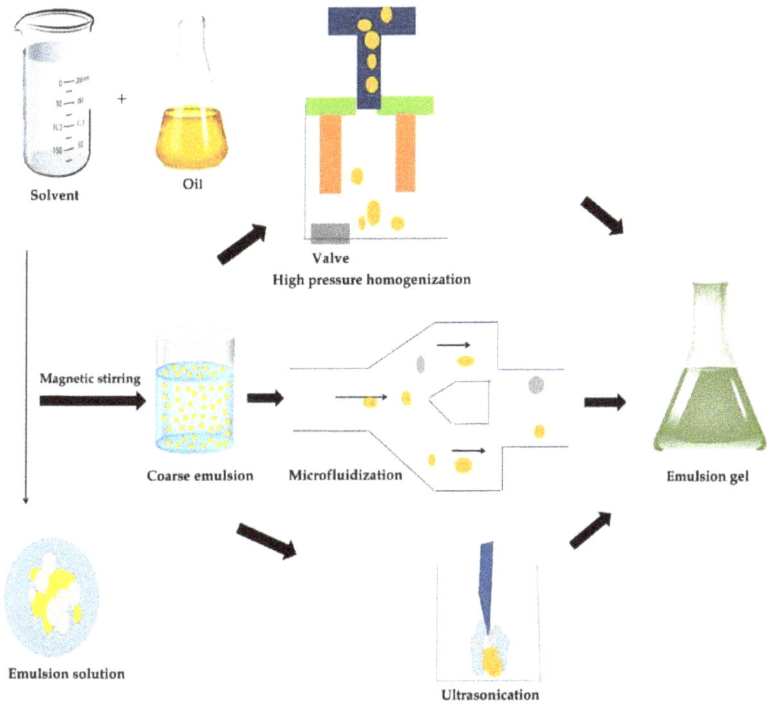

Figure 3. Formation of plant-based emulsion gel.

There are three main processing methods for producing emulsion gels in the food industry, which are HPH, US, and MF. HPH involves flow restriction, which produces greater pressures and shear forces (up to 350 MPa). However, sometimes the restriction of flow generates larger and non-uniform particles due to pressure fluctuation from the higher pressure. This is in contrast to MF, which can only exert a maximum of 275 MPa. The fixed geometry of the MF instrument ensures constant pressure delivery and equal size distribution [68], while ultrasonication works based on the principle of cavitation, which is produced through mechanical vibration. These processing methods are applied to the covert dispersed phase of the emulsion into fine, tiny droplets for better gel characteristics such as water-holding capacity, storage modulus, gel strength, and stability [69].

3.1. High-Pressure Homogenization

High-pressure homogenization (HPH) is an emerging technique with a pressure pump that exerts pressure up to 200 MPa [70]. It is applied for fluid stability, protein and polysaccharide modification, and improving the rheological characteristics of emulsion gels [71,72]. Alvarez-Sabatel et al. [19] evaluated the gelling characteristics of inulin after HPH application at 103, 207, or 296 MPa at an inlet temperature of 3.5 ± 1 °C for 5 min. The HPH treatment lowered the minimum concentration requirement necessary for gel texture and enhanced the crystallization behavior of inulin due to droplet dispersion, while the high pressure of 296 MPa reduced water retention efficiency and hurt the gel structure. Additionally, the inlet temperature was enhanced and reached a maximum (33.51 ± 0.53, 49.99 ± 0.96, and 61.67 ± 0.85 °C) due to the increase in pressure. Moreover, hazelnut beverage samples were subjected to HPH at a pressure of 150 MPa and a concentration of 10 g 100 mL^{-1} at 26–38 °C with a GDL acidification (2 g 100 mL^{-1} at 150 min), and the strength, rheological, and textural properties of cold-set gels were investigated. It increased the viscosity of hazelnut gels and improved the protein structure and gel characteristics with acidification [73].

The impact of thermal treatment on the rheological properties of SPI gels was investigated with acid-induced gelation and further treated with HPH at 400 bar pressure. The results showed that thermal treatment (95 °C, 20 min) improved the mechanical properties, and HPH increased the viscoelastic properties of the gel [74]. Furthermore, the HPH effect on the functional characteristics of SPI emulsion gel was evaluated at different pressures (from 5 MPa to 80 MPa), showing an increase in get strength (G′ = 291 Pa to G′ = 528 Pa at 80 MPa) and water holding capacity (WHC) (87.7% to 91.4%) at 5 to 20 MPa and constant values from 20 to 80 MPa pressure. Overall, the HPH-treated gel has a more uniform network [75]. The potential of using HPH to produce potato protein isolate yogurt alternatives with low and high oil contents has been evaluated at various oil concentrations (1.5, 3, and 10%) and homogenization pressures (0.1 MPa, 30 MPa, and 200 MPa). The HPH decreased the emulsion particle size in comparison to untreated samples. Moreover, the emulsion whiteness index had increased. The greatest value of the whiteness index (76.01 ± 0.50) was obtained at 200 MPa with 10% oil content, while the least (65.33 ± 2.05) was observed at 0.1 MPa with 3% oil content. The creaming velocity at 3% oil concentration was reduced from 10.70 (0.1 MPa) to 0.59 (200 MPa). In the end, gels showed smaller oil droplets and finer constituent distribution [76].

3.2. *Ultrasound*

Emulsion gel structure and functional properties have been significantly affected by homogenization. Ultrasound homogenization (UH), a safe, economical, and environmentally friendly technique, comprises acoustic waves with 10–1000 W/cm^2 power and 20–100 kHz frequency and produces mechanical and shearing effects through cavitation [77]. Various research has proven that the UH enhanced the protein properties such as aggregation size and solubility, thus increasing protein adsorption at the interface [77–79]. Moreover, the UH caused unfolding and an improvement in protein bonds at the interface, improving the gel's properties [77].

For instance, in a study by Aliabbasi et al. [78], the impact of high-intensity ultrasound on pinto bean protein isolate (0, 25, and 50 min at 200 W) was evaluated, and the gelation process was conducted using GDL. The key findings showed that the intrinsic fluorescence and the structure of the gel were improved. Moreover, the WHC and gel strength were increased, and the gel was inoculated with curcumin. Additionally, cross-linking ability decreased the swelling ratio of the gel and changed the curcumin release rate from the emulsion gel [80]. Furthermore, SPI and pectin emulsions were produced by heat treatment, and the effects of ultrasound on the texture, gel characteristics, and emulsion stability were assessed at various powers (0, 150, 300, 450, and 600 W). X-ray diffraction and Fourier transform infrared spectroscopy (FTIR) demonstrated that the interactions between SPI and pectin were improved due to the increase in hydrogen bonding and altered the crystallization of emulsion gels. Moreover, the droplet size was decreased, and WHC was increased by increasing the power (450 W) and denaturation temperature (128.2 °C to 131.9 °C). However, the bioaccessibility (82%) and chemical stability (78.3 ± 2.0%) of β-carotene were improved [81] due to the decrease in the aggregation of oil droplets and degradation of β-carotene during the digestion of the emulsion gel.

In a study by Mozafarpour and Koocheki [82], emulsion gels were produced using grass pea protein isolates (GPPI) at various ultrasonic treatment conditions (amplitudes of 25, 50, and 75% for 5, 10, and 20 min). Emulsion gel formation was stimulated by transglutaminase and stabilized by sonication. The first step, which was called the "fast step," caused weak gel formation, then the second, referred to as the "slow step," increased the gel strength. The results showed uniformity in particle distribution and improved velocities of emulsion gels. A harder emulsion gel was produced at 75% amplitude for 10 min. The WHC and mechanical properties were improved. Ultra-sonicated GPPIs had a fine microstructure compared to untreated GPPIs. Ultrasonication was applied in the formation of emulsion gels for topical drug delivery of metronidazole at 20–25 kHz at 150 W. Sorbitan monostearate (SMS) was used as a stabilizer at the sesame oil/water interface. Emulsion gels were prepared with various proportions of SMS ranging from 2.5–10% (*w*/*w*), while the water proportion ranged from 20% to 80% (*w*/*w*). The emulsion gel's viscosity and firmness had been increased by sonication processing. The drug-loaded gels exhibited antimicrobial efficiency, showing potential as a carrier for drugs [83]. Geng et al. [77] prepared soy protein bulk emulsion gels incorporating $CaCl_2$, GDL, and transglutaminase with the UH treatment (40% amplitude, 20 kHz, 3 min) for β-carotene delivery, showing improved bioaccessibility of β-carotene in bulk emulsion gels. It was noted that β-carotene bioaccessibility was increased when encapsulated in ultrasound-treated emulsion gels (82.39 ± 0.02%) compared to other emulsions (63.37 ± 0.09%). Moreover, findings suggested that WHC (99.81 ± 0.19%) and gel strength were improved (91.02 ± 3.58%) in transglutaminase-induced samples compared to $CaCl_2$ and GDL-induced samples. Overall, it can be suggested that ultrasonication improved WHC and the strength of the gel sample.

3.3. *Microfluidization*

Microfluidization (MF) is a high-energy processing technique in which pressure is applied to force the liquids through micro-channels, thus helping in emulsification by the synergistic effects of shear forces and cavitation [84]. It helps overcome issues such as large particle size and emulsion instability [85]. The MF was applied to the pea protein emulsion (5% pea protein and 50% sunflower oil) at 50 MPa for one pass, which caused the cold-set gel formation. The findings showed that MF increased gel strength, decreased particle size, and improved viscosity with an increase in pressure [86]. In a study by Yang et al. [87], γ-zein from corn was produced into particles to prepare gel-like emulsions. The MF was performed at 0.1 to 120 MPa to assess the rheological properties and structure formation of the emulsions. MF reduced particle size as well as gel network by droplet clusters as shown in microscopy where γ-zein particles provided stabilization, and the excess protein provided particle network. With the increase in pressure, gel strength increased due to the formation of more hydrophobic interactions and disulfide bonds. The results showed that emulsions prepared at 0.1 MPa have a weak gel structure, while emulsions prepared at 120 MPa have a stronger gel structure and higher stability.

Furthermore, MF was applied at 50, 70, and 130 MPa to native pea (NP) and soluble thermally aggregated (SA) pea globulin-based emulsions at neutral pH. Emulsions were assessed for their properties such as protein adsorption ability, charge emulsifying, flocculation, and creaming stability. It showed that NP and SA-based emulsions were more flocculated and had a coarse appearance. The MF pressure reduced the flocculation size when the pressure was increased due to the processing of emulsification. The NP-based emulsion creaming stability was decreased, while the SA-based emulsion creaming stability was improved as the pressure increased due to a decrease in flocculation size and thus the formation of a gel-like structure. The key findings suggested that MF could be used to improve emulsification properties as it reduced particle size, increased physical stability, and improved the viscosity of the emulsion gel [88].

4. Characterization of Plant-Based Food Emulsion Gels

To characterize plant-based food emulsion gels, recent research has mainly focused on these properties: appearance, rheology, texture, microstructure, and stability. Investigations into these properties provide insights into the behavior and overall quality of the emulsion gel, which could optimize the formulation and processing of plant-based emulsion gels. An overview of the characterization techniques of plant-based emulsion gels is presented in Table 1.

Table 1. Techniques used for characterization of plant-based emulsion gel.

Properties	Techniques and Methodology		Characteristics of Emulsion Gel	Ref.
Appearance	Visual inspection	Color, gel fabrication, stability/instability	Color, gel fabrication, stability/instability	[89–95]
	Colorimetry	CIE-LAB color space coordinates	Quantitatively evaluate the color	[96]
		Image converter analysis software (ImageJ®)		[97]
Rheology	Small amplitude oscillatory shear (SAOS)	Frequency sweep	Viscoelasticity	[40,42,82,95,98–100]
		Strain sweep	Linear viscoelastic region (LVR)	
		Time sweep	Viscoelasticity evolution over time	
		Creep-recovery test	Transient viscoelastic behavior	
	Large amplitude oscillatory shear (LAOS)	Stain sweep Lissajous curve Fourier-transform rheology	Gel strength, viscoelastic behavior in the non-linear viscoelastic region	[92,95,101]
Texture	Texture profile analysis (TPA)		Hardness, springiness, gumminess, chewiness, cohesiveness, viscidity, and stiffness, gel strength	[40,42,82,92,93,97, 99,102,103]
Microstructure	Droplet size distribution	Static Laser diffraction	Droplet size distribution	[91,100,104,105]
		Dynamic light scattering (DLS)		[82,92,98]
	Microscopy	Optical microscope (OM)	Droplet size and organization	[89,104,106,107]
		Polarized light microscope (PLM)	Structure of specific material (such as crystals and fibers)	[89,92,105]
		Confocal laser scanning microscope (CLSM)	Droplet size distribution, shape, and behavior	[75,93,97–99]
		Scanning electron microscope (SEM)	Structure of gel network	[40,95,98,102]
		Scanning electron cryo-microscopy (Cryo-SEM)	Structure of a well-maintained gel network	[95]
Stability	Thermal properties	Differential scanning calorimetry (DSC)	The amount of heat required to increase the temperature as a function of temperature	[75,82,92]
		Thermogravimetric analysis (TGA)	the change of the sample mass over time as the temperature changes	[42]
	Zeta potential		Stability	[82,100]
	Water holding capacity (WHC)	Moisture loss from centrifugation	Ability to retain water molecules	[42,75,82,98,100]
	Freeze-thaw cycling	Fluid loss from the freeze-thaw cycle	Stability under extreme temperature stress	[14,90,103]

4.1. Appearance

Visual inspection as a direct method is commonly used in research to rapidly obtain preliminary information on the samples, of which color, gel fabrication (viscoelasticity, compactness, hardness), and stability/instability (phase separation, creaming, aggregation, coalescence) were the most relevant. These results were often discussed with other characteristics to support the conclusion from the perspective of micromorphology [89–95].

Color is recognized as an attribute that mostly affected appearance as perceived by consumers directly. Besides visual inspection, color can also be quantitively characterized by CIELAB color space coordinates (lightness, L*; green/redness, a*; and blue/yellowness, b*) by using a colorimeter [96] or image analysis software (for example, ImageJ® software), which can convert digital images to coordinate data [97].

4.2. Rheological Properties

Rheological properties were measured to characterize the flow and deformation properties of plant-based emulsion gels under applied stress. Rheology measurement is important for understanding the behavior of the gel during processing, storage, and consumption, which is the foundation of optimizing the formulation and processing of the materials. The storage modulus (G′) and the loss modulus (G″) were used to indicate the elasticity (solid-like) and viscosity (liquid-like), respectively. The loss factor (tan δ) was commonly used to indicate the gel-like status of the sample [82,89,95,103]. Usually, a strain sweep test would be used to identify the linear viscoelastic region (LVR). Figure 4 shows an example of research on a starch-based emulsion with different oil volume fractions ranging from 0 to 70%. The transition from a linear to a non-linear viscoelastic region in a strain sweep graph is shown, in which the horizontal-like region was defined as the LVR. The yield stress (σ_c) refers to the stress causing the first non-linear deformation, which was used to reflect the strength of the emulsion gel in many research [91–93].

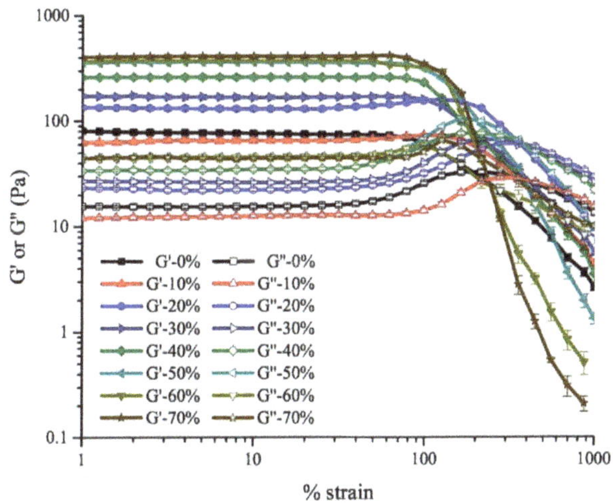

Figure 4. Large-amplitude oscillatory shear behavior as a function of strain amplitude for a novel starch-based emulsion gel with different oil volume fractions (0–70%). Adapted from Ref. [95] with permission from Elsevier.

Small amplitude oscillatory shear (SAOS) within LVR was commonly used in the characterization of the viscoelastic properties of plant-based emulsion gel. A frequency sweep test was used to indicate the fabrication of a gel-like structure (G′ > G″, both frequency independent). The creep-recovery test was used to study transient viscoelastic behavior [95,98,99]. A time sweep test (gelation kinetics) was used to investigate the change in gel texture at a certain frequency and temperature over a period of time [40,42,98]. Moreover, a time sweep has been used to monitor the enzyme performance in terms of gelation induced by crosslinking with transglutaminase in plant-based emulsion gel [82,100].

A large amplitude oscillatory shear (LAOS) test has been used to investigate the rheological behavior while the equilibrium is lost and the intermolecular bonds are (partly) broken down. LAOS can provide more information about the structure evolution under large deformations (e.g., 1–1000%), which is meaningful in industrial applications.

Several research works on plant-based emulsion gel used the LAOS test to gain insight into the texture and structure properties, in which crossover strain, phase angle, Lissajous curves analysis, Fourier-transform, and Chebyshev coefficients analysis were further revealed [92,95,101]. The utilization of the LAOS test is less reported in the current literature. However, this technique has caught considerable interest in recent years as a valuable tool for the characterization of the viscoelastic behavior of plant-based emulsion gel [92,95,101].

4.3. Texture

The texture measurement is important for the sensory experience and overall quality of a plant-based emulsion gel. It can help optimize the formulation and processing of the material to achieve desirable textural characteristics. As reported, texture profile analysis (TPA) is usually performed by a texture analyzer. The penetration test or double-compression tests were usually used to characterize the gel strength of the material. For a penetration test, a sample was penetrated axially at a certain constant speed by a probe, and a force vs. distance plot was obtained. The firmness (FF) was then determined as the initial slope of the penetration profiles [40,82,97,99]. An example is provided in Figure 5, which utilizes the penetration test to study and compare the emulsion gels with different compositions stored under specific conditions for different periods. The results demonstrate significant differences between the various emulsion gel samples, indicating variations in their textures.

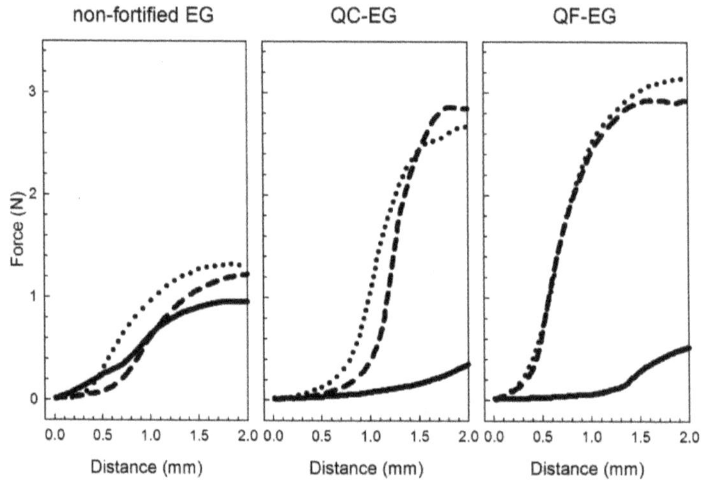

Figure 5. Force vs. distance plots obtained from penetration tests of non-fortified (**left**), QC-fortified (**middle**), and QF-fortified (**right**) soft emulsion gels (EGs) stored at 4 °C for 1, 14, and 28 days (solid, middle dashed, and dotted line styles, respectively) (QF: quinoa flour; QC: quinoa concentrate). Sodium alginate, sodium citrate, $CaCl_2$, and GDL in EG systems used were 10 g/kg, 100 mmol/L, 75 mmol/L, and 30 g/kg, respectively. EGs were prepared by high-speed homogenization (25,000 rpm for 1 min) of the aqueous dispersion and olive oil phases in a mass ratio of 3 g:1 g, respectively. Adapted from Ref. [97] with permission from Elsevier.

For a double-compression test, the samples were subjected to two consecutive deformation cycles over a predetermined distance at a certain constant speed by a probe. A force vs. time plot was obtained, from which parameters such as hardness (N), springiness (mm), gumminess (N), chewiness (N), cohesiveness, viscidity (J), and stiffness (N) [42,93,97,103] can be determined. Additionally, these original parameter data can be normalized to one-dimensional data based on Minkowski distance. Zhang et al. [102] have normalized the three-dimensional data (hardness, springiness, and cohesiveness) into one-dimensional data, which is defined as a comprehensive property index (CPI). This CPI reflected the overall texture properties of the emulsion gel material and was used for further discussion in the relationship between textures and gelation induction methods of emulsion gel.

4.4. Microstructure

4.4.1. Droplet Size Distribution

Droplet size distribution was determined to understand the size distribution of the dispersed phase in plant-based emulsion gels. The droplet size distribution has an impact on the stability, rheology, and sensory properties of the materials [100]. Several techniques were used to measure the droplet size of plant-based emulsion gel, such as laser diffraction [91,94,100,104,105], dynamic light scattering (DLS) [82,92,98], and microscopy (discussed in Section 4.4.2). Additionally, DLS can also be used for the size measurement of protein aggregates. Wang et al. [98] conducted research in which they used DLS to measure the size of soy protein aggregates in a soy oil emulsion.

4.4.2. Microscopy

Microscopy techniques were used to visually inspect the morphological properties of plant-based emulsion gel on a micro level. Some frequently used techniques are optical microscopy (OM), polarized light microscopy (PLM), confocal laser scanning microscopy (CLSM), scanning electron microscopy (SEM), and scanning electron cryo-microscopy (Cryo-SEM).

Optical microscopy (OM), also known as light microscopy, is a technique that has been used to investigate the microstructure of emulsion gel samples. With this technique, it is possible to visualize the size and organization of fat droplets [89,104,107]. A schematic principle of a classical optical microscope is shown in Figure 5 [106]. The polarized light microscope (PLM) was used in some research on plant-based emulsion gel [89,92,105]. This technique can provide additional information about the sample's structure and composition through the use of polarized light, which is affected in specific ways when passing through specific materials, such as crystals, fibers, and starch. Examples of the OM image and the PLM image are shown in Figure 6, from which a styrene-in-water emulsion stabilized by amorphous cellulose was observed. The OM image (a, c, and d) can provide information about the styrene droplet size, shape, and organization. The PLM image (b) can provide information about the cellulose behavior in the system.

Confocal laser scanning microscopy (CLSM) uses a focused laser beam to illuminate a sample, and only the in-focus light emitted from the sample is collected, resulting in 3D images with a high resolution [108]. According to reported research, the lipid phase was usually stained by Nile blue, and proteins were usually stained by Nile red or rhodamine B. The use of CLSM in studies on plant-based emulsion gel can reveal information about lipid droplets and protein behavior. For example, it can determine the lipid droplet size and distribution, the existence of aggregation, and the affinity of proteins for the interface. This information can provide insights into the structure, strength, and unification of the emulsion gel [75,93,97–99]. An example of a CLSM image can be found in Figure 2.

Scanning electron microscopy (SEM) uses a focused electron beam to scan the sample surface, producing high-resolution images by detecting secondary electrons emitted from the sample. This technique can provide more detailed information about the network structure of emulsion gels, such as pore size and distribution, wall thickness, and oil droplet distribution [40,95,102]. Freeze drying is a common method for sample dehydration. However, dehydration by ethanol was also reported in the research by Wang et al. [98]. Scanning electron cryo-microscopy (Cryo-SEM) is another similar microscopy technique in which a pre-freezing by liquid nitrogen is performed to maintain the structure of emulsion gels. The frozen samples are maintained at very low temperatures (e.g., −70 °C) while imaging [95]. An example of SEM images and Cryo-SEM images is shown in Figure 7. For the same sample, the image from SEM was blurred due to some alterations in their morphologies, resulting from the high oil content. While from Cryo-SEM, the original structures were well maintained.

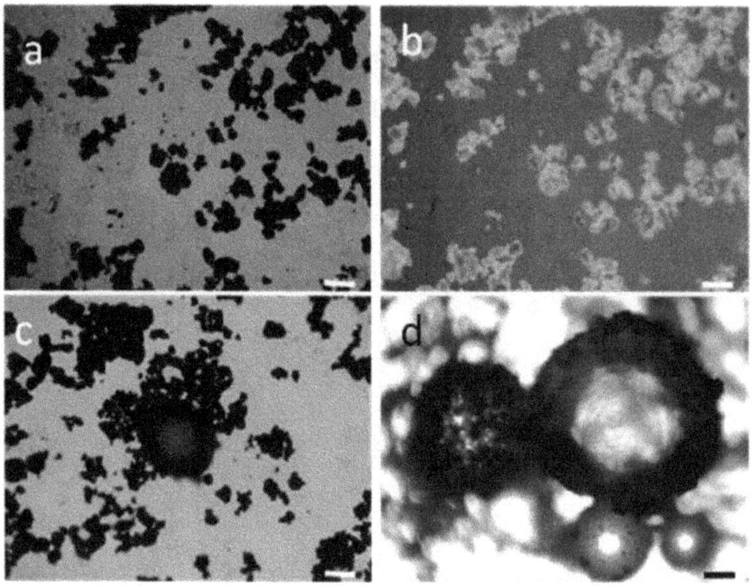

Figure 6. Typical bright-field optical (**a**,**c**,**d**) and polarized optical micrographs (**b**) of air-dried amorphous cellulose (**a**,**b**), and styrene Pickering emulsion (**c**,**d**) stabilized by amorphous cellulose and polymerized using AIBN (azobisisobutyronitrile) as initiator. The scare bar is 100 mm. Adapted from Ref. [89] with permission from Elsevier.

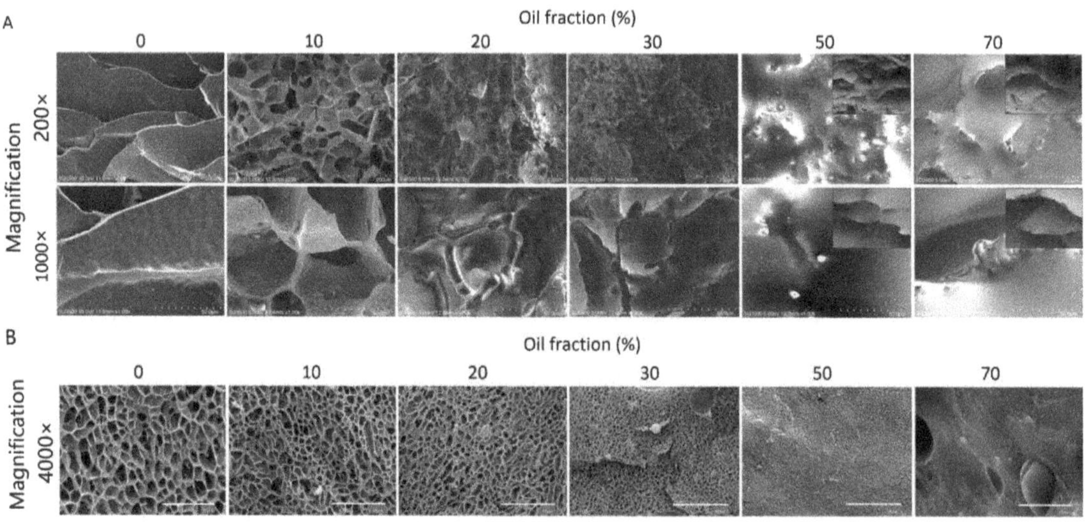

Figure 7. (**A**) SEM micrographs of the emulsion gels with 0–70% oil fractions and the corresponding Cryo-SEM micrographs of the emulsion gels with 50% and 70% oil fractions (located in the upper right corner of the SEM micrographs): The blue arrows in the 10% oil sample indicate the oil droplets; the Cryo-SEM micrographs of 50% and 70% oil emulsion gels were taken because these samples were blurred under SEM observation due to the high content of oil. (**B**) Cryo-SEM micrographs of the emulsion gels with 0–70% oil fractions. Adapted from Ref. [95] with permission from Elsevier.

4.5. Stability

The stability study of plant-based emulsion gels is important to maintain their characteristics over time. This study can provide information on the factors that affect gel stability, which can be used to optimize the formulation and processing of the materials. The stability of the plant-based emulsion gel can be reflected in multiple aspects, such as the properties mentioned previously on appearance, rheological properties, texture, and microstructure (Sections 4.1–4.4). Thermal properties and the zeta potential were measured to quantitatively characterize the gel stability, while freeze-thaw cycling tests and centrifugation tests were used for stability study in the reported study.

4.5.1. Thermal Properties

The thermal stability of plant-based emulsion gel has been reported as having important implications for sensory and physical properties, as well as processing and storage characteristics. Differential scanning calorimetry (DSC) is a commonly used technique to characterize the thermal properties of plant-based emulsion gel, from which the amount of heat required to increase the temperature of a sample is measured as a function of temperature. Thermal information such as the on-set and maximum temperatures of phase transition, the temperature for protein denaturation, and their respective enthalpies can be obtained from the DSC curves. Research conducted by Liu et al. [92] has used this technique to monitor rapeseed oil solidification in emulsion gel formation (Figure 8). The behavior of the intermolecular interaction, small aggregate disruption, and protein denaturation were also investigated in the reported research on plant-based emulsion gel [75,82]. Thermogravimetric analysis (TGA) is another common method to characterize the thermal properties of a plant-based emulsion gel, in which the mass of a sample is measured over time as the temperature changes. Thermal information, such as the thermal degradation temperature of emulsion gels, can be obtained to indicate the gel's thermal stability [42].

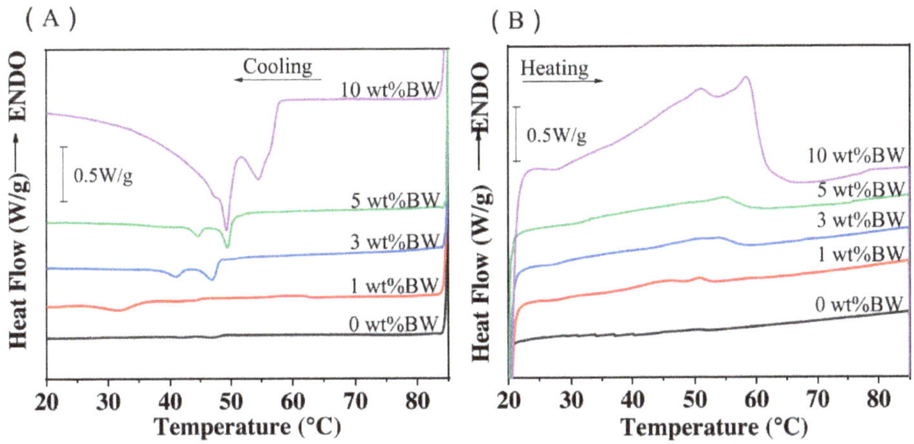

Figure 8. DSC thermographs of low-oil emulsion gels with 0, 1, 3, 5, and 10% (*w/w*) Beeswax during the cooling (**A**) cycle at −5 °C/min and the heating (**B**) cycle at 5 °C/min. Adapted from Ref. [92] with permission from Elsevier.

4.5.2. Zeta Potential

The use of zeta potential measurement by an electrophoresis instrument in the characterization of plant-based emulsion gel stability has been reported. The zeta potential is the electrostatic potential at the slipping plane, where the continuous phase (typically aqueous) begins to flow at a small distance from the droplet surface. Zeta potential measurements can indicate the interaction between droplets. Generally, a larger absolute value of zeta potential reflects a more stable emulsion system. This value can be influenced by the oil-weight fractions (Figure 9), droplet size, the amount of attached protein on the interface, and the denaturation and unfolding of proteins as reported in [82,100].

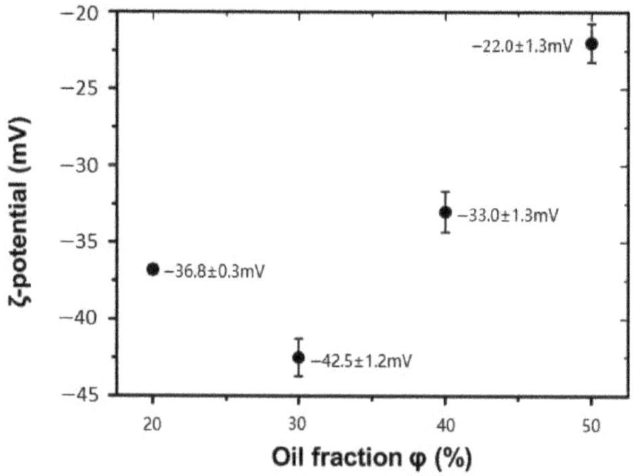

Figure 9. Zeta potential of PPI-stabilized emulsions with different oil-weight fractions (φ). Adapted from Ref. [100] with permission from John Wiley and Sons.

4.5.3. Water Holding Capacity (WHC)

The pores in the gel network can provide space for additional water. WHC measurement can provide insights into the microstructure and functional properties of emulsion gels [42,75,82,98,100]. It is generally accepted that a higher value of WHC implies the formation of a gel network with higher strength and a more uniform structure, resulting in a stronger ability to retain water molecules.

4.5.4. Freeze-Thaw Stability

Freeze-thaw cycling was reported as a method to assess the stability of plant-based emulsion gel under temperature stress [14,90,103]. Briefly, this method involves subjecting the emulsion gel to repeated freezing and thawing cycles, with the fluid loss being measured after each cycle. Freezing destabilizes an emulsion by promoting flocculation and coalescence as water and lipid crystalize [109]. The structural change is then made apparent as the gel is thawed, where phase separation will be observed. The formulation, including the type of oil used, greatly influences freeze-thaw loss and the emulsifier used [109].

5. Potential Applications of Plant-Based Emulsion Gels in the Food Industry

The prevalence of emulsion-based foods in the everyday diet has created immense opportunities for research and development in replacing current animal-based foods with plant-based mimics. Even though annual sales of plant-based food had grown by 54% from US$4.8 billion to US$7.4 billion in 2021, plant-based milk remains the largest sector of sales [110]. Although other plant-based dairy products make up the second largest sector, the contribution of other key areas, such as plant-based meat and egg alternatives, remained minor [110]. This may be due to the challenges involved in creating plant-based alternatives that can effectively replicate the flavor, texture, and mouthfeel of animal-based food products such as meat and dairy products. Therefore, emulsion gels, in whole or as an ingredient, were thought to be able to improve some of these limitations. A summary of selected studies on possible plant-based emulsion gel application is presented in Table 2.

Table 2. Summary of studies in food application of plant-based emulsion gel.

Target Food	Formulation	Summary	Ref.
Yogurt (Dairy)	Lentil protein isolate (LPI), sunflower oil, Yoflex® Acidifix™, sucrose	Similar cohesiveness and viscosity but higher firmness and consistency than dairy yogurt. The fermented gel shows pale pink color. Low FODMAP content, suitable for irritable bowel syndrome.	[111]
	PPI, canola oil, GDL, Custom starter culture (MEGAN, VEGAN, ExECO)	VEGAN strain produced yogurt less associated with "cut herb" and "woody" and better associated with "coffee" and "smoked". Reduction in volatiles associated with "grassy" odor.	[112]
	Potato protein isolate, sunflower oil, glucose, starter culture	High-pressure homogenization created highly stable emulsions at 1.5–10% oil. The whiteness of emulsion gel may be increased through higher pressure and oil content. Possible application in low-fat or Greek-style yogurt.	[76]
Cheese (Dairy)	Yellow pea/fava bean protein, canola oil, carrageenans (κ-, ι-) and xanthan gum, nutritional yeast, calcium sulfate	17.5% boiled fava bean flour and 1% κ- carrageenan were found to have indifferent springiness and chewiness to Gouda cheese but were also harder and less cohesive. Boiled yellow pea flour has the closest color resemblance to Gouda cheese.	[113]
	Zein, starches (corn, tapioca), sunflower oil	30% protein sample showed similar extensibility to the textural properties (hardness, chewiness, and gumminess) of cheddar cheese. Reduction in texture parameters at 50 °C is seen in both zein and cheddar cheese.	[114]
	Pea protein, pea fiber, potato fiber, sunflower oil/coconut fat, shea stearin, lactic acid, salt	Pea protein slurry with 15% emulsion gel showed a similar spreadability index to dairy cheese. Adding oregano and rosemary essential oils was able to reduce the perception of the grassy odor	[96]
Pork fat (Animal fat analog	SPI, fully hydrogenated canola oil, canola oil, transglutaminase oil	Softer than pork fat tissue. Thermal properties of animal fat may be achieved by blending solid plant fat with oil.	[115]
	SPI, KGM, coconut oil, transglutaminase	Similar hardness to pork fat (5% SPI, 4% KGM, and 10% oil (*w*/*w*)) Similar in color space to L* and b*	[50]
	DKG, MC, canola/coconut oil	MC and DKG complement each other in thermal properties at different temperatures. At 80 °C, coconut oil emulsion gel with a high MC better emulates the textural properties of pork fat.	[38]
Pork/beef fat (Animal fat analog)	Lecithin, potato starch (PS), inulin, soybean oil/coconut oil	At 12.8% (*w*/*w*) PS, soybean oil emulsion gel meltability is similar to that of pork fat. At the same PS content, coconut oil emulsion gel was similar to beef fat. Soybean/coconut oil emulsion gel with different PS contents may have a similar hardness to pork fat. Emulsion gels were consistently softer than beef fat.	[41]
Butter (Fat replacer)	Extra virgin olive oil, inulin, soy, lecithin	Harder product compared to control. Increasing emulsion gel content decreased spread during baking. Emulsion gel lacks plasticity and is therefore unable to create a porous structure.	[116]
Mayonnaise (Egg yolk)	Sunflower oil, chickpea protein/fava bean protein/yellow split lentils protein, xanthan gum, vinegar, sugar, salt, mustard powder	Xanthan gum is required to increase the viscosity and stability of the mayonnaise analog. Chickpea protein mayonnaise is indifferent to egg mayonnaise control. (Extrudability, compression texture analysis, color, and sensory analysis).	[117]
	Citrus fruit fiber, corn peptides, sunflower oil	HIPE has been characterized by above 90% thixotropic recovery. High heat and freeze-thaw stability are seen in HIPE. The creaminess and thickness of HIPE may be altered by fiber content.	[118]

Table 2. *Cont.*

Target Food	Formulation	Summary	Ref.
Functional food	β-Carotene, zein, glycerol, and corn oil	β-Carotene loaded in the oil phase showed higher retention after UV treatment than in the aqueous phase. Additional β-carotene as an antioxidant in glycerol further increases retention in the oil phase.	[93]
	Rhamnogalacturonan-I enriched pectin, soybean oil, Tween 20, curcumin	Active filler pectin gel showed high thermal stability. The release of curcumin may be modified through gel structure, but no difference in bioavailability is seen.	[61]
Gummy candy (Functional foods)	Vitamin B_{12} and D_3, gum Arabic, inulin, flaxseed oil, pectin	No reduction in vitamin activity after 30 days of storage. No unpleasant taste was noted by panelists after the addition of the strawberry aroma.	[59]

5.1. Dairy Alternatives

Emulsion gels formed by dairy proteins such as whey and casein were typically described as model emulsion gels [5]. However, the key textural characteristics of casein gels, such as those seen in yogurt and cheese, have been difficult to replicate by plant protein alone [18]. This is due to differences in plant protein structure, namely the lack of random coils and phosphate groups that could be linked by the presence of calcium ions [18]. Thus, to create convincing analogs of dairy products, emulsion gels of various formulations have been proposed.

The development of plant-based yogurt has seen some success in emulating dairy yogurt in both research and commercial products. [26,111,112,119]. However, to design a successful plant-based yogurt-like product, further insights may be drawn from the literature to determine critical factors in formulation and processing. Several studies have indicated that the sensory attributes (taste, aroma, texture, and appearance) of plant-based emulsion gel for yogurt purposes depend on the starter culture used [112,120]. In a previous study for pea protein fermentation, Ben-Harb et al. [112] used a microbial consortium design and found a reduction in undesirable odors from the use of legume-proteins after fermentation, as "smoked" and "coffee" notes were better identified by panelists than "cut grass" notes. Aside from full-fat yogurt, HPH treatment of plant protein yogurt may be used to create plant-based low-fat and Greek-style yogurt [76]. Rheological and textural characteristics were shown to be tunable at different pressures and oil contents without changing the formulation [76]. Thus, as studies into interactions between yogurt components are further advanced, a successful yogurt-like product may be created using plant-based emulsion gel.

On the other hand, emulsion gel was also employed to create plant-based cheese analogs of various types, including soft, semi-hard, and spreadable cheeses [96,113,114,121,122]. In this regard, several important properties of plant-based cheese analogs were identified by Grossmann and McClements [121], such as texture, meltability, shreddability, and aroma. The recreation of texture in plant-based cheese analog broadly follows other similar applications that require texturization. Ferawati et al. [113] trailed the creation of a semi-hard cheese analog using various pulse proteins, canola oil, and κ-carrageenan. In their experiments, fava bean flour was most suitable to mimic the texture of Gouda cheese with comparable characteristics that may be further improved with optimization of processing conditions and protein ratios [113]. In terms of meltability, recent advances in zein protein-based cheese formulation with xanthan gum and starches (tapioca and corn) have shown promising results in creating cheese analogs with similar characteristics to Cheddar cheese [114]. Zein protein gel was shown to better trace cheddar cheese than PPI-based analog, gluten-based analog, and commercial plant-based cheese at 30% (*w/w*) protein content [114]. Nevertheless, as pointed out by others, a detailed sensory analysis would be required to provide a holistic analysis of the formulation's suitability as an equivalent of a given cheese type [123]. This may be especially true as plant-based cheese analogs were only able to achieve comparable meltability at high protein and low oil content, which may impact the mouthfeel of a product.

5.2. Meat Alternatives

Emulsion gel is often designed as an "animal fat analog" (AFA) to mimic animal fat tissue. Emulsion gel-based AFA has seen use as the primary fat component in alternative meat products or as fat replacers in animal meat products to improve the nutrition profile of these products [56]. In the current commercial plant-based meat products, the fat component is often represented using unstructured plant fat, such as coconut oil or cocoa butter [124]. Although unstructured fat may be present as fat marbling in a raw and chilled state, naked plant fat often lacks texture, especially after cooking [115]. This is further compounded by possible processing difficulties during extrusion and loss of appeal due to smearing if mishandled [125,126]. Hence, emulsion gel was proposed as a potential solution.

Emulsion gels for AFA had been created using various gelation methods in both the protein-only gel and mixed gel forms. For this application, qualities such as appearance, texture, and thermal performance are parameters that were widely measured. Dreher et al. [115] created a soy-based emulsion gel crosslinked by transglutaminase with a plant fat and oil blend. Although the hardness observed was lower compared to pork fat tissue, the introduction of solid fat content enabled tunable characteristics in crosslinked emulsion that may imitate the melting characteristic of fat in animal tissue [115]. A transglutaminase crosslinked soy protein and KGM gel using coconut oil were proposed by Huang et al. [50]. Emulsion gel at 5% SPI (*w*/*w*), 4% KGM (*w*/*w*), and 10% (*w*/*w*) oil was found to have similar hardness and springiness values to pork back fat. The colorimetry of emulsion gel samples was also found to be able to mimic pork back fat in terms of lightness (L^*) and blue/yellow (b^*) [50]. Both characteristics demonstrated promising results in recreating animal fat using plant-based emulsion gel. Furthermore, the stability of emulsion gel during and after cooking was demonstrated by several studies [38,41,50]. Thermal irreversibility and cooking profiles may be modified by the selection of polysaccharides or the introduction of additives [38,41]. However, a good balance between each characteristic, including additional parameters such as health and palatability, is yet to be found at present.

5.3. Egg Yolk Alternatives and Baked Goods

Egg yolk protein has an extensive role in food as an emulsifier and thickener in condiments and confectionaries [127]. Notably, in mayonnaise analogs, statistically indifferent textural, extrusion, and sensory properties were seen in chickpea protein stabilized mayonnaise compared to egg-based mayonnaise at 70% oil content [117]. In other formulations, particularly in HIPE, it was shown that the textural and rheological characteristics of gels are tunable based on biopolymer content [118,128]. Such a characteristic allows for the formulation of low-fat mayonnaise, which maintains similar textural characteristics.

Similarly, an emulsion gel-based fat replacer was also used in baked goods to deliver a better nutritional profile. Inulin and extra virgin olive oil emulsion gel were used as a partial replacement for butter in shortbreads [116]. However, the lower plasticity of emulsion gels compared to butter was deemed problematic as it was not able to entrap gas released by the leavening agent. Along with a lower spread during heating and limited gluten formation, a harder product was obtained [116]. The observation revealed that additional processing may be required for emulsion gel, such as the incorporation of air within the gel matrix, for better uses as a fat replacement in baked goods, especially for short doughs.

5.4. Functional Foods

As previously indicated, emulsion gels may be used as a controlled-release regimen to protect and deliver nutrients or other active ingredients into the human body. Emulsion gels have been proven to protect active ingredients against ultraviolet light, free radicals, and temperature [61,62,93]. Both the lipid and aqueous phases could act as carriers of ingredients. Ghialdi et al. [59] developed an emulsion gel-based confectionary using inulin, pectin, and gum Arabic to protect and deliver vitamin B_{12} and D_3 in their aqueous and lipid phases, respectively. No difference in active vitamin was found between the day of fabrication and after 30 days of storage, indicating that the gel was effective in limiting light and oxygen exposure. Moreover, the panelists showed a positive reception, with no unpleasant taste being noticed after the strawberry flavoring was added to the candies [59].

5.5. Consumer Acceptance and Sensory Properties of Plant-Based Emulsion Gels in Food

Consumer acceptance is one of the challenges hindering the mass adoption of plant-based emulsion gel in food. One major issue with plant-based products named by consumers lies with the heavy use of legume proteins, giving the product an undesirable "grassy" or "beany" odor and taste [129]. The addition of volatile additives enhanced consumer acceptance of plant-based emulsion gel. In spreadable plant-based cheese based on pea protein, inulin, and olive oil, the incorporation of essential oils leads to better odor perception in the sensory analysis [96]. In addition, fermentation may also reduce "beany" perception to consumers [112,130]. Volatile analysis of fermented pea gel revealed that volatiles responsible for "grassy" and "earthy" odors are substantially reduced after inoculation and fermentation [130].

Despite the structural importance that polysaccharides provide to emulsion gels, the incorporation of polysaccharides may lead to adverse sensory properties. Insights may be drawn from primarily plant-based fat substitutes developed to improve the fatty acid profile in deli meats. Incorporating KGM emulsion gel into chorizo sausage led to a loss in the perception of juiciness and firmness, with an increasing degree at higher KGM content [131]. The lower perceived juiciness was attributed to KGM's high WHC as less fluid was released from chewing. Similar results were seen across various plant-based polysaccharides and in mixed gels [50,132]. In a similar vein, oil release and mouthfeel of emulsion gel are also ingredient dependent. For instance, Hu et al. [133] demonstrated that at 5% oil content, the oiliness perception of modified starch/gellan gum gel was higher than that of whey protein/gellan gum gel at 20% oil content. The authors attribute this observation to the onset of enzymatic digestion of starch during mastication [133]. Therefore, consumer perception is shown to be formulation dependent on the types of biopolymers used and their respective ratios. To achieve desired product characteristics, plant-based emulsion gel requires extensive optimization.

6. Conclusions

The recent boost in the popularity of plant-based foods has been the primary driver in the development and understanding of plant-based emulsion gel as an ingredient or an entire food matrix. Plant-based ingredients are often perceived as healthier, more sustainable, and more environmentally friendly than food from animal sources. Plant proteins, polysaccharides, and lipids were shown in research to function synergistically in various configurations to deliver desirable qualities for a food product. As evidenced in this review, altering the types and amounts of protein, oil, and polysaccharide could change the emulsion gels' rheology, texture, and thermal performance. Processing techniques, such as high-pressure homogenization and ultrasonication, have widened the scope for plant-based emulsion gel design. Essential qualities such as droplet size, stability, and bioavailability in functional emulsion gels could be substantially improved because of the choice of treatment. On the other hand, characterization techniques have allowed researchers to quantify and evaluate the properties of plant-based emulsion gel in detail. These techniques helped elucidate the properties of plant-based components (e.g., microscopy and DSC) and, critically, their applicability as a food analog (e.g., TPA).

Potential applications and possible future research were described in this study. Research has indicated that shortcomings in using plant-based ingredients may be improved using additives. Some food types, such as yogurt and mayonnaise, have been more successful in overcoming the challenges of using plant-based ingredients compared to other types of food. The development of functional foods using emulsion gel was also shown to be feasible in principle. Given the prevalence of emulsions and analogous structures in food, it is believed that creating a plant-based emulsion gel that could closely mimic the physical properties of their designated animal-based counterparts will be instrumental for the success of a plant-based alternative food product.

Author Contributions: Conceptualization, C.C.-Y.Y., Y.W. and C.S.; data curation, C.C.-Y.Y., S.W.L., K.M. and W.K.; formal analysis, C.C.-Y.Y., S.W.L. and K.M.; investigation, C.C.-Y.Y., S.W.L., K.M., W.K. and Y.W.; visualization, C.C.-Y.Y., S.W.L. and K.M.; methodology, C.C.-Y.Y., S.W.L., K.M. and Y.W.; writing—original draft, C.C.-Y.Y., S.W.L., K.M. and Y.W.; writing—review & editing, C.C.-Y.Y., W.K., Y.W. and C.S.; project administration, Y.W.; resources, C.S. supervision, Y.W. and C.S.; funding acquisition, C.S. All authors have read and agreed to the published version of the manuscript.

Funding: This research received no external funding.

Institutional Review Board Statement: Not applicable.

Informed Consent Statement: Not applicable.

Data Availability Statement: Not applicable.

Conflicts of Interest: The authors declare no conflict of interest.

References

1. Farjami, T.; Madadlou, A. An overview on preparation of emulsion-filled gels and emulsion particulate gels. *Trends Food Sci. Technol.* **2019**, *86*, 85–94. [CrossRef]
2. Guo, J.; Cui, L.; Meng, Z. Oleogels/emulsion gels as novel saturated fat replacers in meat products: A review. *Food Hydrocoll.* **2023**, *137*, 108313. [CrossRef]
3. Domínguez, R.; Munekata, P.E.S.; Pateiro, M.; López-Fernández, O.; Lorenzo, J.M. Immobilization of oils using hydrogels as strategy to replace animal fats and improve the healthiness of meat products. *Curr. Opin. Food Sci.* **2021**, *37*, 135–144. [CrossRef]
4. Wang, Y.; Selomulya, C. Food rheology applications of large amplitude oscillation shear (LAOS). *Trends Food Sci. Technol.* **2022**, *127*, 221–244. [CrossRef]
5. Dickinson, E. Emulsion gels: The structuring of soft solids with protein-stabilized oil droplets. *Food Hydrocoll.* **2012**, *28*, 224–241. [CrossRef]
6. Tan, C.; McClements, D.J. Application of Advanced Emulsion Technology in the Food Industry: A Review and Critical Evaluation. *Foods* **2021**, *10*, 812. [CrossRef]
7. Yu, J.; Wang, X.Y.; Li, D.; Wang, L.J.; Wang, Y. Development of soy protein isolate emulsion gels as extrusion-based 3D food printing inks: Effect of polysaccharides incorporation. *Food Hydrocoll.* **2022**, *131*, 107824. [CrossRef]
8. Zhao, X.; Li, D.; Wang, L.J.; Wang, Y. Role of gelation temperature in rheological behavior and microstructure of high elastic starch-based emulsion-filled gel. *Food Hydrocoll.* **2023**, *135*, 108208. [CrossRef]
9. Mao, L.; Lu, Y.; Cui, M.; Miao, S.; Gao, Y. Design of gel structures in water and oil phases for improved delivery of bioactive food ingredients. *Crit. Rev. Food. Sci. Nutr.* **2020**, *60*, 1651–1666. [CrossRef]
10. Lu, Y.; Mao, L.; Hou, Z.; Miao, S.; Gao, Y. Development of Emulsion Gels for the Delivery of Functional Food Ingredients: From Structure to Functionality. *Food Eng. Rev.* **2019**, *11*, 245–258. [CrossRef]
11. Poore, J.; Nemecek, T. Reducing food's environmental impacts through producers and consumers. *Science* **2018**, *360*, 987–992. [CrossRef] [PubMed]
12. Nikbakht Nasrabadi, M.; Sedaghat Doost, A.; Mezzenga, R. Modification approaches of plant-based proteins to improve their techno-functionality and use in food products. *Food Hydrocoll.* **2021**, *118*, 106789. [CrossRef]
13. Gao, K.; Rao, J.; Chen, B. Unraveling the mechanism by which high intensity ultrasound improves the solubility of commercial pea protein isolates. *Food Hydrocoll.* **2022**, *131*, 107823. [CrossRef]
14. Yu, J.; Wang, Y.; Li, D.; Wang, L.-J. Freeze-thaw stability and rheological properties of soy protein isolate emulsion gels induced by NaCl. *Food Hydrocoll.* **2022**, *123*, 107113. [CrossRef]
15. Lin, D.; Kelly, A.L.; Miao, S. Preparation, structure-property relationships and applications of different emulsion gels: Bulk emulsion gels, emulsion gel particles, and fluid emulsion gels. *Trends Food Sci. Technol.* **2020**, *102*, 123–137. [CrossRef]

16. Zhu, Y.; Gao, H.; Liu, W.; Zou, L.; McClements, D.J. A review of the rheological properties of dilute and concentrated food emulsions. *J. Texture. Stud.* **2020**, *51*, 45–55. [CrossRef]
17. Ma, K.K.; Greis, M.; Lu, J.; Nolden, A.A.; McClements, D.J.; Kinchla, A.J. Functional Performance of Plant Proteins. *Foods* **2022**, *11*, 594. [CrossRef]
18. McClements, D.J.; Grossmann, L. The science of plant-based foods: Constructing next-generation meat, fish, milk, and egg analogs. *Compr. Rev. Food Sci. Food Saf.* **2021**, *20*, 4049–4100. [CrossRef]
19. Alvarez-Sabatel, S.; Marañón, I.M.D.; Arboleya, J.-C. Impact of high pressure homogenisation (HPH) on inulin gelling properties, stability and development during storage. *Food Hydrocoll.* **2015**, *44*, 333–344. [CrossRef]
20. Clark, A.H.; Kavanagh, G.M.; Ross-Murphy, S.B. Globular protein gelation—Theory and experiment. *Food Hydrocoll.* **2001**, *15*, 383–400. [CrossRef]
21. Shand, P.J.; Ya, H.; Pietrasik, Z.; Wanasundara, P.K.J.P.D. Physicochemical and textural properties of heat-induced pea protein isolate gels. *Food Chem.* **2007**, *102*, 1119–1130. [CrossRef]
22. Ferreira, L.S.; Brito-Oliveira, T.C.; Pinho, S.C. Emulsion-filled gels of soy protein isolate for vehiculation of vitamin D3: Effect of protein solubility on their mechanical and rheological characteristics. *Food Biosci.* **2022**, *45*, 101455. [CrossRef]
23. Kim, K.H.; Renkema, J.M.S.; van Vliet, T. Rheological properties of soybean protein isolate gels containing emulsion droplets. *Food Hydrocoll.* **2001**, *15*, 295–302. [CrossRef]
24. Ben-Harb, S.; Panouillé, M.; Huc-Mathis, D.; Moulin, G.; Saint-Eve, A.; Irlinger, F.; Bonnarme, P.; Michon, C.; Souchon, I. The rheological and microstructural properties of pea, milk, mixed pea/milk gels and gelled emulsions designed by thermal, acid, and enzyme treatments. *Food Hydrocoll.* **2018**, *77*, 75–84. [CrossRef]
25. Zha, F.; Rao, J.; Chen, B. Plant-based food hydrogels: Constitutive characteristics, formation, and modulation. *Curr. Opin. Colloid Interface Sci.* **2021**, *56*, 101505. [CrossRef]
26. Klost, M.; Drusch, S. Structure formation and rheological properties of pea protein-based gels. *Food Hydrocoll.* **2019**, *94*, 622–630. [CrossRef]
27. Wang, X.; Luo, K.; Liu, S.; Zeng, M.; Adhikari, B.; He, Z.; Chen, J. Textural and Rheological Properties of Soy Protein Isolate Tofu-Type Emulsion Gels: Influence of Soybean Variety and Coagulant Type. *Food Biophys.* **2018**, *13*, 324–332. [CrossRef]
28. Wang, X.; Luo, K.; Liu, S.; Adhikari, B.; Chen, J. Improvement of gelation properties of soy protein isolate emulsion induced by calcium cooperated with magnesium. *J. Food Eng.* **2019**, *244*, 32–39. [CrossRef]
29. Glusac, J.; Davidesko-Vardi, I.; Isaschar-Ovdat, S.; Kukavica, B.; Fishman, A. Gel-like emulsions stabilized by tyrosinase-crosslinked potato and zein proteins. *Food Hydrocoll.* **2018**, *82*, 53–63. [CrossRef]
30. Drusch, S.; Klost, M.; Kieserling, H. Current knowledge on the interfacial behaviour limits our understanding of plant protein functionality in emulsions. *Curr. Opin. Colloid Interface Sci.* **2021**, *56*, 101503. [CrossRef]
31. McClements, D.J.; Gumus, C.E. Natural emulsifiers—Biosurfactants, phospholipids, biopolymers, and colloidal particles: Molecular and physicochemical basis of functional performance. *Adv. Colloid Interface Sci.* **2016**, *234*, 3–26. [CrossRef] [PubMed]
32. Keerati-u-rai, M.; Corredig, M. Heat-induced changes in oil-in-water emulsions stabilized with soy protein isolate. *Food Hydrocoll.* **2009**, *23*, 2141–2148. [CrossRef]
33. Peng, W.; Kong, X.; Chen, Y.; Zhang, C.; Yang, Y.; Hua, Y. Effects of heat treatment on the emulsifying properties of pea proteins. *Food Hydrocoll.* **2016**, *52*, 301–310. [CrossRef]
34. Kim, W.; Wang, Y.; Selomulya, C. Dairy and plant proteins as natural food emulsifiers. *Trends Food Sci. Technol.* **2020**, *105*, 261–272. [CrossRef]
35. Amine, C.; Dreher, J.; Helgason, T.; Tadros, T. Investigation of emulsifying properties and emulsion stability of plant and milk proteins using interfacial tension and interfacial elasticity. *Food Hydrocoll.* **2014**, *39*, 180–186. [CrossRef]
36. Gumus, C.E.; Decker, E.A.; McClements, D.J. Formation and Stability of ω-3 Oil Emulsion-Based Delivery Systems Using Plant Proteins as Emulsifiers: Lentil, Pea, and Faba Bean Proteins. *Food Biophys.* **2017**, *12*, 186–197. [CrossRef]
37. Pang, S.; Shao, P.; Sun, Q.; Pu, C.; Tang, W. Relationship between the emulsifying properties and formation time of rice bran protein fibrils. *LWT* **2020**, *122*, 108985. [CrossRef]
38. Jeong, H.; Lee, J.; Jo, Y.-J.; Choi, M.-J. Thermo-irreversible emulsion gels based on deacetylated konjac glucomannan and methylcellulose as animal fat analogs. *Food Hydrocoll.* **2023**, *137*, 108407. [CrossRef]
39. Xu, Q.; Qi, B.; Han, L.; Wang, D.; Zhang, S.; Jiang, L.; Xie, F.; Li, Y. Study on the gel properties, interactions, and pH stability of pea protein isolate emulsion gels as influenced by inulin. *LWT* **2021**, *137*, 110421. [CrossRef]
40. Zhang, Y.; Yang, H.; Zheng, H.; Yuan, D.; Mao, L. Physical properties and salt release of potato starch-based emulsion gels with OSA starch-stabilized oil droplets. *LWT* **2021**, *141*, 110929. [CrossRef]
41. Wen, Y.; Che, Q.T.; Kim, H.W.; Park, H.J. Potato starch altered the rheological, printing, and melting properties of 3D-printable fat analogs based on inulin emulsion-filled gels. *Carbohydr. Polym.* **2021**, *269*, 118285. [CrossRef] [PubMed]
42. Feng, L.; Jia, X.; Yan, J.; Yan, W.; Yin, L. Mechanical, thermal stability and microstructural properties of emulsion-filled gels: Effect of sugar beet pectin/soy protein isolate ratio. *LWT* **2021**, *141*, 110917. [CrossRef]
43. Lin, D.; Kelly, A.L.; Miao, S. The role of mixing sequence in structuring O/W emulsions and emulsion gels produced by electrostatic protein-polysaccharide interactions between soy protein isolate-coated droplets and alginate molecules. *Food Hydrocoll.* **2021**, *113*, 106537. [CrossRef]

44. Tavernier, I.; Patel, A.R.; Van der Meeren, P.; Dewettinck, K. Emulsion-templated liquid oil structuring with soy protein and soy protein: κ-carrageenan complexes. *Food Hydrocoll.* **2017**, *65*, 107–120. [CrossRef]
45. McClements, D.J. Recent progress in hydrogel delivery systems for improving nutraceutical bioavailability. *Food Hydrocoll.* **2017**, *68*, 238–245. [CrossRef]
46. Yu, J.; Li, D.; Wang, L.-J.; Wang, Y. Improving freeze-thaw stability and 3D printing performance of soy protein isolate emulsion gel inks by guar & xanthan gums. *Food Hydrocoll.* **2023**, *136*, 108293. [CrossRef]
47. Wang, Y.; Liu, J.; Li, Q.; Wang, Y.; Wang, C. Two natural glucomannan polymers, from Konjac and Bletilla, as bioactive materials for pharmaceutical applications. *Biotechnol. Lett.* **2015**, *37*, 1–8. [CrossRef] [PubMed]
48. Sala, G.; van Vliet, T.; Cohen Stuart, M.; van de Velde, F.; van Aken, G.A. Deformation and fracture of emulsion-filled gels: Effect of gelling agent concentration and oil droplet size. *Food Hydrocoll.* **2009**, *23*, 1853–1863. [CrossRef]
49. Li, R.; Xue, H.; Gao, B.; Liu, H.; Han, T.; Hu, X.; Tu, Y.; Zhao, Y. Physicochemical properties and digestibility of thermally induced ovalbumin–oil emulsion gels: Effect of interfacial film type and oil droplets size. *Food Hydrocoll.* **2022**, *131*, 107747. [CrossRef]
50. Huang, L.; Ren, Y.; Li, H.; Zhang, Q.; Wang, Y.; Cao, J.; Liu, X. Create Fat Substitute From Soybean Protein Isolate/Konjac Glucomannan: The Impact of the Protein and Polysaccharide Concentrations Formulations. *Front. Nutr.* **2022**, *9*, 843832. [CrossRef]
51. Jiang, Y.; Liu, L.; Wang, B.; Yang, X.; Chen, Z.; Zhong, Y.; Zhang, L.; Mao, Z.; Xu, H.; Sui, X. Polysaccharide-based edible emulsion gel stabilized by regenerated cellulose. *Food Hydrocoll.* **2019**, *91*, 232–237. [CrossRef]
52. Zhou, P.; Wen, L.; Ai, T.; Liang, H.; Li, J.; Li, B. A novel emulsion gel solely stabilized by the hot water extracted polysaccharide from psyllium husk: Self-healing plays a key role. *Food Hydrocoll.* **2022**, *130*, 107718. [CrossRef]
53. Martellet, M.C.; Majolo, F.; Ducati, R.G.; Volken de Souza, C.F.; Goettert, M.I. Probiotic applications associated with Psyllium fiber as prebiotics geared to a healthy intestinal microbiota: A review. *Nutrition* **2022**, *103–104*, 111772. [CrossRef]
54. Marangoni, A.G.; Acevedo, N.; Maleky, F.; Co, E.; Peyronel, F.; Mazzanti, G.; Quinn, B.; Pink, D. Structure and functionality of edible fats. *Soft Matter* **2012**, *8*, 1275–1300. [CrossRef]
55. de Oliveira Otto, M.C.; Mozaffarian, D.; Kromhout, D.; Bertoni, A.G.; Sibley, C.T.; Jacobs, D.R., Jr.; Nettleton, J.A. Dietary intake of saturated fat by food source and incident cardiovascular disease: The Multi-Ethnic Study of Atherosclerosis. *Am. J. Clin. Nutr.* **2012**, *96*, 397–404. [CrossRef] [PubMed]
56. Ren, Y.; Huang, L.; Zhang, Y.; Li, H.; Zhao, D.; Cao, J.; Liu, X. Application of Emulsion Gels as Fat Substitutes in Meat Products. *Foods* **2022**, *11*, 1950. [CrossRef]
57. Ribeiro, A.P.B.; Grimaldi, R.; Gioielli, L.A.; Gonçalves, L.A.G. Zero trans fats from soybean oil and fully hydrogenated soybean oil: Physico-chemical properties and food applications. *Food Res. Int.* **2009**, *42*, 401–410. [CrossRef]
58. Gu, X.; Campbell, L.J.; Euston, S.R. Effects of different oils on the properties of soy protein isolate emulsions and gels. *Food Res. Int.* **2009**, *42*, 925–932. [CrossRef]
59. Ghiraldi, M.; Franco, B.G.; Moraes, I.C.F.; Pinho, S.C. Emulsion-Filled Pectin Gels for Vehiculation of Vitamins D3 and B12: From Structuring to the Development of Enriched Vegan Gummy Candies. *ACS Food Sci. Technol.* **2021**, *1*, 1945–1952. [CrossRef]
60. Yi, J.; Gan, C.; Wen, Z.; Fan, Y.; Wu, X. Development of pea protein and high methoxyl pectin colloidal particles stabilized high internal phase pickering emulsions for β-carotene protection and delivery. *Food Hydrocoll.* **2021**, *113*, 106497. [CrossRef]
61. Zhang, L.; Zheng, J.; Wang, Y.; Ye, X.; Chen, S.; Pan, H.; Chen, J. Fabrication of rhamnogalacturonan-I enriched pectin-based emulsion gels for protection and sustained release of curcumin. *Food Hydrocoll.* **2022**, *128*, 107592. [CrossRef]
62. Lu, Y.; Mao, L.; Cui, M.; Yuan, F.; Gao, Y. Effect of the Solid Fat Content on Properties of Emulsion Gels and Stability of beta-Carotene. *J. Agric. Food Chem.* **2019**, *67*, 6466–6475. [CrossRef]
63. Himanath, G.; Shruthy, R.; Preetha, R.; Sreejit, V. Nanoemulsion with Coconut Oil and Soy Lecithin as a Stable Delivery System for Lycopene and Its Incorporation into Yogurt to Enhance Antioxidant Properties and Maintain Quality. *ACS Food Sci. Technol.* **2021**, *1*, 1538–1549. [CrossRef]
64. Jiang, Z.Q.; Wang, J.; Stoddard, F.; Salovaara, H.; Sontag-Strohm, T. Preparation and Characterization of Emulsion Gels from Whole Faba Bean Flour. *Foods* **2020**, *9*, 755. [CrossRef] [PubMed]
65. Kakran, M.; Antipina, M.N. Emulsion-based techniques for encapsulation in biomedicine, food and personal care. *Curr Opin Pharm.* **2014**, *18*, 47–55. [CrossRef] [PubMed]
66. Gao, H.; Ma, L.; Cheng, C.; Liu, J.; Liang, R.; Zou, L.; Liu, W.; McClements, D.J. Review of recent advances in the preparation, properties, and applications of high internal phase emulsions. *Trends Food Sci. Technol.* **2021**, *112*, 36–49. [CrossRef]
67. Wu, D.; Tu, M.; Wang, Z.; Wu, C.; Yu, C.; Battino, M.; El-Seedi, H.R.; Du, M. Biological and conventional food processing modifications on food proteins: Structure, functionality, and bioactivity. *Biotechnol Adv* **2020**, *40*, 107491. [CrossRef]
68. Ozturk, O.K.; Turasan, H. Applications of microfluidization in emulsion-based systems, nanoparticle formation, and beverages. *Trends Food Sci. Technol.* **2021**, *116*, 609–625. [CrossRef]
69. Aswathanarayan, J.B.; Vittal, R.R. Nanoemulsions and Their Potential Applications in Food Industry. *Front. Sustain. Food Syst.* **2019**, *3*, 95. [CrossRef]
70. Codina-Torrella, I.; Guamis, B.; Zamora, A.; Quevedo, J.M.; Trujillo, A.J. Microbiological stabilization of tiger nuts' milk beverage using ultra-high pressure homogenization. A preliminary study on microbial shelf-life extension. *Food Microbiol.* **2018**, *69*, 143–150. [CrossRef]

71. Saricaoglu, F.T.; Gul, O.; Besir, A.; Atalar, I. Effect of high pressure homogenization (HPH) on functional and rheological properties of hazelnut meal proteins obtained from hazelnut oil industry by-products. *J. Food Eng.* **2018**, *233*, 98–108. [CrossRef]
72. Zhou, L.; Guan, Y.; Bi, J.; Liu, X.; Yi, J.; Chen, Q.; Wu, X.; Zhou, M. Change of the rheological properties of mango juice by high pressure homogenization. *LWT—Food Sci. Technol.* **2017**, *82*, 121–130. [CrossRef]
73. Gul, O.; Saricaoglu, F.T.; Atalar, I. Effect of high pressure homogenization on microstructure and rheological properties of hazelnut beverage cold-set gels induced glucono-δ-lactone. *Lwt.* **2021**, *143*, 111154. [CrossRef]
74. Huang, Z.-G.; Wang, X.-Y.; Zhang, J.-Y.; Liu, Y.; Zhou, T.; Chi, S.-Y.; Gao, F.; Li, J.; Tian, B.; Shi, W.-T.; et al. Effect of heat treatment on the nonlinear rheological properties of acid-induced soy protein isolate gels modified by high-pressure homogenization. *LWT* **2022**, *157*, 113094. [CrossRef]
75. Bi, C.-H.; Wang, P.-L.; Sun, D.-Y.; Yan, Z.-M.; Liu, Y.; Huang, Z.-G.; Gao, F. Effect of high-pressure homogenization on gelling and rheological properties of soybean protein isolate emulsion gel. *J. Food Eng.* **2020**, *277*, 109923. [CrossRef]
76. Levy, R.; Okun, Z.; Shpigelman, A. Utilizing high-pressure homogenization for the production of fermented plant-protein yogurt alternatives with low and high oil content using potato protein isolate as a model. *Innov. Food Sci. Emerg. Technol.* **2022**, *75*, 102909. [CrossRef]
77. Geng, M.; Wang, Z.; Qin, L.; Taha, A.; Du, L.; Xu, X.; Pan, S.; Hu, H. Effect of ultrasound and coagulant types on properties of β-carotene bulk emulsion gels stabilized by soy protein. *Food Hydrocoll.* **2022**, *123*, 107146. [CrossRef]
78. Li, K.; Fu, L.; Zhao, Y.-Y.; Xue, S.-W.; Wang, P.; Xu, X.-L.; Bai, Y.-H. Use of high-intensity ultrasound to improve emulsifying properties of chicken myofibrillar protein and enhance the rheological properties and stability of the emulsion. *Food Hydrocoll.* **2020**, *98*, 105275. [CrossRef]
79. Fu, Q.-Q.; Liu, R.; Zhou, L.; Zhang, J.-W.; Zhang, W.-G.; Wang, R.-R. Effects of psyllium husk powder on the emulsifying stability, rheological properties, microstructure, and oxidative stability of oil-in-water emulsions. *Food Control.* **2022**, *134*, 108716. [CrossRef]
80. Aliabbasi, N.; Djomeh, Z.E.; Askari, G.; Salami, M.J. Technology. Pinto bean protein ultrasonicated cold-set emulsion gels catalyzed by transglutaminase/glucono-δ-lactone: Development, characterization and in vitro release characteristics. *J. Drug Deliv. Sci. Technol.* **2023**, *81*, 104239. [CrossRef]
81. Zhang, X.; Chen, X.; Gong, Y.; Li, Z.; Guo, Y.; Yu, D.; Pan, M. Emulsion gels stabilized by soybean protein isolate and pectin: Effects of high intensity ultrasound on the gel properties, stability and β-carotene digestive characteristics. *Ultrason. Sonochemistry.* **2021**, *79*, 105756. [CrossRef] [PubMed]
82. Mozafarpour, R.; Koocheki, A. Effect of ultrasonic pretreatment on the rheology and structure of grass pea (*Lathyrus sativus* L.) protein emulsion gels induced by transglutaminase. *Ultrason. Sonochemistry.* **2023**, *92*, 106278. [CrossRef] [PubMed]
83. Singh, V.K.; Behera, B.; Pramanik, K.; Pal, K. Ultrasonication-assisted preparation and characterization of emulsions and emulsion gels for topical drug delivery. *J Pharm Sci.* **2015**, *104*, 1035–1044. [CrossRef] [PubMed]
84. Baydin, T.; Arntsen, S.W.; Hattrem, M.N.; Draget, K.I. Physical and functional properties of plant-based pre-emulsified chewable gels for the oral delivery of nutraceuticals. *Food Res. Int.* **2022**, *2*, 100225. [CrossRef]
85. He, X.-H.; Luo, S.-J.; Chen, M.-S.; Xia, W.; Chen, J.; Liu, C.-M. Effect of industry-scale microfluidization on structural and physicochemical properties of potato starch. *Innov. Food Sci. Emerg. Technol.* **2020**, *60*, 102278. [CrossRef]
86. McCarthy, N.A.; Kennedy, D.; Hogan, S.A.; Kelly, P.M.; Thapa, K.; Murphy, K.M.; Fenelon, M.A. Emulsification properties of pea protein isolate using homogenization, microfluidization and ultrasonication. *Food Res. Int.* **2016**, *89*, 415–421. [CrossRef]
87. Yang, Q.; Zheng, Q.; Jin, M.; Chen, Y.; Guo, L.; Lin, J.; Zou, Y. Fabrication of gel-like emulsions with γ-zein particles using microfluidization: Structure formation and rheological properties. *Food Res. Int.* **2022**, *158*, 111514. [CrossRef] [PubMed]
88. Oliete, B.; Potin, F.; Cases, E.; Saurel, R. Microfluidization as Homogenization Technique in Pea Globulin-Based Emulsions. *Food and Bioprocess Technology.* **2019**, *12*, 877–882. [CrossRef]
89. Jia, X.; Xu, R.; Shen, W.; Xie, M.; Abid, M.; Jabbar, S.; Wang, P.; Zeng, X.; Wu, T. Stabilizing oil-in-water emulsion with amorphous cellulose. *Food Hydrocoll.* **2015**, *43*, 275–282. [CrossRef]
90. Jiang, W.-X.; Qi, J.-R.; Liao, J.-S.; Yang, X.-Q. Acid/ethanol induced pectin gelling and its application in emulsion gel. *Food Hydrocoll.* **2021**, *118*, 106774. [CrossRef]
91. Li, S.; Zhang, B.; Tan, C.P.; Li, C.; Fu, X.; Huang, Q. Octenylsuccinate quinoa starch granule-stabilized Pickering emulsion gels: Preparation, microstructure and gelling mechanism. *Food Hydrocoll.* **2019**, *91*, 40–47. [CrossRef]
92. Liu, C.; Zheng, Z.; Shi, Y.; Zhang, Y.; Liu, Y. Development of low-oil emulsion gel by solidifying oil droplets: Roles of internal beeswax concentration. *Food Chem.* **2021**, *345*, 128811. [CrossRef] [PubMed]
93. Lu, Y.; Ma, Y.; Zhang, Y.; Gao, Y.; Mao, L. Facile synthesis of zein-based emulsion gels with adjustable texture, rheology and stability by adding β-carotene in different phases. *Food Hydrocoll.* **2022**, *124*, 107178. [CrossRef]
94. Song, X.; Pei, Y.; Qiao, M.; Ma, F.; Ren, H.; Zhao, Q. Preparation and characterizations of Pickering emulsions stabilized by hydrophobic starch particles. *Food Hydrocoll.* **2015**, *45*, 256–263. [CrossRef]
95. Zhao, X.; Li, D.; Wang, L.-J.; Wang, Y. Rheological properties and microstructure of a novel starch-based emulsion gel produced by one-step emulsion gelation: Effect of oil content. *Carbohydr. Polym.* **2022**, *281*, 119061. [CrossRef] [PubMed]
96. Mefleh, M.; Pasqualone, A.; Caponio, F.; De Angelis, D.; Natrella, G.; Summo, C.; Faccia, M. Spreadable plant-based cheese analogue with dry-fractioned pea protein and inulin–olive oil emulsion-filled gel. *J. Sci. Food Agric.* **2022**, *102*, 5478–5487. [CrossRef] [PubMed]

97. Ingrassia, R.; Busti, P.A.; Boeris, V. Physicochemical and mechanical properties of a new cold-set emulsion gel system and the effect of quinoa protein fortification. *LWT* **2022**, *156*, 113048. [CrossRef]
98. Wang, X.; He, Z.; Zeng, M.; Qin, F.; Adhikari, B.; Chen, J. Effects of the size and content of protein aggregates on the rheological and structural properties of soy protein isolate emulsion gels induced by CaSO4. *Food Chem.* **2017**, *221*, 130–138. [CrossRef] [PubMed]
99. Yang, X.; Gong, T.; Li, D.; Li, A.; Sun, L.; Guo, Y. Preparation of high viscoelastic emulsion gels based on the synergistic gelation mechanism of xanthan and konjac glucomannan. *Carbohydr. Polym.* **2019**, *226*, 115278. [CrossRef]
100. Zhan, F.; Tang, X.; Sobhy, R.; Li, B.; Chen, Y. Structural and rheology properties of pea protein isolate-stabilised emulsion gel: Effect of crosslinking with transglutaminase. *Int. J. Food Sci. Technol.* **2022**, *57*, 974–982. [CrossRef]
101. Li, Q.; Xu, M.; Xie, J.; Su, E.; Wan, Z.; Sagis, L.M.C.; Yang, X. Large amplitude oscillatory shear (LAOS) for nonlinear rheological behavior of heterogeneous emulsion gels made from natural supramolecular gelators. *Food Res. Int.* **2021**, *140*, 110076. [CrossRef] [PubMed]
102. Zhang, M.; Yin, L.; Yan, W.; Gao, C.; Jia, X. Preparation and Characterization of a Novel Soy Protein Isolate-Sugar Beet Pectin Emulsion Gel and Its Application as a Multi-Phased Nutrient Carrier. *Foods* **2022**, *11*, 469. [CrossRef] [PubMed]
103. Zhang, X.; Chen, D.; Zhao, Z.; Wan, J.; Prakash, S. Rheological and textural properties of emulsion-filled gel based on enzymatically hydrolyzed rice starch. *Food Hydrocoll.* **2022**, *126*, 107463. [CrossRef]
104. Neckebroeck, B.; Verkempinck, S.H.E.; Van Audenhove, J.; Bernaerts, T.; de Wilde d'Estmael, H.; Hendrickx, M.E.; Van Loey, A.M. Structural and emulsion stabilizing properties of pectin rich extracts obtained from different botanical sources. *Food Res. Int.* **2021**, *141*, 110087. [CrossRef] [PubMed]
105. Sjöö, M.; Emek, S.C.; Hall, T.; Rayner, M.; Wahlgren, M. Barrier properties of heat treated starch Pickering emulsions. *J. Colloid Interface Sci.* **2015**, *450*, 182–188. [CrossRef]
106. Ghafoor, M. Synthesis of High Refractive Index Materials for Manufacturing Apochromatic Lens by 3D Printing. Ph.D. Thesis, University of Eastern Finland, Kuopio, Finland, 2017.
107. Leal-Castañeda, E.J.; García-Tejeda, Y.; Hernández-Sánchez, H.; Alamilla-Beltrán, L.; Téllez-Medina, D.I.; Calderón-Domínguez, G.; García, H.S.; Gutiérrez-López, G.F. Pickering emulsions stabilized with native and lauroylated amaranth starch. *Food Hydrocoll.* **2018**, *80*, 177–185. [CrossRef]
108. Houérou, V. Rayabilité des Verres Silico-Sodo-Calciques. Ph.D. Thesis, Université Rennes, Rennes, France, 2005.
109. Degner, B.M.; Chung, C.; Schlegel, V.; Hutkins, R.; McClements, D.J. Factors Influencing the Freeze-Thaw Stability of Emulsion-Based Foods. *Compr. Rev. Food Sci. Food Saf.* **2014**, *13*, 98–113. [CrossRef]
110. Ignaszewski, E. *2021 U.S. Retail Market Insights for Plant Based Food*; Good Food Institute: Washington, DC, USA, 2022.
111. Boeck, T.; Zannini, E.; Sahin, A.W.; Bez, J.; Arendt, E.K. Nutritional and Rheological Features of Lentil Protein Isolate for Yoghurt-Like Application. *Foods* **2021**, *10*, 1692. [CrossRef]
112. Ben-Harb, S.; Irlinger, F.; Saint-Eve, A.; Panouillé, M.; Souchon, I.; Bonnarme, P. Versatility of microbial consortia and sensory properties induced by the composition of different milk and pea protein-based gels. *LWT* **2020**, *118*, 108720. [CrossRef]
113. Ferawati, F.; Hefni, M.; Ostbring, K.; Witthoft, C. The Application of Pulse Flours in the Development of Plant-Based Cheese Analogues: Proximate Composition, Color, and Texture Properties. *Foods* **2021**, *10*, 2208. [CrossRef]
114. Mattice, K.D.; Marangoni, A.G. Physical properties of plant-based cheese products produced with zein. *Food Hydrocoll.* **2020**, *105*, 105746. [CrossRef]
115. Dreher, J.; Blach, C.; Terjung, N.; Gibis, M.; Weiss, J. Formation and characterization of plant-based emulsified and crosslinked fat crystal networks to mimic animal fat tissue. *J. Food Sci.* **2020**, *85*, 421–431. [CrossRef] [PubMed]
116. Paciulli, M.; Littardi, P.; Carini, E.; Paradiso, V.M.; Castellino, M.; Chiavaro, E. Inulin-based emulsion filled gel as fat replacer in shortbread cookies: Effects during storage. *LWT* **2020**, *133*, 109888. [CrossRef]
117. Armaforte, E.; Hopper, L.; Stevenson, G. Preliminary investigation on the effect of proteins of different leguminous species (Cicer arietinum, Vicia faba and Lens culinarius) on the texture and sensory properties of egg-free mayonnaise. *LWT* **2021**, *136*, 110341. [CrossRef]
118. Ruan, Q.; Yang, X.; Zeng, L.; Qi, J. Physical and tribological properties of high internal phase emulsions based on citrus fibers and corn peptides. *Food Hydrocoll.* **2019**, *95*, 53–61. [CrossRef]
119. Greis, M.; Sainio, T.; Katina, K.; Nolden, A.A.; Kinchla, A.J.; Seppa, L.; Partanen, R. Physicochemical Properties and Mouthfeel in Commercial Plant-Based Yogurts. *Foods* **2022**, *11*, 941. [CrossRef] [PubMed]
120. Boeck, T.; Ispiryan, L.; Hoehnel, A.; Sahin, A.W.; Coffey, A.; Zannini, E.; Arendt, E.K. Lentil-Based Yogurt Alternatives Fermented with Multifunctional Strains of Lactic Acid Bacteria-Techno-Functional, Microbiological, and Sensory Characteristics. *Foods* **2022**, *11*, 2013. [CrossRef]
121. Grossmann, L.; McClements, D.J. The science of plant-based foods: Approaches to create nutritious and sustainable plant-based cheese analogs. *Trends Food Sci. Technol.* **2021**, *118*, 207–229. [CrossRef]
122. Michel, S.E.S.; Scheermeijer, R.; Ambühl, M.; Fernández Farrés, I. Novel plant-based cream cheese: A tribology perspective. *J. Food Eng.* **2022**, *335*, 111172. [CrossRef]
123. Short, E.C.; Kinchla, A.J.; Nolden, A.A. Plant-Based Cheeses: A Systematic Review of Sensory Evaluation Studies and Strategies to Increase Consumer Acceptance. *Foods* **2021**, *10*, 725. [CrossRef]

124. Bohrer, B.M. An investigation of the formulation and nutritional composition of modern meat analogue products. *Food Sci. Hum. Wellness* **2019**, *8*, 320–329. [CrossRef]
125. Kyriakopoulou, K.; Keppler, J.K.; van der Goot, A.J. Functionality of Ingredients and Additives in Plant-Based Meat Analogues. *Foods* **2021**, *10*, 600. [CrossRef] [PubMed]
126. Tangkham, W.; LeMieux, F. The Effects of Replacing Pork Fat with Cold-Pressed Coconut Oil on the Properties of Fresh Sausage. *J. Food Res.* **2017**, *6*, 83. [CrossRef]
127. McClements, D.J.; Grossmann, L. Eggs and Egg Products. In *Next-Generation Plant-based Foods*; Springer: Berlin/Heidelberg, Germany, 2022; pp. 341–388.
128. Yin, W.-J.; Chen, X.-W.; Ma, C.-G.; Wang, J.-M. Fabrication and Characterization of Tunable High Internal Phase Emulsion Gels (HIPE-Gels) Formed by Natural Triterpenoid Saponin and Plant Soy Protein. *ACS Food Sci. Technol.* **2022**, *2*, 1103–1113. [CrossRef]
129. Ouyang, H.; Kilcawley, K.N.; Miao, S.; Fenelon, M.; Kelly, A.; Sheehan, J.J. Exploring the potential of polysaccharides or plant proteins as structuring agents to design cheeses with sensory properties focused toward consumers in East and Southeast Asia: A review. *Crit. Rev. Food Sci. Nutr.* **2022**, *62*, 4342–4355. [CrossRef] [PubMed]
130. El Youssef, C.; Bonnarme, P.; Fraud, S.; Peron, A.C.; Helinck, S.; Landaud, S. Sensory Improvement of a Pea Protein-Based Product Using Microbial Co-Cultures of Lactic Acid Bacteria and Yeasts. *Foods* **2020**, *9*, 349. [CrossRef] [PubMed]
131. Jiménez-Colmenero, F.; Triki, M.; Herrero, A.M.; Rodríguez-Salas, L.; Ruiz-Capillas, C. Healthy oil combination stabilized in a konjac matrix as pork fat replacement in low-fat, PUFA-enriched, dry fermented sausages. *LWT—Food Sci. Technol.* **2013**, *51*, 158–163. [CrossRef]
132. Koç, H.; Drake, M.; Vinyard, C.J.; Essick, G.; van de Velde, F.; Foegeding, E.A. Emulsion filled polysaccharide gels: Filler particle effects on material properties, oral processing, and sensory texture. *Food Hydrocoll.* **2019**, *94*, 311–325. [CrossRef]
133. Hu, X.; Karthik, P.; Chen, J. Enhanced oral oil release and mouthfeel perception of starch emulsion gels. *Food Res. Int.* **2021**, *144*, 110356. [CrossRef]

gels

MDPI

Review

Pectin Hydrogels: Gel-Forming Behaviors, Mechanisms, and Food Applications

Nurul Saadah Said [1,†], Ibukunoluwa Fola Olawuyi [1,2,†] and Won Young Lee [1,2,*]

[1] School of Food Science and Technology, Kyungpook National University, Daegu 41566, Republic of Korea; nurulsaadah.said@gmail.com (N.S.S.); ifolawuyi@knu.ac.kr (I.F.O.)
[2] Research Institute of Tailored Food Technology, Kyungpook National University, Daegu 41566, Republic of Korea
* Correspondence: wonyoung@knu.ac.kr
† These authors contributed equally to this work.

Abstract: Pectin hydrogels have garnered significant attention in the food industry due to their remarkable versatility and promising properties. As a naturally occurring polysaccharide, pectin forms three-dimensional (3D) hydrophilic polymer networks, endowing these hydrogels with softness, flexibility, and biocompatibility. Their exceptional attributes surpass those of other biopolymer gels, exhibiting rapid gelation, higher melting points, and efficient carrier capabilities for flavoring and fat barriers. This review provides an overview of the current state of pectin gelling mechanisms and the classification of hydrogels, as well as their crosslinking types, as investigated through diverse research endeavors worldwide. The preparation of pectin hydrogels is categorized into specific gel types, including hydrogels, cryogels, aerogels, xerogels, and oleogels. Each preparation process is thoroughly discussed, shedding light on how it impacts the properties of pectin gels. Furthermore, the review delves into the various crosslinking methods used to form hydrogels, with a focus on physical, chemical, and interpenetrating polymer network (IPN) approaches. Understanding these crosslinking mechanisms is crucial to harnessing the full potential of pectin hydrogels for food-related applications. The review aims to provide valuable insights into the diverse applications of pectin hydrogels in the food industry, motivating further exploration to cater to consumer demands and advance food technology. By exploiting the unique properties of pectin hydrogels, food formulations can be enhanced with encapsulated bioactive substances, improved stability, and controlled release. Additionally, the exploration of different crosslinking methods expands the horizons of potential applications.

Keywords: pectin; gels; gelling mechanism; hydrogel; application

Citation: Said, N.S.; Olawuyi, I.F.; Lee, W.Y. Pectin Hydrogels: Gel-Forming Behaviors, Mechanisms, and Food Applications. *Gels* **2023**, *9*, 732. https://doi.org/10.3390/gels9090732

Academic Editors: Baskaran Stephen Inbaraj, Kandi Sridhar and Minaxi Sharma

Received: 18 August 2023
Revised: 6 September 2023
Accepted: 7 September 2023
Published: 9 September 2023

1. Introduction

In recent times, there has been a growing interest in developing innovative biomaterials derived from natural biopolymers that could potentially revolutionize the food sector by improving product quality and providing functional advantages to customers. Among these biomaterials, carbohydrate-based polymers such as starch, cellulose, chitosan, and pectin have emerged as versatile and valuable assets in food technology [1–5]. These naturally derived polymers offer a wide range of applications as stabilizers, fat replacers, emulsifiers, etc., in the food science industry, and they contribute distinct functionalities to food products. In addition, they are applicable for ingredient enhancement, additive integration, and food packaging [6,7]. For instance, starch and cellulose can contribute as dietary fiber, act as an effective thickening and stabilizing agent [8,9], or be employed as materials for food packaging [10,11]. Chitosan, with its potent antimicrobial properties, is used as a preservative packaging material for extending the shelf life of perishable goods, particularly fruits and vegetables [12]. On the other hand, pectin, renowned for its remarkable gelling properties, frequently serves as a crucial gelling agent, transforming liquid formulations into stable gels, imparting desirable textures, and enhancing product quality [13]. Besides these diverse applications, the preparation of food hydrogels using these biopolymers has recently garnered significant interest in the food industry.

Hydrogels are polymeric materials with a 3D network capable of absorbing and retaining water, making them highly hydrophilic [14]. Besides their controlled release properties, ability to provide structural/textural stability, and ability to mimic desired food textures, they are indispensable in modern food technology, further extending their applications to other fields, including cosmetics, drug delivery, and tissue engineering [15–17]. Hydrogels are employed for tasks such as enabling intricate shapes in 3D printing, serving as fat substitutes, and promoting satiety with smaller portions [18,19]. They also serve as thickeners and stabilizers due to their capacity to retain a substantial quantity of water or polar solvents while maintaining a solid-like structure, achieved through the physical or chemical crosslinking of hydrophilic polymer chains.

Pectin, a naturally occurring polysaccharide found in plant cell walls, is primarily composed of repeating units of α-(1-4)-linked D-galacturonic acid units [20]. Depending on the plant source, a pectin structure may consist of homogalacturonan (HG), rhamnogalacturonan I (RG-I), and rhamnogalacturonan II (RG-II) domains [21]. This diversity in composition accounts for the unique properties exhibited by different pectins, such as their gelation, solubility, and rheological behavior. Additionally, pectin can form 3D networks of hydrophilic polymer chains, making it ideal for preparing hydrogels. Gels prepared using pectin are advantageous over other biopolymer gels, such as gelatin, in terms of their ability to form gels rapidly, elevated thermal stability, and exceptional capacity for encapsulating flavors and creating fat barriers [22]. Compared to other natural biopolymers such as starch, cellulose, chitosan, collagen, protein, and agarose, pectin distinguishes itself through its exceptional gelling properties, allowing for the creation of stable hydrogels under milder conditions. Moreover, pectin provides the advantage of controllable gelation and interactions through its adjustability by modifying its degree of methoxylation and acetylation [23–25]. Also, its amphiphilic nature, with both polar and non-polar sites within its structure [26], enables effective interaction with water and oil, making it versatile for encapsulating hydrophobic bioactives.

In application, pectin hydrogels are currently being investigated for their potential in the development of innovative food products and the fabrication of structured food with specific textures for specific purposes while elevating sensory properties. Also, there is evidence that pectin hydrogels can be employed for effective encapsulation and targeted release of bioactives to specific digestive tract regions [27]. Therefore, this paper aims to provide valuable insights through a systematic review of the present status of research and advancements in pectin hydrogels. This review could stimulate further exploration and utilization of pectin hydrogels, aligning with the evolving demands of modern consumers and promoting advancements in food technology.

2. Pectin Extraction and Characterization

2.1. Extraction of Pectin from Various Sources

Pectin can be extracted from a diverse range of sources, each possessing its own distinctive composition and properties. Among these sources, pectin is commonly obtained from citrus fruits like oranges, lemons, and limes due to their substantial peel content and higher yield of pectic polysaccharides (Table 1). Also, apples offer a pectin-rich pomace that includes both peels and cores, contributing to their relatively high pectin yield. Other alternative sources, such as bananas, potato pulp, pumpkin peels, watermelon rinds, cocoa husks, and soy hulls, have also been investigated for their pectin content. These unconventional sources hold promise for diversifying pectin extraction options and utilizing agricultural byproducts effectively.

Table 1. Sources and extraction methods of pectin.

Source	Pectin Yield (%)	Extraction Methods	References
Lime peel	17.70–26.30	EAE	[28]
Blood orange peel	19.24	MAE	[29]
Sour orange peel	29.10	MAE	[30]
Navel orange peel	15.47–20.44	CE, MAE, UHP	[31]
Apple pomace	3.63–14.50	SW, EAE	[32,33]
Citrus peel	0.15–28.82	CE, SW, UAE	[32,34–36]
Grapefruit peel	23.50–27.34	CE, UAE	[37]
Pineapple peel	1.02–2.12	CE, MAE	[38]
Apple peel	3.60–6.40	CE	[39]
Plantain peel	6.20–13.40	CE, EAE	[40]
Pumpkin peel	8.08–10.03	EAE	[41,42]
Mango peel	1.55–21.82	CE, UAE	[43,44]
Watermelon rind peel	11.25–25.79	CE, MAE	[45–48]
Dragon fruit peel	7.50	MAE	[49]
Cocoa husk	8.00–11.31	CE	[50,51]
Soy hull	26.00–28.00	CE	[52]
Potato pulp	14.34	CE	[53]
Banana peel	15.89–24.08	CE	[54]
Strawberry	4.10–9.00	CE, UAE, EAE	[55]
Redcurrent	2.20–8.80	CE, UAE, EAE	[55]
Blackberry	4.30–9.10	CE, UAE, EAE	[55]
Raspberry	8.70–12.20	CE, UAE, EAE	[55]

Recently, pectin extraction studies have progressively focused on the upcycling of byproducts from the fruit industry. By valorizing the value of peel waste and husks, which are often discarded, these byproducts offer an eco-friendly and sustainable avenue for pectin extraction. This approach aligns with the principles of the circular economy, turning what was once considered waste into a valuable resource.

Pectin is typically extracted through aqueous methods (Table 1), including conventional heating (CE) [31,34], microwave heating (MAE) [29,30], ultrasonic (UAE) [37,43], and enzymatic extraction (EAE) methods [28]. Unconventional techniques like ultra-high pressure (UHP) and subcritical water (SW) extraction have also been explored, but they can lead to some degree of pectin quality limitation and degradation [31,56]. It is worth noting that pectin yield is also influenced by factors like temperature, extraction time, pH, and raw material characteristics, along with extraction parameters [57].

2.2. *Pectin Structure and Characterization*

Pectin, being a complex polysaccharide, has an extensive variety of physical and chemical configurations, which affect its properties and functionality in a diverse array of food applications. Pectin is mostly made of α-(1,4)-linked D-galacturonic acid units, which form the polysaccharide's backbone [58]. Moreover, pectin molecules exhibit pendant groups that encompass both hydrophilic functional units, including hydroxyl and carboxyl groups, and hydrophobic functional units, such as carboxylic ester and amide groups [59]. The polysaccharide chain found in pectin is hydrophilic [60], while proteins, feruloylated groups, and methyl and acetyl groups in pectin molecules are hydrophobic, which gives pectin good amphiphilic properties [59–61]. Among the pendant groups in pectin's backbone structure, acetyl groups (CH_3CO–) and methoxy groups (CH_3O–) are the most prevalent, as they play essential roles in defining the functionality of pectin. The acetyl groups are typically attached to the hydroxyl (-OH) groups of galacturonic acid units within the pectin structure. This is typically presented as the degree of acetylation (DA), which is the number of acetyl groups present per galacturonic acid unit [62]. The presence of acetyl groups profoundly influences pectin's solubility, gelation characteristics, and interactions with other molecules. Higher acetylation levels can impede molecular interactions among pectin molecules, leading to a reduced ability to form gels [63,64].

Conversely, lower acetylation levels encourage stronger interactions, enhancing pectin's gelling properties and potential applications. In contrast, methoxy groups are linked to the carbon atoms of galacturonic acid units, and the degree of methoxylation (DM) signifies the number of methoxy groups per galacturonic acid unit [65–67]. Methoxylation considerably affects the gelation behavior of pectin [68], especially in the presence of calcium ions, which will be further explained in subsequent sections that focus on the pectin gelling mechanism.

Based on the degree of esterification (DE) of these groups, pectin is classified either as low-methoxyl (LMP; DE 50%) or high-methoxyl (HMP; DE > 50%) [69]. The DE influences the gelation behavior, solubility, and engagement with other constituents within food matrices, making it a critical parameter in tailoring pectin for specific applications. The presence of side branches and neutral sugar side chains, such as rhamnose and arabinose residues, also has a substantial impact on pectin's gelling property, influencing its overall structure and activity [70–72]. Pectin consists of three main regions (Figure 1): homogalacturonan (HG), which is regarded as the "smooth" region and made up of α-(1,4)-linked D-galacturonic acid residues with methyl and acetyl esterification; rhamnogalacturonan I (RG-I), which is termed the "hairy" region with alternating L-Rha and D-GalA residues and variable side chains; and rhamnogalacturonan II (RG-II), a complex structure containing up to 13 different sugars and 21 glycosidic linkages [73].

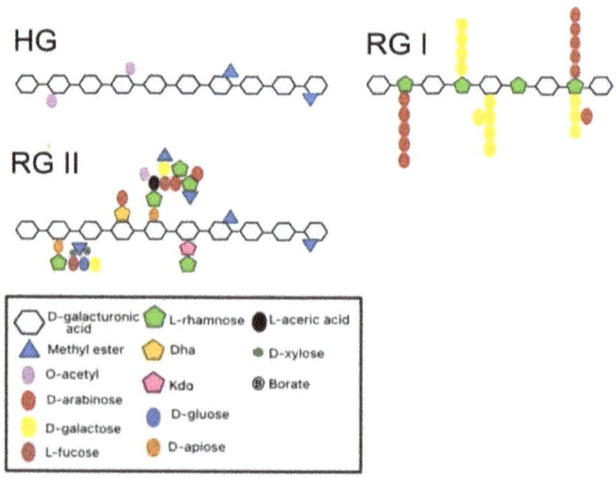

Figure 1. Schematic diagram of various pectin structures.

HG, a linear chain of α-(1,4)-linked galacturonic acid residues, is the primary structural component of pectin [74]. HG plays a pivotal role in establishing the gel matrix of pectin hydrogels. Commercial pectin, with a higher proportion of the HG domain, is renowned for its superior gelling ability, making it a widely used and stable gel-forming agent in the food and pharmaceutical industries [75]. The gelation mechanism includes divalent cations, such as calcium ions, interacting with the carboxyl groups present in the galacturonic acid residues. This results in the creation of "egg-box" structures in which calcium ions bridge adjacent HG chains, resulting in a 3D gel network with improved strength and stability. Also, for pectin consisting of both HG and RG-I structures, its ability to form gels is influenced by the ratio of HG:RG-I and the length of the HG domain. In addition, it has been reported that hydrogels exhibit optimal properties when prepared using pectin with a high content of galacturonic acid (GalA), a high molecular weight, and a suitable proportion of side chains (>15.8%) [76].

RG-I is a complex side chain composed of alternating galacturonic acid and rhamnose residues that is connected to the HG backbone [77]. The length of the RG-I backbone can range between 20 and 300 repeating units. At the C-4 position of rhamnose residues in pectin, there are side chains composed of galactose and/or arabinose residues. These side chains can form individual sugar units or combine to create chains of arabinans, galactans, or arabinogalactans. The branched rhamnose units in these chains account for 20–80% of the total structure [67,78]. A study by Zheng et al. [70] demonstrated that RG-I-enriched pectin may produce gels under both cation-induced and acid-induced circumstances. The existence of numerous arabinose sugar side chains in RG-I contributes significantly to the gel network's strength by generating entanglements and stabilizing both chain–chain and dimer–dimer structures. These side-chain entanglements provide a denser gel network, limiting network chain mobility, and strengthening hydrophobic and hydrogen bonding in the HG region, which results in enhanced gel strength.

RG-II is the most complex and structurally unique component of pectin. It consists of a branched backbone of alternating galacturonic acid and rhamnose residues, with various side chains containing arabinan, apiose, and xylose residues [67,74,77]. RG-II has great crosslinking capabilities [79], which help to generate robust and stable pectin hydrogels. Because of its structural complexity, RG-II may interact with a wide spectrum of macromolecules in food systems, making it a crucial element in improving the functionality and performance of pectin hydrogels. However, there are yet few studies on hydrogels prepared with RG-II pectins or pectin structures containing a higher proportion of RG-II.

3. Gelling Mechanism of Pectin

The gelling properties and mechanisms of pectin have primarily been categorized and discussed in various studies, with a focus on factors such as its degree of esterification. This section discusses the impact of the presence of esters and other molecules in the pectin network on its gelation behavior and structural characteristics.

3.1. High-Ester Pectin

High-methoxyl (HM) pectins, with a DE ranging from 50% and above, predominantly form gels through the cohesion of hydrophobic forces and the formation of hydrogen bonds under specific environmental circumstances. These conditions include low pH levels (around 2.5 to 3.5) and the presence of soluble solids like sucrose (55% to 75%) or similar co-solutes for the gelling process to occur [74,80,81]. Sugar plays a crucial role in gel formation by reducing the amount of available water, which stabilizes the junction zones through hydrophobic interactions. As a result, the formed gel exhibits a two-dimensional (2D) network of interconnected pectin molecules with water and co-solutes trapped inside, contributing to its ability to resist deformation. The 3D network of HM pectin gels is established through junction zones that are stabilized by hydrogen bonding between carboxyl and secondary alcohol groups, as well as hydrophobic interactions involving methyl esters. These gels exhibit thermal reversibility, meaning they can undergo gel-to-sol transitions with changes in temperature. When exposed to hot water, HM pectin gels are soluble, and to prevent lumping, they are often used with a dispersion agent like dextrose [74]. The gelling mechanism of HM pectin is depicted in Figure 2.

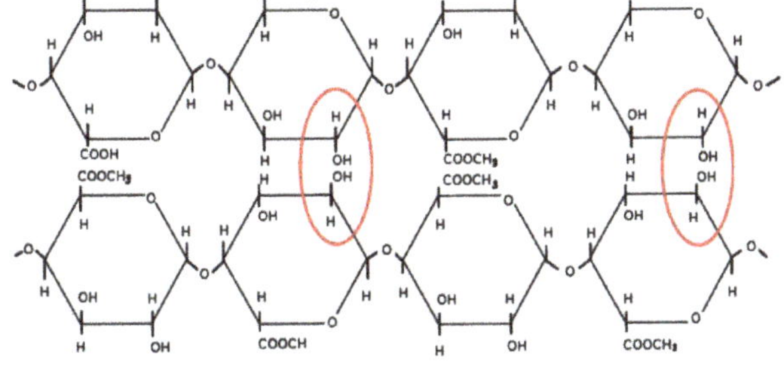

Figure 2. Gelling mechanism of high-methoxyl pectin. Red circle represents hydrogen bond formation between the pectin chains.

The gel formation of high-ester pectins is a complex process affected by various factors beyond just the DE. These pectins typically create gels by linking polymer chains at junction zones, facilitated by hydrogen bonding and hydrophobic interactions among methyl-ester groups. In cases where esters are grouped together, allowing some parts of the molecule to remain as free acids, calcium bridges might also contribute to the gelation process. The factors influencing the gelation process and gel structure of high-ester pectins include the concentration of pectin, its molecular weight, degree of acetylation, branching pattern, pH, ionic strength, water content, type of sugar present, cooling rate, and storage temperature [63]. These parameters play crucial roles in determining the properties and characteristics of the pectin gels [63]. During the gelation process, 3D networks are established, effectively entrapping water and solute molecules within the gel structure. The gel strength of high-ester pectins increases with higher pectin concentration, creating more junction zones and elastic chains [74,82]. The DE determines pH and temperature range for gelling, with higher DE pectins gelling at higher pH and temperature [83]. Acetylation reduces pectin's gelling ability by hindering interactions [84,85], while neutral sugars can either hinder or enhance gel cohesion through hydrophobic interactions [63,74]. For sugar composition in high-ester pectin, the effect is dependent on the molecular geometry of the sugar interacting with neighboring water molecules [80]. The other factor affecting pectin gelling properties is pH level. Lower pH promotes gel formation by facilitating interactions between pectin molecules [63]. The carboxyl groups on galacturonic acid residues are less dissociated in acidic circumstances, resulting in less electrostatic repulsion between pectin chains. This permits the chains to generate additional hydrogen bonds and form a gel network. On the other hand, an excessively low pH level might cause rapid gelling without sufficient organization, resulting in a weak and poorly organized gel.

In addition, pectic polysaccharides, such as pectins, are polyelectrolytes whose gelation behavior and interactions with ions are affected by the solution's ionic strength [86]. Higher ionic strength can change the pH range for gel formation and enhance the creation of stronger gel networks by binding divalent cations such as calcium (Ca^{2+}), which function as bridges between pectin molecules [63]. This can also enhance the creation of junction zones between pectin molecules, resulting in stronger gel networks. These cations operate as bridges between the negatively charged carboxyl groups of pectin, further solidifying the gel structure. Meanwhile, lowering the water activity by increasing the concentration of soluble solids to around 65 wt% is preferred, as it speeds up the gelling process and enhances the strength of the resulting gel [87]. When water activity is reduced, gelation occurs faster because there is less water available to hydrate the pectin molecules. As a result, the molecules are forced closer together, resulting in a denser and closer woven gel network. As a result, the final gel strength increases. The cooling rate of gels has been indicated as another factor influencing the pectin gelation rate [63]. Intermediate cooling rates and temperatures are favorable for gel formation as they promote the formation of a network with the highest elasticity. During cooling, hydrogen bonding and hydrophobic interactions between pectin molecules play an important role in gel structure stabilization. A slower cooling rate gives these interactions more time to occur, resulting in a stronger and more elastic gel. Therefore, to explicitly determine the gelling behavior and consequent gel characteristics of high-ester pectin obtained from different sources, the interactions between these parameters and their optimal conditions for desirable gel formation must be investigated.

3.2. Low-Ester Pectin

Low-ester pectins have historically served as the go-to option for gelling food products in scenarios where high-ester pectins may not be as effective in forming a gel. In addition, low-ester pectins have demonstrated their usefulness in forming stable gels in low-to-moderate sugar and acidic environments. Amidated pectins, which are a subcategory of low-ester pectins, exhibit distinct gelling characteristics, providing a wide array of functional properties in the food industry. These unique characteristics make them versatile choices for various food applications.

Low-ester pectins with a low degree of methylation (DM) have more free carboxylic acid groups, which interact with Ca^{2+} ions to create a continuous gel network through the "egg-box" paradigm [68]. Figure 3 presents the "egg-box" model, elucidating the gel formation mechanism of low-methoxyl pectin. However, it remains disputed how many contiguous non-methoxylated galacturonic acid residues are necessary for cooperative egg-box formation (6 to 20 residues) [88–91]. The capacity of the gel to produce stable egg-box junction zones is governed by the presence of extended blocks of non-methoxylated galacturonic acid residues for cooperative Ca^{2+} ion binding [92]. The amount of calcium required for gelation is determined by the DE, size, and distribution of non-methylesterified galacturonic acid, as well as the process parameters [92,93]. Calcium excess can cause pre-gelation or the formation of pectin precipitates.

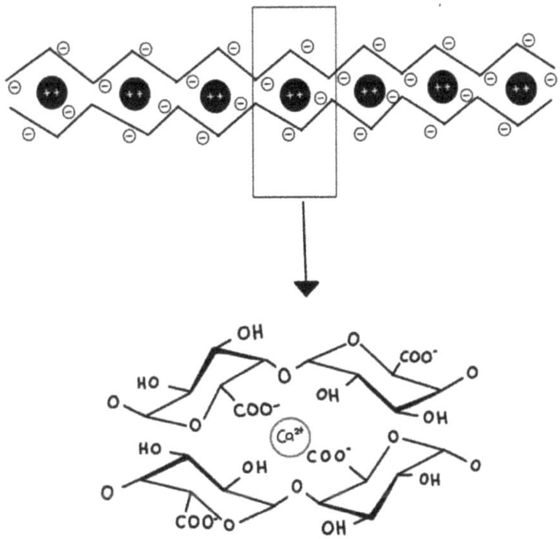

Figure 3. The "egg-box" model illustrating the gelling mechanism of low-methoxyl pectin.

According to Flutto [63] and Vriesmann [64], the presence of ester, acetyl, or amide groups in pectin disrupts the stabilization of polar groups in the junction zones between neighboring pectic chains, resulting in hindered gel formation. The side chains of pectin may also influence the flexibility of the molecule, preventing aggregation through steric hindrance. Hydrogen bonds and hydrophobic interactions can also impact the ultimate texture of low-ester pectin gels, particularly in conditions of low pH and high soluble solid concentrations. Various parameters influence the gelling of low-ester pectins, including the number and distribution of ester and amide groups, molecular weight, pH, ionic strength, and water activity of the gelling system [63].

Several variables determine the gel strength of low-ester pectins. As calcium bonds can only form in esterification-free areas, lower esterification levels result in greater gel strength. Pectin amidation enhances gelling power, promotes hydrogen bonding, and leads to tougher and thermo-reversible gels with reduced calcium requirements. Amidation increases the gelling ability of low-methoxy pectins by requiring less calcium to gel and making them less likely to precipitate at high calcium levels [94]. The molecular weight of low-ester pectins governs gelation by influencing the number of required connection zones [63]. Pectins with higher molecular weights have more junction zones, leading to faster gelation and reduced syneresis. The length of the pectin chain directly relates to its molecular weight, creating a greater number of junction zones and resulting in stronger gel formation [95,96].

The pH of the pectin solution influences gel texture as well as calcium needs. Studies on LM-pectin gel properties [97,98] revealed that lowering the pH < 3 weakened the Ca^{2+} -induced gel for non-amidated pectin but strengthened it for amidated pectin. Interestingly, LM pectin can form gels even at low pH without Ca^{2+} [99]. They proposed that below a certain pH, a conformational transition induces pectin aggregation and gelation when $G' > G''$. Additionally, gelation of LM pectin is favored at high pH, as Ca^{2+} bridges require an adequate number of dissociated carboxyl groups [92]. These findings provide valuable insights into the complex gelation behavior of LM-pectin under different pH and Ca^{2+} conditions. Managing calcium requirements is another important parameter influencing gelation, which can be achieved by adjusting the water activity through the addition of sugar or by varying the concentration of soluble solids in the solution [63]. Increasing the solid level reduces the amount of calcium needed while accelerating the gelling process, elevating the setting temperature, and enhancing the final gel strength. However, this approach also leads to a narrower optimal calcium window, prompting practical applications to favor pectin with higher esterification at higher solid levels.

4. Types of Pectin Gels

Pectin gels exhibit diverse forms, comprising hydrogels, cryogels, aerogels, xerogels, and oleogels, each of which will be explored in detail in the subsequent sections. A visual representation of these distinct pectin gel types is presented in Figure 4, offering a schematic diagram for better understanding.

Figure 4. Schematic diagram illustrating various pectin-based gel types according to preparation conditions and varying levels of drying conditions.

4.1. Hydrogels

Hydrogels are 3D porous materials made from crosslinked hydrophilic polymers, whether natural or synthetic. They possess the ability to absorb substantial quantities of water or biological fluids without dissolution [100,101]. Hydrogels are created by either physically or chemically crosslinking polymer chains, which can be synthetic or naturally derived. The control of hydrogel formation and the enhancement of interactions depend significantly on factors such as pH and charge balance. Most hydrogels, especially those based on polysaccharides, exhibit desirable biocompatibility, biodegradability, tunable structures, and stable physicochemical properties. These unique features of pectin hydrogels enable their wide application in wound healing [102,103], tissue engineering [104], drug delivery [59,105,106], strain sensors [107], supercapacitors [108], aqueous batteries [108], and various other fields. However, hydrogels formed through single chemically or physically cross-linked methods may have poor mechanical properties and weak energy dissipation during deformation. This brittleness limits their potential applications. To enhance their properties, various techniques in preparing hydrogels are utilized, such as ionic gelation, ionotropic gelation, casting, and filtration, offering versatile routes to tailor hydrogels for specific needs, as demonstrated in Table 2.

Table 2. Types of pectin gels and their applications.

Types of Pectin Gel	Composite Material	Methods	Outcomes	Applications	References
Hydrogel	Pectin/chitosan/essential oils	Ionic gelation (Dripping method)	Good antimicrobial activity against six types of microorganism	-	[109]
Hydrogel	Pectin/chitosan	Ionic charge interaction	Good antibacterial and wound healing properties	Tissue regeneration	[110]
Hydrogel	LM apple pectin	-	Low toxicity, improved stability towards elastic and plastic deformation, ability to adhere to macrophages and the non-specific adsorption of blood plasma proteins	Scaffold for tissue engineering	[13]
Hydrogel	LM apple pectin/LM hogweed pectin	Ionotropic	Increased gel strength	-	[111]
Hydrogel	HM apple pectin/Glucono-δ-lactone	-	Great mechanical strength, stronger thermo-reversibility, and higher pH stability		[112]
Hydrogel (membrane layer)	Banana peel pectin/Water hyacinth carboxymethyl cellulose	Casting	Increased hydrophobicity of hydrogel membrane	-	[113]
Hydrogel	LM pectin/Resistant starch/Lactobacillus bulgaricus	Filtration	High storage ability and protective effects on *L. bulgaricus*	Synbiotic encapsulation, protection, and delivery of probiotics	[114]
Aerogel	Citrus pectin/cellulose nanofiber	Freeze drying	Improved tensile and compressive properties	Edible fungus moisture-regulating packaging	[115]
Aerogel	LM pectin/alginate	Freeze drying	Strong antioxidant activity with good controlled released of proanthocyanidins	Matrix for the controlled release of proanthocyanidin compound	[116]
Aerogel	1.Citrus pectin 2. Watermelon rind pectin	Supercritical drying with CO_2	High specific surface and low bulk density	Matrix for the controlled release of vanillin compound	[117]
Aerogel	Citrus pectin/PLA	Supercritical drying	Increased swelling and simulated body fluid (SBF) uptake	Active wound-healing materials	[118]
Aerogel	Pectin/TiO_2	Supercritical CO_2 drying	Great mechanical, thermal, and antimicrobial properties	Temperature-sensitive food	[119]
Aerogel	Citrus pectin	Supercritical CO_2 drying	Low density with high porosity and pore volume resulted in small pores size, mainly mesopores and small macropores	Matrix for the controlled release of theophylline compound	[120]
Oleogel	Citrus pectin/camellia oil/tea polyphenol-palmitate particles	Freeze drying	Improved oil binding capacity and gel strength	-	[121]
Oleogel	Citrus pectin/ ovotransferrin fibrils	Homogenizing	Better stability, smaller droplet size, more prominent gel-like structure, high viscosity, and superior texture properties	Matrix for the controlled release of curcumin compound	[122]
Oleogel	Citrus pectin/tea polyphenol ester	Freeze drying	Increased stability and viscoelasticity of emulsions, improved oil binding capacity and gel strength of the oleogels	Fat replacer in cookies product	[123]
Cryogel	1. Apple pectin/chitosan 2. Heracleum pectin/chitosan	Cryotropic gelation (Freeze drying)	Possessed biocompatibility, biodegradability, and low toxicity	Potential medical purposes	[124]
Cryogel	LM pectin/sucrose	Freeze drying	Reduction of ice crystal in gel	-	[125]
Cryogel	Citrus pectin	Freeze drying	High loading efficiency of theophylline compound	Matrix for the controlled release of theophylline compound	[120]
Cryogel	LM, MM and HM pectin/polyvinyl alcohol	Film drying	Ability to keep the enrofloxacin antibiotic inside the matrix and control of the cargo amount in the gel	Can be used for different infectious pathologies and/or treatments	[126]

Table 2. *Cont.*

Types of Pectin Gel	Composite Material	Methods	Outcomes	Applications	References
Xerogel	Citrus pectin	Oven drying	High density, low porosity, low pore volume and compact morphology	Matrix for the controlled release of theophylline compound	[120]
Xerogel	Sugar beet pectin	Air drying	Improved stability and reusability of the gels with good sorption capability of metal compounds	Heavy metal removal	[127]
Xerogel	Sugar beet pectin	Air drying	Good mechanical strength with high continuous biosorption and desorption of copper	Biosorbent for copper removal in a fixed-bed column	[128]
Xerogel	LM pectin/brea gum	Oven drying	Showed good compatibility between both polymers with high gel strength, while also able to respond to the changes in pH of the medium and modify dye release	Matrix for the controlled release of methylene blue dye	[129]

Hydrogel Preparation

Ionic gelation is a commonly employed technique for producing hydrogels, especially those derived from natural polymers such as pectin, alginate, or chitosan. This method is widely used due to its effectiveness and versatility in creating hydrogel networks. This method involves crosslinking polymer chains through ionic interactions with multivalent ions, typically divalent cations such as calcium (Ca^{2+}) or zinc (Zn^{2+}) [109,130,131]. In this process, the polymer solution is mixed with the crosslinking ion solution, resulting in the creation of a cohesive gel structure. The gelation process occurs when the divalent ions interact with the functional groups (e.g., carboxylate or sulfate groups) on the polymer chains, forming strong ionic bonds. The crosslinking of polymer chains creates a 3D network that traps water molecules, giving rise to the hydrogel structure. The gelation can be triggered by changing pH, temperature, or simply mixing the polymer and ion solutions. In research conducted by Torpol et al. [109], a pectin hydrogel was developed using the ionic gelation method, which involved the combination of chitosan and essential oils with the addition of $CaCl_2$. In another study, the ionic gelation method was utilized for microencapsulation-based gelation, enabling the crosslinking of polyelectrolytes (pectin and chitosan) with multivalent ions, including calcium (Ca^{2+}), to create the hydrogel bead [132]. The study findings indicated that the hydrogel developed exhibited significant inhibition activity against various harmful bacteria, such as *B. cereus*, *C. perfringens*, *E. coli*, *P. fluorescens*, *L. monocytogenes*, and *S. aureus*. This suggests the potential of the pectin–chitosan hydrogel for antimicrobial applications.

Ionotropic gelation is a specific type of ionic gelation that relies on the capacity of polyelectrolytes to undergo crosslinking when exposed to counterions, leading to the formation of hydrogels [133]. In this method, the polymer solution is mixed with a solution containing metal cations, such as calcium (Ca^{2+}) or aluminum (Al^{3+}). The metal cations interact with the carboxylate or sulfate groups on the polymer chains, leading to gel formation [134]. The formation of the hydrogel structure through ionotropic gelation is influenced by factors such as the polymer concentration, crosslinking ions, and ionic strength of the surrounding medium. By adjusting the concentration and types of metal cations used, the crosslinking process can be customized. Ionotropic gelation finds widespread use in various applications, including controlled drug release, encapsulation of bioactive substances, and tissue engineering purposes. In another study by Popov et al. [111], the ionotropic gelation method was employed to create hydrogels using low-methyl apple and hogweed pectin samples with the addition of calcium gluconate. Ionotropic hydrogels are formed when polymers gel in the presence of metal cations. Pectin, with its carboxylate groups, readily forms gels in the presence of metal cations like Ca^{2+}. Calcium gluconate, a divalent metal cation, can crosslink with pectin and contribute to its gelling properties. The use of calcium gluconate was preferred over calcium chloride due to its milder taste, improving consumer acceptance. The presence of sucrose was observed to positively influence the creation of pectin gels by stabilizing the crosslinks between pectin and calcium ions. Moreover, sucrose forms hydrogen bonds with water molecules, resulting in the immobilization of free water and promoting the concentration of the polymer environment, thereby facilitating gelation. The study demonstrated that a mixture of both apple and hogweed pectin showed a synergistic effect, contributing to higher gel strength in the hydrogels formed through ionotropic gelation [111].

The casting method involves the preparation of hydrogels by casting a solution or dispersion of hydrogel precursors into a mold or container of the desired shape [135]. Natural polymers, synthetic polymers, or a combination of both can be used as hydrogel precursors. Precursor gelation happens by physical or chemical crosslinking, depending on the formulation. Intermolecular forces like hydrogen bonding and hydrophobic interactions cause gelation in physical hydrogels. Conversely, chemical hydrogels are created through covalent bonding of polymer chains, in which chemical processes generate permanent links between polymer chains. The casting approach provides for exact control over the shape and size of the hydrogel, making it suited for applications such as drug delivery systems, wound dressings, and tissue engineering scaffolds. The casting method of developing hydrogel was demonstrated by Elma et al. [113], who showed good compatibility between CMC and pectin from banana peels that led to stabilization of cross-linking the hydrogel membrane synthesis. The study also showed an increase in the hydrophobicity of the hydrogel membrane due to the addition of banana peel pectin.

The filtering process includes extruding a mixture of hydrogel precursors through a membrane or filter with precise pore sizes to create hydrogel beads [114]. The extrusion procedure results in the creation of uniformly sized hydrogel beads. Polymers containing crosslinking functional groups, such as alginate or chitosan, can be used as hydrogel precursors [132]. During the extrusion process, the hydrogel precursors interact and create a gel network [136]. The addition of crosslinking agents or ions can improve the gelation process even more. The filtration method is commonly used for the encapsulation of drugs, enzymes, or bioactive compounds, as well as the delivery of therapeutic agents and the immobilization of cells for various biomedical applications. A study by Lee et al. [114] demonstrated the filtration method for preparing hydrogels. The study showed that adding pectic oligosaccharide (POS) resulted in smooth hydrogel beads with fewer surface imperfections. The smoothness was achieved through hydrogen bonding between resistant starch (RS) beads and POS. The study suggests that RS-POS (1.2%) hydrogel beads could be used as an effective carrier for encapsulating *L. bulgaricus* probiotics, offering protection and controlled delivery.

4.2. Cryogels

Cryogels are a type of pectin gel formed using a cryotropic gelation process. In this method, pectin solutions are frozen at sub-zero temperatures, and the ice crystals formed act as templates for the gelation process [137]. As the frozen gel is thawed, the ice crystals melt, leaving behind a porous network of interconnected pectin chains [138]. Cryogels have a highly open and porous structure, making them ideal for applications requiring high surface area and rapid mass transfer, such as in food packaging, adsorption, and filtration processes. Cryogels are 3D porous materials formed by a process called cryotropic gelation. The preparation of cryogels involves two main methods: freeze-drying (also known as lyophilization) and film drying, as stated in Table 1. A study by Konovalova et al. [124] demonstrated polymeric cryogels formed through freeze-drying, which involves freezing and thawing the initial solutions to create the gel. In this study, low-methyl-esterified pectin from apples and Heracleum were used as the main components, capable of forming a gel with Ca^{2+} ions. The cryogels are prepared by diffusing pectin into a frozen chitosan solution, resulting in the formation of a pectin/chitosan polyelectrolyte complex. The freeze-drying process shapes the unique macroporous structure of the cryogels, contributing to their special properties and applications. However, the freeze-dying process has been revealed to affect the microstructure and mechanical properties of pectin cryogels [125]. This method creates a cellular, less dense structure with a smooth surface and homogeneous honeycomb-like pores. The freezing temperature influences porosity, with higher temperatures leading to decreased porosity. The process also impacts mechanical properties, reducing density and increasing porosity. Slow freezing produces larger ice crystals, resulting in shorter drying times and lower hardness in the cryogels. Findings by Groult et al. [120] showed that freeze-dried pectin cryogels undergo limited sample shrinkage (10–13 vol%) due to water freezing and ice crystal growth within the sample. However, this creates large pores and a damaged morphology with cracks and macropores. The resulting freeze-dried cryogels have a very low density (0.07 g/cm^3), high porosity (95%), and high pore volume (13 cm^3/g). Compared to hydrogel, aerogel, and xerogel, pectin cryogels have the lowest bulk density (0.073 $\pm$ 0.003) and volume shrinkage.

Meanwhile, pectin cryogels formed using the film drying method have been demonstrated in a previous study [126]. The advantage of film drying properties over pectin cryogel properties is that they allow for the modulation of the releasing rate of drugs, such as enrofloxacin in this case. By incorporating high-methoxylated pectin into the cryogel film, the release rate of the antibiotic enrofloxacin was significantly slowed down [126]. Additionally, the two-layer film system, with the top film equilibrated with different NaCl concentrations, further controlled the release rate of enrofloxacin. This demonstrates the potential of film drying to tailor the drug release properties of pectin cryogels, making them suitable for transcutaneous antibiotic delivery applications.

4.3. *Aerogels*

Aerogels are solid structures composed of colloidal or polymeric networks, known for their extremely low weight and exceptionally high porosity, reaching up to 99.9% (v/v) open spaces [139]. Aerogels are fabricated through a drying process where the liquid within the gel's pores is substituted with air. These materials can exhibit a wide range of pore sizes, from macro to micro, resulting in high surface areas and low thermal conductivities. Due to their tunable properties, aerogels have garnered attention as versatile nanomaterials. Their unique attributes, such as ultra-low density, high specific surface area, and remarkable acoustic, mechanical, and thermal insulation properties, make them a special class of advanced materials with great potential for bioactive encapsulation and controlled release applications [116]. In addition, bio-aerogel is a remarkable material characterized by its low density, extensive surface area, and porous structure, offering ample opportunities for functionalization. This exceptional feature arises from the abundant hydroxyl groups present on the polymer backbone, enabling straightforward modification and customization of the material's properties [140].

Aerogels have emerged as highly desirable materials for supporting single or multiple component nanoparticles, owing to their adjustable characteristics like pore size, surface area, and density. The narrow distribution of pore sizes and substantial surface areas enable excellent dispersion of nanoparticles. This dispersion leads to enhanced control over the rates of reactant and product diffusion to and from catalytic sites composed of nanoparticles. By combining robust sol–gel chemistry with various preparation methods, such as supercritical deposition, researchers have successfully developed aerogel-supported nanoparticles with exceptional catalytic properties tailored for specific targeted reactions [141]. The drying step is the most crucial stage in the production of aerogels. The majority of research efforts focused on fabricating polymer aerogels have employed supercritical CO_2 drying and freeze-drying techniques, as evidenced by the data presented in Table 1. It is worth noting that the properties of the final products vary depending on the drying method employed.

Pectin-based aerogels have been extensively studied using supercritical CO_2 drying. However, this method has some drawbacks compared to freeze-drying, including complexity, expensive raw materials, and high energy and CO_2 consumption. A study by Méndez et al. [117] investigated pectin hydrogels prepared through a sol–gel process followed by supercritical drying. They analyzed how pectin composition affected the aerogel structure and release properties when impregnated with vanillin. The developed aerogel particles exhibited high specific surface areas and low bulk density. Pectin's affinity with vanillin influences shrinkage during aerogel formation and the release profile of vanillin, making it a promising carrier for active compounds in food and biomedical applications. Horvat et al. [118] described the synthesis of biodegradable hybrid aerogels using pectin and polylactic acid as wound-dressing materials. These aerogels were loaded with model drugs and oxygen-generating compounds to assess their drug-release properties. Pectin's high water uptake and swelling ability make it attractive for wound-dressing applications. The addition of polylactic acid improved the material's stability in simulated body fluid, which is crucial for wound healing. The resulting hybrid material exhibited a highly porous structure with a large surface area, making it advantageous for drug delivery applications. A study by Groult et al. [120] observed that changing the drying method from freeze-drying (cryogels) to supercritical drying (aerogels) creates noticeable structural differences, particularly concerning specific surface area and pore sizes. Supercritical fluids, like CO_2, exhibit properties between liquids and gases, allowing for a gentle drying process without damaging the network structure. This results in low-density aerogels with high porosity and pore volume, similar to pectin cryogels. However, aerogels have smaller pores, mainly mesopores and small macropores (50–150 nm in diameter), leading to a significantly higher specific surface area (SBET) of 360 m^2/g, compared to cryogels with a SBET of 10–20 m^2/g [120].

Recent studies have shown that freeze drying is a cost-effective method for producing polymer aerogels, comparable to supercritical drying. A study by Wu et al. [115] investigated composite citrus pectin combined with cellulose nanofiber to create aerogels for thymol release. During freeze drying, the emulsion structure around oil droplets is destroyed, leaving oil droplet-shaped pores as a template for the aerogel structure. This aerogel maintained thymol activity, reduced susceptibility to oxygen, and provided slow-release properties. The aerogel was tested on fresh edible mushrooms (*Agaricus bisporus*), extending their storage time up to 5 days by adjusting the humidity in the packaging to 97%. In another study, biopolymer aerogel microspheres were fabricated using alginate and pectin crosslinked with divalent cations (Ca^{2+}) via the sol–gel method followed by freeze drying [116]. In this study, as the pectin ratio in the aerogels increased, greater porosity and pore size were observed. Moreover, the encapsulated proanthocyanidins within these aerogel microspheres exhibited controlled release behaviors, conforming to both the first-order and Korsmeyer–Peppas models. Notably, aerogels with higher pectin content exhibited stronger antioxidant activity based on radical scavenging and ferric-reducing antioxidant power results.

4.4. Xerogels

Xerogels are a specific category of gels that are formed into solid structures by slow drying at room temperature, allowing them to shrink freely during the process [142,143]. Pectin xerogel can be obtained through the removal of the solvent from the hydrogel by evaporation at room temperature or under vacuum conditions. During this process, the solvent is gradually removed, causing the gel structure to collapse, resulting in a solid material with a high content of interconnected pectin chains. Xerogels have a lower water content compared to hydrogels but retain their 3D network structure. They are commonly used in food applications for the encapsulation and controlled release of bioactive compounds and flavors. Xerogels are a type of hydrogel that is prepared by drying the gel at low temperatures to remove the solvent and water, leaving behind a solid porous material. Evaporative drying commonly results in pore collapse due to elevated capillary pressure, leading to materials with high density and low porosity [144]. There are two common methods used to prepare xerogels: oven drying and air drying, as shown in Table 1.

Oven drying, also known as conventional drying, is another method for preparing xerogels. In this process, the wet hydrogel is placed in an oven at a controlled temperature to facilitate the removal of solvent and water. The controlled environment ensures more precise drying conditions compared to air drying, but it still requires a longer drying time than freeze drying. Oven drying offers a cost-effective approach that can be easily implemented using standard laboratory equipment. The study conducted by Groult et al. [120] investigated evaporative drying in the development of citrus pectin xerogel under vacuum conditions at 60 °C. This process led to a significant shrinkage of over 90% in the volume of the material. Consequently, the drying method created a compact morphology with a high density (~1 g/cm^3), a relatively low porosity (approximately 30%), and a low pore volume (0.3 cm^3/g). The pectin xerogels exhibited some degree of porosity when observed through a scanning electron microscope (SEM), but the specific surface area could not be measured due to possible closed pores. Notably, loading efficiency was high in pectin xerogels (94%), indicating that the impregnation time was sufficient to fully load the pectin alcogel. Another study demonstrated the usage of oven drying at 40 °C to produce xerogel consisting of low-methyl pectin and brea gum [129]. The study found the xerogel exhibited a compact and dense structure with good compatibility between pectin and brea gum, and its swelling and erosion behavior were influenced by the external pH, reaching equilibrium states for water absorption and erosion. These properties of the xerogel have implications for its potential applications in medical, food, and industrial uses, given its response to changes in pH and controlled release behavior.

An alternative method was employed by Mata et al. [127,128], which utilized air drying to remove the solvent from the gels and obtain the desired xerogel structure. These studies developed sugar-beet pectin xerogels, which were later found to be effective in removing heavy metals (cadmium, lead, and copper) from effluents and wastewater in continuous systems. The xerogels show promising potential as a biosorbent for metal recovery due to their high adsorption capacity and stability. In addition, xerogels also exhibit excellent reusability after multiple batch sorption–desorption cycles. The biosorption capacity and mass of the xerogel beads remain largely unchanged even after multiple reuse cycles, making them suitable for metal remediation technologies.

4.5. Oleogels

An oleogel is a type of gel that is formed by structuring liquid oil using a gelling agent. Typically, the gelling agent is a hydrophilic material, such as a polymer or a surfactant, that can interact with the oil molecules to create a 3D network or structure [145]. This network traps and immobilizes the oil, transforming it into a gel-like consistency. Oleogels are often used as fat replacers in various food products to reduce the amount of solid fats like butter or margarine while maintaining desirable texture and sensory properties [146]. They offer potential benefits by reducing saturated fat content and improving the nutritional profile of food products. Pectin oleogels have been developed using two methods: freeze-drying and homogenizing, as demonstrated in Table 1.

In the freeze-drying method, a stable emulsion of pectin and oil is first formed. The emulsion is then frozen, and the water in it is removed by sublimation, leaving behind a porous structure of pectin and oil. The oil is trapped within the pores, creating an oleogel. This method preserves the original emulsion structure and results in a highly porous and stable oleogel, but it can be time-consuming and requires specialized equipment. A study by Luo et al. [121] investigated the preparation and application of oleogels made with camellia oil, tea polyphenol-palmitate particles, and citrus pectin using the emulsion-templated method. The concentration of citrus pectin had a significant impact on the physical properties of the emulsions, dried products, and oleogels. Higher pectin concentrations led to more stable and viscoelastic emulsions, as well as dried products with a denser structure and increased hardness. The oleogels exhibited enhanced oil binding capacity and gel strength, with a high gel strength ($G' > 17{,}000$ Pa) observed when the citrus pectin concentration exceeded 1.5% (m/v). These polyphenol-rich oleogels also demonstrated strong antioxidant activity. When used as a replacement for butter in cakes, the oleogels achieved a satisfactory overall quality with hedonic scores ranging from 21.49 to 27.58, compared to a score of 32.03 for cakes made with butter. In addition, Pan et al. [123] developed pectin oleogels combined with tea polyphenol ester particles of different fatty acid chain lengths, which were further used in cookie production as a fat replacer. The study found that the fatty acid chain length influenced the characteristics of the oleogels, including appearance, firmness, and gel intensity. When using pectin oleogels as a butter replacement in cookies, the texture and sensory qualities of the cookies changed. At certain replacement levels, cookies made with specific fatty acid chain lengths in the oleogels showed similar qualities to traditional butter cookies, making them a potential alternative for fat replacement in cookies.

In the homogenizing method, pectin and oil are mixed to form an emulsion using a homogenizer. The pectin molecules stabilize the oil droplets within the water phase. The emulsion is then allowed to cool and set to form the oleogel structure. This method is relatively simple and scalable, allowing for controlled manipulation of the gel structure by adjusting homogenization conditions. However, the resulting oleogel may have a lower porosity and specific surface area compared to freeze-dried oleogels. Dong et al. [122] explored the effect of the interaction between ovotransferrin fibrils (OVTFs) and citrus pectin on the properties of oleogel-based pickering emulsions. OVTF–citrus pectin complexes with better stability were obtained at a mass ratio of 3:1 and pH 5.0, exhibiting pearl chain-like structures. Subsequently, oleogel-based OVTF-stabilized pickering emulsions (OEs) and oleogel-based OVTF–CP complex-stabilized pickering emulsions (OCPEs) were developed. In comparison to OE, the combination of OVTFs with citrus pectin in OCPE resulted in greater stability, smaller droplet sizes, a more noticeable gel-like structure, higher viscosity, and superior textural qualities. The OCPE was also employed as a curcumin delivery method, with superior curcumin preservation, a higher rate of lipolysis, and improved bioaccessibility. This novel strategy sheds new insight on how to customize the characteristics of oleogel-based pickering emulsions by leveraging the interaction between protein fibrils and polysaccharides, which might lead to the precise production of emulsions with preferred shapes and properties.

5. Crosslinking in Hydrogel

The crosslinking of hydrogels encompasses three main processes: physical, chemical, and interpenetrating polymer networks (IPNs), as depicted in Figure 5. Each process imparts distinct characteristics to the hydrogel, making it suitable for specific applications. The choice of crosslinking methods plays a crucial role in determining the hydrogel's properties, and different crosslinking approaches are employed based on the desired characteristics and applications.

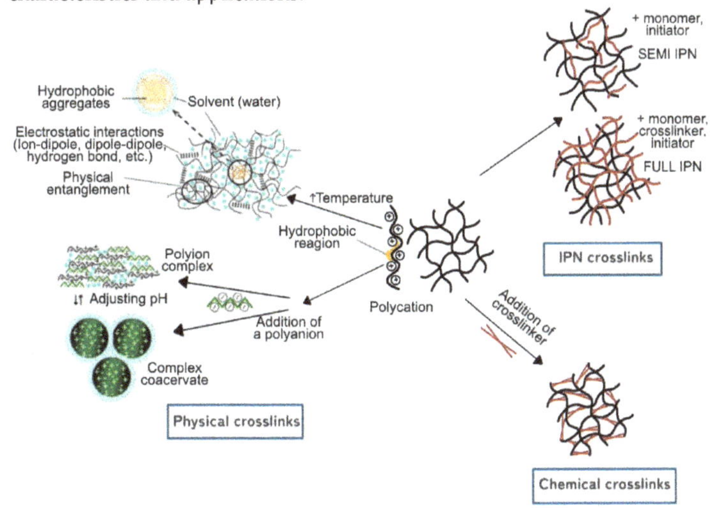

Figure 5. Crosslinking methods, including physical interactions, chemical bonds, and interpenetrating polymer network (IPN) formation, for the preparation of pectin hydrogels. Illustrative figure modified and permitted by [106,147].

5.1. Physical Crosslink Hydrogels

Physical hydrogels are formed through non-covalent interactions, such as electrostatic, hydrogen bonding, and hydrophobic forces, between oppositely charged biopolymers [18,147,148]. These interactions allow for the creation of polyion complexes, where multiple macromolecules come together to form a stable network. The polymer chains in physical hydrogels have strong inter-chain interactions, leading to a cohesive molecular network. At the same time, these hydrogels possess a high affinity for water, encouraging water molecules to access and reside within the gel structure. Due to their reversible and water-sensitive nature, physically crosslinked hydrogels have a short lifespan, typically lasting from a few days to a month when exposed to physiological conditions [18]. This property makes them advantageous for applications where short-term drug release is required, especially in clinical settings, as they do not rely on toxic covalent crosslinking molecules for gelation. In the case of physical crosslinking, the modification process involves interactions that are reversible and do not involve the formation of new covalent bonds. Instead, existing forces such as electrostatic interactions, hydrogen bonding, or hydrophobic interactions are utilized to create the network structure. As an example, a study [149] found that the combination of gelatin and low-methoxyl pectin leads to the formation of a physical co-gel. Electrostatic forces between gelatin and pectin facilitate the interactions, resulting in a reversible physical polyion complex. Gelatin forms the primary network, while pectin is dispersed within. Electrostatic forces facilitate interactions between gelatin and pectin molecules, forming a reversible physical polyion complex with enhanced performance [147]. Furthermore, the study also incorporated glutaraldehyde to achieve 3D crosslinking, which involves the formation of strong and enduring connections through covalent bonds between polymer chains. Upon introduction to the physical polyion complex of gelatin and pectin, glutaraldehyde reacts with specific polymer functional groups, generating new covalent bonds. This process establishes a stable and enduring hydrogel structure characterized by enhanced mechanical strength and water resistance. While physically crosslinked hydrogels may exhibit reduced strength compared to chemically crosslinked ones, they can offer limited stability and durability.

5.2. Chemical Crosslink Hydrogels

In contrast to physical hydrogels, chemical hydrogels are formed through the covalent crosslinking of biopolymers at specific sites [147,150]. This crosslinking is achieved using crosslinkers, which act as bridges between polymer chains, resulting in a stable and homogenous network. Unlike physical hydrogels, the synthesis and properties of chemical hydrogels are not solely dependent on pH but can be easily controlled by manipulating the crosslinking process. Chemical crosslinking allows for the modification of various hydrogel properties, including swelling behavior, biodegradability, and mechanical strength. Different approaches, such as the inclusion of small molecules, ionizing radiation, and free radical mechanisms, can be employed for covalent crosslinking [18]. Chemically crosslinked hydrogels offer enhanced stability and durability, making them suitable for longer-term applications. The process of modifying pectin to create its derivatives also encompasses ionic gelation between pectin and another polymer. A study on the combination of pectin and chitosan was investigated by Maciel et al. [151] and Shishir et al. [152]. The study prepared pectin-chitosan hydrogel through the formation of a polyelectrolyte complex. This complex arises from the electrostatic interaction between the negatively charged carboxyl groups (COOH) of pectin and the positively charged amino groups (NH_2) of chitosan, resulting in the development of a chemically stable hydrogel. In addition, the prepared hydrogel exhibited remarkable moisturizing properties, was biocompatible, and provided a protective effect on skin wounds [110]. Furthermore, an investigation into the synergy of pectin and cellulose integration was undertaken by Chen et al. [153] using an ionic liquid approach, resulting in the development of a chemically crosslinked hydrogel. This research synthesized a natural composite hydrogel by combining flexible pectin and cellulose within an ionic liquid environment. Ionic liquids are often used as solvents to dissolve cellulose due to their ability to disrupt the strong hydrogen bonding network in cellulose [154]. When pectin, which carries negative charges on its carboxyl groups, is combined with cellulose in the ionic liquid, an electrostatic attraction occurs between the positive charges on the cellulose and the negative charges on the pectin. This interaction leads to the formation of a stable composite hydrogel through chemical crosslinking, which involves the creation of ester bonds between cellulose and pectin molecules. This process leads to the development of a hydrogel with a dense network structure and enhanced properties, as described in the study [153].

5.3. IPN (Interpreting Polymer Network) Crosslink Hydrogels

IPNs are a unique type of hydrogel that involves the physical entanglement of two or more polymer networks, each with its own distinct properties [147,150]. These networks are interlaced at a molecular scale but not covalently bonded to each other, and they cannot be separated unless chemical bonds are broken. IPNs can be semi-IPNs or full-IPNs, depending on the level of crosslinking between the polymers [18]. In semi-IPNs, one polymer network is crosslinked, while the other is physically associated with the crosslinked network. On the other hand, full-IPNs occur when both polymer networks are crosslinked [104]. IPNs provide a way to combine different polymers, such as natural polysaccharides, proteins, or synthetic hydrophilic polymers, to complement each other's deficiencies. By utilizing this entangled structure, IPNs offer unique mechanical, swelling, and biocompatible properties, making them valuable in various applications, including drug delivery and tissue engineering. IPNs can be prepared through different routes by combining natural and synthetic polymers, offering versatility and control in hydrogel design. Yan et al. [155] explored the potential use of IPNs consisting of soy protein isolate (SPI) and sugar beet pectin as carriers for probiotic delivery. The researchers employed an enzymatic approach to create the IPN hydrogels and investigated the influence of laccase's amount as well as the concentrations of SPI and sugar beet pectin on the swelling, textural, and rheological properties of the hydrogels. The authors observed that by altering the laccase quantity and the concentrations of SPI and sugar beet pectin, it was possible to regulate the swelling, texture, and rheological characteristics of the IPN hydrogels [155].

6. Potential Application of Pectin Hydrogel in Food Industry

Pectin hydrogels present a wide range of potential applications in the food industry, offering innovative solutions to address various challenges. One of the characteristics of hydrogels is their ability to retain a subsequent amount of water, making them appealing for innovative use in the food industry. Their eco-friendly nature, coupled with inherent biocompatibility, positions pectin hydrogels as an attractive choice for applications focused on minimizing environmental impact and addressing consumer demand for cleaner and healthier food products. The application of pectin hydrogels as agent carriers, fat replacers, 3D-printed food, and food packaging and coating material is depicted in Figure 6 and further explained in this section.

6.1. Carrier for Active Compound

One of the primary uses of food hydrogels is the encapsulation of bioactive molecules, including food ingredients, additives, antioxidants, vitamins, probiotics, and drugs. Since hydrogels offer regulated release by virtue of their 3D network, they assure the safety and stability of bioactives during food preparation and storage. Peng et al. [156] achieved the encapsulation of vitamin C in citrus peel pectin hydrogel conjugated with bovine serum albumin. The study observed a 65.31% encapsulation efficiency for vitamin C in pectin hydrogel as a carrier. Another study conducted by Zhou et al. [157] investigated nanohydrogel development involving the combination of pectin with low-density lipoprotein as a carrier for curcumin. The nanogels withstood the challenges posed by stomach acid and various digestive enzymes and facilitated an efficient, controlled release of curcumin over a period of time, enhancing its bioavailability and targeted delivery. In another study, Jung et al. [158] explored the potential of various hydrogel sources derived from low-methoxyl citrus pectins and citrus pectin methylesterase (PME)-modified pectin as carriers for the drug indomethacin. Impressively, the study achieved favorable results in terms of drug encapsulation efficiency, particularly for applications in a drug delivery system targeting the colon through oral administration. Additionally, a novel pH-responsive biopolymer mixture known as Al-P, comprising alginate and pectin, was designed to form a hydrogel at pH levels below 3.0. This innovative approach was demonstrated in the study by Guo and Kaletunç [159]. Notably, the production of disc-shaped particles using this approach was innovative and had the potential to enhance adhesion within the intestines. The hydrogel's dissolution characteristics adapt to changes in pH within the environment, enabling the controlled and efficient release of bioactive compounds that align with specific physiological conditions. The study aimed to elucidate the factors impacting the dissolution kinetics of Al-P hydrogel and to create mathematical models describing the degradation behavior of these hydrogels under conditions similar to product storage and the lower gastrointestinal tract. Overall, it is evident from the aforementioned studies that pectin hydrogel has the potential to serve as an effective mechanism for delivering active bioingredients into food delivery systems. This property is very effective for increasing the bioavailability of nutrients and functional components, potentially offering consumers health benefits.

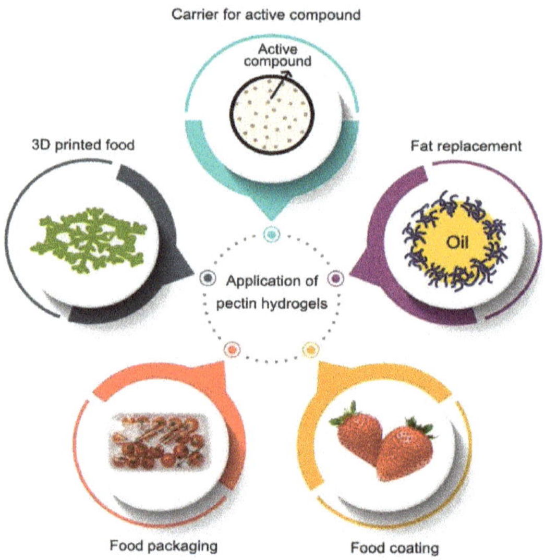

Figure 6. Application of pectin hydrogels in the food industry.

6.2. Fat Replacement and Emulsifiers

Pectin hydrogel particles emulate the texture and deformability of fat particles, effectively mimicking the sensory and physical properties of emulsified fats. Notably, pectin-based fat substitutes have emerged, employing various pectin variants with distinct degrees of esterification. For instance, low-methoxyl pectin (LMP), harnessed through calcium gelation, has found application as a fat mimic in products like mayonnaise [160]. On the other hand, high-methoxyl pectin (HMP) played a pivotal role in crafting oil-filled hydrogel granules through controlled phase separation via hydrophobic interactions and hydrogen bonding. This strategic use of HMP serves the purpose of both fat substitutes and emulsifiers [161]. Their flexible and soft nature makes them a healthier alternative to typical fats without compromising taste or texture. Pectin hydrogels can also be used to improve the nutritional profile of meals by increasing the mouthfeel of low-fat products and developing fat-barrier functions. A study by Kavya et al. [162] demonstrated the utilization of pectin sourced from passion fruit rind to produce an emulsion with varying oil content (20–40%

oil v/v). This was carried out to investigate the transformation process from emulsion to emulgel and its consequent impact on the structural and rheological properties. Passion fruit rind pectin demonstrated impressive emulsifying capabilities by significantly lowering the interfacial tension between water and oil. Furthermore, the study highlights passion fruit rind pectin emulgel as a sustainable fat substitute for commercial use, showcasing the potential of pectin hydrogel as a viable fat replacer. In all, the application of pectin hydrogel as a fat replacer and emulsifier is yet to be fully explored and requires more studies for a consistent report.

6.3. Three-Dimensionally-Printed Food

Another useful property of pectin is its gelation ability, which allows for the creation of structured meals with certain textures and uses. This feature not only enhances sensory characteristics but also allows for the creation of distinctive meals tailored to consumer preferences. Pectin hydrogels are a significant combination for innovative technologies such as 3D printing. Their structural stability makes them suitable for 3D printing applications in food design, enabling the precise fabrication of complex shapes and customized food products. Among the pectin types, LMP has been proposed by a few studies as a suitable food-ink material for the 3D printing of customizable food simulants. A study by Lu et al. [163] formulated polysaccharide-based hydrogel food inks using ionic crosslinked LMP and cellulose nanocrystalline (CNC). LMP, characterized by its lower degree of methoxylation, typically forms gels through electrostatic interactions with cations like Ca^{2+}. The formation of a polymeric network by crosslinking LMP with calcium ions contributed to maintaining the 3D structure of the hydrogels formulated for the food ink in this study. In another investigation conducted by Vancauwenberghe et al. [164], the adjustment of pectin, sugar syrup, and bovine serum albumin (BSA) concentrations was explored to manipulate the desired texture and structural properties of the printed food. The results showed that the viscosity and mechanical properties of the printed food were primarily influenced by pectin and sugar concentrations, while BSA enhanced the gel's porosity.

6.4. Food Packaging

In food packaging, hydrogels are applicable due to their unique properties, such as water retention and controlled release. They prolong the shelf life of perishables, notably fruits and vegetables, by regulating moisture and gas exchange, reducing food waste, and ensuring fresher products. Moreover, hydrogels can also be tailored to release antimicrobial agents or antioxidants, enhancing food preservation and safety. Importantly, they contribute to sustainability by reducing single-use plastics. This section summarizes their various film- and coating-based approaches.

6.4.1. Film-Based Applications

Pectin-based films incorporated with essential oils and plant extracts, such as clove essential oil [165], copaiba oil [166], marjoram [167], and tea polyphenols [168], have been demonstrated to exhibit good antioxidant and antimicrobial activity while also enhancing the film's water barrier properties, which led to longer preservation of intended food products. In another instance, Torpol et al. [109] successfully encapsulated antimicrobial compounds like garlic and holy basil essential oils in chitosan-pectin hydrogel beads, combating various pathogens. The beads demonstrated the capacity to hinder the growth of *Bacillus cereus*, *Clostridium perfringens*, *Escherichia coli*, *Pseudomonas fluorescens*, *Listeria monocytogenes*, and *Staphylococcus aureus*. Another finding by Nešić et al. [119] demonstrated the promising potential of pectin-TiO_2 nanocomposite aerogels as an environmentally friendly and effective material for food packaging. These aerogels, prepared through a sol–gel process and supercritical drying, exhibit improved mechanical, thermal, and antimicrobial properties compared to traditional pectin aerogels. Notably, their thermal conductivity is lower than that of air, which is a valuable attribute for temperature-sensitive food storage. The study by Otálora González et al. [169] successfully developed functional composite edible films based on pectin with beetroot and red cabbage powder fillers. These films exhibited favorable physico-chemical, mechanical, and thermal properties and demonstrated color stability over a 30-day storage period, suggesting their potential as smart indicators for edible food packaging applications. Another study by Dudnyk et al. [170] developed a pectin-based sensor incorporating red cabbage as a food-derived material, which represents an innovative and edible solution for food packaging. This sensor operates as a colorimetric indicator of food freshness, demonstrating high sensitivity to gaseous amines. It effectively detects degradation in various food samples, including beef, chicken, shrimp, and fish, with colorimetric changes aligning well with standard degradation markers. The sensor's ability to correlate visual and measured changes with established freshness indicators like total volatile basic nitrogen and aerobic colony counting highlights its potential as a smart indicator for food packaging, offering both safety and utility.

6.4.2. Coating Applications

Hydrogel coatings have the ability to protect fresh food from deterioration by providing semi-permeable barriers against harmful factors, reducing enzymatic browning and water loss, and can be fortified with minerals, antioxidants, nutrients, vitamins, or probiotics. A study by Muñoz-Labrador et al. [171] investigated the potential use of citrus pectin gels applied as edible coatings for fresh strawberries. The results demonstrated that these pectin gels effectively enhanced the quality of strawberries during storage, reducing moisture loss, changes in acidity, and alterations in color. Furthermore, the utilization of pectin derived from crude cacao shells as a coating for tomatoes demonstrated the capability to postpone quality deterioration, thereby extending the shelf life of the coated samples to 27 days at 4 °C [172]. This underscores the potential of pectin-based coatings to extend the shelf life and preserve the quality of perishable food products like fresh produce and fruits. Additionally, pectin-based coatings enriched with essential oils have been studied to exhibit both antioxidant and antimicrobial effects. These coatings preserve food quality and safety by preventing oxidative degradation and inhibiting microbial growth, offering a natural and eco-friendly alternative to synthetic preservatives. Pectin-based coatings, enriched with essential oils like oregano, rosemary, Mentha piperita, and lemon, have demonstrated efficacy in enhancing the shelf life of various food items, including broccoli, shrimp, and rainbow trout fillets, by mitigating the growth of spoilage microorganisms [173–175]. Similarly, research by Nisar et al. [176] also highlighted the remarkable potential of pectin-based coatings enriched with clove essential oil as potent edible coatings for preserving bream fillets during refrigeration. These coatings, with their demonstrated antimicrobial properties, effectively extend the shelf life of the fillets by inhibiting lipid oxidation and suppressing bacterial growth while simultaneously improving the weight loss, water holding capacity, and textural and color attributes of the bream samples. In addition, research also indicates that active compounds can migrate from pectin-based packaging, influencing sensory characteristics. For example, coating carrots with pectin reduced the accumulation of substances such as lignin precursors and flavonoids, which can contribute to undesirable flavors, resulting in improved overall taste and sensory qualities of the carrots [177]. This demonstrates how pectin coatings can positively impact the way food tastes and feels when consumed.

7. Conclusions

This comprehensive review uncovered the distinct gelling mechanisms of pectin, classified into high-ester and low-ester pectins. High-methoxyl pectins form gels through hydrophobic interactions and hydrogen bonding under specific conditions, while low-methoxyl pectins create continuous gel networks through calcium-mediated "egg-box" formations. Both types of pectin hydrogels offer unique properties with vast potential for various food applications. The review has highlighted that pectin's gelling behavior is influenced by several factors, including degree of esterification (DE), molecular weight, acetylation, pH, ionic strength, and water activity. Understanding these factors and their impact on gel properties is crucial for optimizing the applications of pectin hydrogel in food design.

Furthermore, the review explored the wide array of pectin gel types, including cryogels, aerogels, xerogels, and oleogels, each offering distinct characteristics with vast potential in diverse fields. Cryogels and aerogels, characterized by their high surface area and porous structures, demonstrate considerable potential for drug delivery and wound dressing applications. Xerogels, with reduced water content while retaining the 3D network, are valuable for encapsulating and releasing bioactive compounds in food applications. On the other hand, oleogels, formed by structuring liquid oil with pectin, serve as fat substitutes in food items, contributing to formulations that are both healthier and nutritionally enhanced. The review also highlighted the significant influence of processing factors, such as ionic interactions, ionotropic gelation, filtration, and drying methods, on the properties of pectin gels. Understanding and optimizing these factors is essential for tailoring gel properties to specific applications and enhancing the efficiency of gel preparation techniques. However, there are still knowledge gaps, particularly in optimizing preparation methods and functionalizing gels with nanoparticles or bioactive compounds. Interdisciplinary collaborations and eco-friendly approaches are recommended to advance the field and unleash the full potential of pectin-based gels in diverse industries, benefiting both consumers and the environment.

In addition to pectin gel exploration, this paper also discussed the crosslinking mechanisms in hydrogels, including physical, chemical, and interpenetrating polymer networks (IPNs). Physical hydrogels are formed through non-covalent interactions and are suitable for short-term drug release, while chemical hydrogels, formed through covalent crosslinking, offer enhanced stability and control over properties for longer-term applications. IPNs combine different polymer networks to achieve unique properties, but gaps in understanding cooperative gelation mechanisms and the influence of amidation on gel properties remain. Innovative methods, interdisciplinary collaboration, and synergy between different hydrogel preparation techniques offer potential avenues for advancing the field and unlocking new applications in targeted drug delivery, tissue engineering, and food design.

Author Contributions: N.S.S.: Writing—original draft preparation, review and editing, data curation, and visualization. I.F.O.: Conceptualization, Writing—original draft preparation, review and editing, validation, and supervision. W.Y.L.: Review and editing, project administration, resources. All authors have read and agreed to the published version of the manuscript.

Funding: This work was supported by the Basic Research Program (2021R1I1A3046987) of the National Research Foundation (NRF), Ministry of Education, Korea.

Institutional Review Board Statement: Not applicable.

Informed Consent Statement: Not applicable.

Data Availability Statement: Not applicable.

Conflicts of Interest: The authors declare no conflict of interest.

References

1. Carvalho, A.J. Starch: Major sources, properties and applications as thermoplastic materials. *Monomers Polym. Compos. Renew. Resour.* **2008**, 321–342. [CrossRef]
2. Mellinas, C.; Ramos, M.; Jiménez, A.; Garrigós, M.C. Recent trends in the use of pectin from agro-waste residues as a natural-based biopolymer for food packaging applications. *Materials* **2020**, *13*, 673. [CrossRef]
3. Farooq, A.; Patoary, M.K.; Zhang, M.; Mussana, H.; Li, M.; Naeem, M.A.; Mushtaq, M.; Farooq, A.; Liu, L. Cellulose from sources to nanocellulose and an overview of synthesis and properties of nanocellulose/zinc oxide nanocomposite materials. *Int. J. Biol. Macromol.* **2020**, *154*, 1050–1073. [CrossRef]
4. Saberi Riseh, R.; Gholizadeh Vazvani, M.; Hassanisaadi, M.; Skorik, Y.A. Micro-/nano-carboxymethyl cellulose as a promising biopolymer with prospects in the agriculture sector: A review. *Polymers* **2023**, *15*, 440. [CrossRef]
5. Elieh-Ali-Komi, D.; Hamblin, M.R. Chitin and chitosan: Production and application of versatile biomedical nanomaterials. *Int. J. Adv. Res.* **2016**, *4*, 411.

6. Garavand, F.; Rouhi, M.; Razavi, S.H.; Cacciotti, I.; Mohammadi, R. Improving the integrity of natural biopolymer films used in food packaging by crosslinking approach: A review. *Int. J. Biol. Macromol.* **2017**, *104*, 687–707. [CrossRef] [PubMed]
7. Tang, X.; Kumar, P.; Alavi, S.; Sandeep, K. Recent advances in biopolymers and biopolymer-based nanocomposites for food packaging materials. *Crit. Rev. Food Sci. Nutr.* **2012**, *52*, 426–442. [CrossRef]
8. Wang, X.; Fang, J.; Cheng, L.; Gu, Z.; Hong, Y. Interaction of starch and non-starch polysaccharides in raw potato flour and their effects on thickening stability. *Int. J. Biol. Macromol.* **2023**, *242*, 124702. [CrossRef] [PubMed]
9. Fangwei, L.; Junyi, Y.; Shaoping, N. Classification, properties and development trend of polysaccharides thickeners. *China Food Addit.* **2023**, *34*, 37–45.
10. Chen, Z.; Aziz, T.; Sun, H.; Ullah, A.; Ali, A.; Cheng, L.; Ullah, R.; Khan, F.U. Advances and applications of cellulose bio-composites in biodegradable materials. *J. Polym. Environ.* **2023**, *31*, 2273–2284. [CrossRef]
11. Casalini, S.; Giacinti Baschetti, M. The use of essential oils in chitosan or cellulose-based materials for the production of active food packaging solutions: A review. *J. Sci. Food Agric.* **2023**, *103*, 1021–1041. [CrossRef]
12. Wang, Z.; Ng, K.; Warner, R.D.; Stockmann, R.; Fang, Z. Application of cellulose-and chitosan-based edible coatings for quality and safety of deep-fried foods. *Compr. Rev. Food Sci. Food Saf.* **2023**, *22*, 1418–1437. [CrossRef]
13. Markov, P.A.; Krachkovsky, N.S.; Durnev, E.A.; Martinson, E.A.; Litvinets, S.G.; Popov, S.V. Mechanical properties, structure, bioadhesion, and biocompatibility of pectin hydrogels. *J. Biomed. Mater. Res. Part A* **2017**, *105*, 2572–2581. [CrossRef]
14. Ahmed, E.M. Hydrogel: Preparation, characterization, and applications: A review. *J. Adv. Res.* **2015**, *6*, 105–121. [CrossRef] [PubMed]
15. Singh, S.K.; Dhyani, A.; Juyal, D. Hydrogel: Preparation, characterization and applications. *Pharma Innov.* **2017**, *6*, 25.
16. Gulrez, S.K.; Al-Assaf, S.; Phillips, G.O. Hydrogels: Methods of preparation, characterisation and applications. In *Progress in Molecular and Environmental Bioengineering—From Analysis and Modeling to Technology Applications*; Intech Open: London, UK, 2011; Volume 117150.
17. Ullah, F.; Othman, M.B.H.; Javed, F.; Ahmad, Z.; Akil, H.M. Classification, processing and application of hydrogels: A review. *Mater. Sci. Eng. C* **2015**, *57*, 414–433. [CrossRef] [PubMed]
18. Bashir, S.; Hina, M.; Iqbal, J.; Rajpar, A.; Mujtaba, M.; Alghamdi, N.; Wageh, S.; Ramesh, K.; Ramesh, S. Fundamental concepts of hydrogels: Synthesis, properties, and their applications. *Polymers* **2020**, *12*, 2702. [CrossRef] [PubMed]
19. Nath, P.C.; Debnath, S.; Sridhar, K.; Inbaraj, B.S.; Nayak, P.K.; Sharma, M. A Comprehensive Review of Food Hydrogels: Principles, Formation Mechanisms, Microstructure, and Its Applications. *Gels* **2022**, *9*, 1.
20. Olawuyi, I.F.; Kim, S.R.; Hahn, D.; Lee, W.Y. Influences of combined enzyme-ultrasonic extraction on the physicochemical characteristics and properties of okra polysaccharides. *Food Hydrocoll.* **2020**, *100*, 105396. [CrossRef]
21. Harholt, J.; Suttangkakul, A.; Vibe Scheller, H. Biosynthesis of pectin. *Plant Physiol.* **2010**, *153*, 384–395. [CrossRef]
22. Schrieber, R.; Gareis, H. *Gelatine Handbook: Theory and Industrial Practice*; John Wiley & Sons: Hoboken, NJ, USA, 2007.
23. Chen, J.; Liu, W.; Liu, C.-M.; Li, T.; Liang, R.-H.; Luo, S.-J. Pectin modifications: A review. *Crit. Rev. Food Sci. Nutr.* **2015**, *55*, 1684–1698. [CrossRef]
24. Belkheiri, A.; Forouhar, A.; Ursu, A.V.; Dubessay, P.; Pierre, G.; Delattre, C.; Djelveh, G.; Abdelkafi, S.; Hamdami, N.; Michaud, P. Extraction, characterization, and applications of pectins from plant by-products. *Appl. Sci.* **2021**, *11*, 6596.
25. Cao, L.; Lu, W.; Mata, A.; Nishinari, K.; Fang, Y. Egg-box model-based gelation of alginate and pectin: A review. *Carbohydr. Polym.* **2020**, *242*, 116389.
26. Humerez-Flores, J.N.; Verkempinck, S.H.; De Bie, M.; Kyomugasho, C.; Van Loey, A.M.; Moldenaers, P.; Hendrickx, M.E. Understanding the impact of diverse structural properties of homogalacturonan rich citrus pectin-derived compounds on their emulsifying and emulsion stabilizing potential. *Food Hydrocoll.* **2022**, *125*, 107343. [CrossRef]
27. Morales-Medina, R.; Drusch, S.; Acevedo, F.; Castro-Alvarez, A.; Benie, A.; Poncelet, D.; Dragosavac, M.M.; Tesoriero, M.V.D.; Löwenstein, P.; Yonaha, V. Structure, controlled release mechanisms and health benefits of pectins as an encapsulation material for bioactive food components. *Food Funct.* **2022**, *13*, 10870–10881. [CrossRef] [PubMed]
28. Dominiak, M.; Søndergaard, K.M.; Wichmann, J.; Vidal-Melgosa, S.; Willats, W.G.; Meyer, A.S.; Mikkelsen, J.D. Application of enzymes for efficient extraction, modification, and development of functional properties of lime pectin. *Food Hydrocoll.* **2014**, *40*, 273–282.
29. Maran, J.P.; Sivakumar, V.; Thirugnanasambandham, K.; Sridhar, R. Optimization of microwave assisted extraction of pectin from orange peel. *Carbohydr. Polym.* **2013**, *97*, 703–709. [CrossRef]
30. Hosseini, S.S.; Khodaiyan, F.; Yarmand, M.S. Optimization of microwave assisted extraction of pectin from sour orange peel and its physicochemical properties. *Carbohydr. Polym.* **2016**, *140*, 59–65. [CrossRef]
31. Guo, X.; Han, D.; Xi, H.; Rao, L.; Liao, X.; Hu, X.; Wu, J. Extraction of pectin from navel orange peel assisted by ultra-high pressure, microwave or traditional heating: A comparison. *Carbohydr. Polym.* **2012**, *88*, 441–448.
32. Wang, X.; Chen, Q.; Lü, X. Pectin extracted from apple pomace and citrus peel by subcritical water. *Food Hydrocoll.* **2014**, *38*, 129–137. [CrossRef]
33. Wikiera, A.; Mika, M.; Grabacka, M. Multicatalytic enzyme preparations as effective alternative to acid in pectin extraction. *Food Hydrocoll.* **2015**, *44*, 156–161. [CrossRef]
34. Kurita, O.; Fujiwara, T.; Yamazaki, E. Characterization of the pectin extracted from citrus peel in the presence of citric acid. *Carbohydr. Polym.* **2008**, *74*, 725–730. [CrossRef]

35. Panwar, D.; Panesar, P.S.; Chopra, H.K. Green extraction of pectin from *Citrus limetta* peels using organic acid and its characterization. In *Biomass Conversion and Biorefinery*; Springer: Berlin/Heidelberg, Germany, 2022.
36. Panwar, D.; Panesar, P.S.; Chopra, H.K. Ultrasound-assisted extraction of pectin from Citrus limetta peels: Optimization, characterization, and its comparison with commercial pectin. *Food Biosci.* **2023**, *51*, 102231. [CrossRef]
37. Wang, W.; Ma, X.; Xu, Y.; Cao, Y.; Jiang, Z.; Ding, T.; Ye, X.; Liu, D. Ultrasound-assisted heating extraction of pectin from grapefruit peel: Optimization and comparison with the conventional method. *Food Chem.* **2015**, *178*, 106–114. [CrossRef] [PubMed]
38. Rodsamran, P.; Sothornvit, R. Preparation and characterization of pectin fraction from pineapple peel as a natural plasticizer and material for biopolymer film. *Food Bioprod. Process.* **2019**, *118*, 198–206. [CrossRef]
39. Cho, E.-H.; Jung, H.-T.; Lee, B.-H.; Kim, H.-S.; Rhee, J.-K.; Yoo, S.-H. Green process development for apple-peel pectin production by organic acid extraction. *Carbohydr. Polym.* **2019**, *204*, 97–103. [PubMed]
40. Xie, J.; Zhang, Y.; Klomklao, S.; Simpson, B.K. Pectin from plantain peels: Green recovery for transformation into reinforced packaging films. *Waste Manag.* **2023**, *161*, 225–233. [CrossRef]
41. Ptichkina, N.; Markina, O.; Rumyantseva, G. Pectin extraction from pumpkin with the aid of microbial enzymes. *Food Hydrocoll.* **2008**, *22*, 192–195. [CrossRef]
42. Cui, S.W.; Chang, Y.H. Emulsifying and structural properties of pectin enzymatically extracted from pumpkin. *LWT-Food Sci. Technol.* **2014**, *58*, 396–403. [CrossRef]
43. Wang, M.; Huang, B.; Fan, C.; Zhao, K.; Hu, H.; Xu, X.; Pan, S.; Liu, F. Characterization and functional properties of mango peel pectin extracted by ultrasound assisted citric acid. *Int. J. Biol. Macromol.* **2016**, *91*, 794–803. [CrossRef]
44. Ribeiro, A.C.B.; Cunha, A.P.; da Silva, L.M.R.; Mattos, A.L.A.; de Brito, E.S.; de Souza, M.d.S.M.; de Azeredo, H.M.C.; Ricardo, N.M.P.S. From mango by-product to food packaging: Pectin-phenolic antioxidant films from mango peels. *Int. J. Biol. Macromol.* **2021**, *193*, 1138–1150. [CrossRef] [PubMed]
45. Jiang, L.N.; Shang, J.J.; He, L.B.; Dan, J.M. Comparisons of microwave-assisted and conventional heating extraction of pectin from seed watermelon peel. *Adv. Mater. Res.* **2012**, *550*, 1801–1806. [CrossRef]
46. Petkowicz, C.; Vriesmann, L.; Williams, P. Pectins from food waste: Extraction, characterization and properties of watermelon rind pectin. *Food Hydrocoll.* **2017**, *65*, 57–67. [CrossRef]
47. Maran, J.P.; Sivakumar, V.; Thirugnanasambandham, K.; Sridhar, R. Microwave assisted extraction of pectin from waste Citrullus lanatus fruit rinds. *Carbohydr. Polym.* **2014**, *101*, 786–791. [CrossRef] [PubMed]
48. Hartati, I.; Subekti, E. Microwave assisted extraction of watermelon rind pectin. *Int. J. ChemTech Res.* **2015**, *8*, 163–170.
49. Thirugnanasambandham, K.; Sivakumar, V.; Maran, J.P. Process optimization and analysis of microwave assisted extraction of pectin from dragon fruit peel. *Carbohydr. Polym.* **2014**, *112*, 622–626. [CrossRef]
50. Mollea, C.; Chiampo, F.; Conti, R. Extraction and characterization of pectins from cocoa husks: A preliminary study. *Food Chem.* **2008**, *107*, 1353–1356. [CrossRef]
51. Hennessey-Ramos, L.; Murillo-Arango, W.; Vasco-Correa, J.; Paz Astudillo, I.C. Enzymatic extraction and characterization of pectin from cocoa pod husks (*Theobroma cacao* L.) using Celluclast® 1.5 L. *Molecules* **2021**, *26*, 1473. [CrossRef]
52. Kalapathy, U.; Proctor, A. Effect of acid extraction and alcohol precipitation conditions on the yield and purity of soy hull pectin. *Food Chem.* **2001**, *73*, 393–396. [CrossRef]
53. Yang, J.-S.; Mu, T.-H.; Ma, M.-M. Extraction, structure, and emulsifying properties of pectin from potato pulp. *Food Chem.* **2018**, *244*, 197–205. [CrossRef]
54. Khamsucharit, P.; Laohaphatanalert, K.; Gavinlertvatana, P.; Sriroth, K.; Sangseethong, K. Characterization of pectin extracted from banana peels of different varieties. *Food Sci. Biotechnol.* **2018**, *27*, 623–629. [CrossRef]
55. Muñoz-Almagro, N.; Ruiz-Torralba, A.; Méndez-Albiñana, P.; Guerra-Hernández, E.; García-Villanova, B.; Moreno, R.; Villamiel, M.; Montilla, A. Berry fruits as source of pectin: Conventional and non-conventional extraction techniques. *Int. J. Biol. Macromol.* **2021**, *186*, 962–974. [CrossRef] [PubMed]
56. Eisenmenger, M.J.; Reyes-De-Corcuera, J.I. High hydrostatic pressure increased stability and activity of immobilized lipase in hexane. *Enzym. Microb. Technol.* **2009**, *45*, 118–125. [CrossRef]
57. Spinei, M.; Oroian, M. The influence of extraction conditions on the yield and physico-chemical parameters of pectin from grape pomace. *Polymers* **2022**, *14*, 1378. [CrossRef]
58. Ropartz, D.; Ralet, M.-C. Pectin structure. In *Pectin: Technological and Physiological Properties*; Kontogiorgos, V., Ed.; Springer: Cham, Switzerland, 2020; Volume 1, pp. 17–36.
59. Li, D.-Q.; Li, J.; Dong, H.-L.; Li, X.; Zhang, J.-Q.; Ramaswamy, S.; Xu, F. Pectin in biomedical and drug delivery applications: A review. *Int. J. Biol. Macromol.* **2021**, *185*, 49–65. [CrossRef]
60. Niu, H.; Chen, X.; Luo, T.; Chen, H.; Fu, X. Relationships between the behavior of three different sources of pectin at the oil-water interface and the stability of the emulsion. *Food Hydrocoll.* **2022**, *128*, 107566. [CrossRef]
61. Shahin, L.; Zhang, L.; Mohnen, D.; Urbanowicz, B.R. Insights into Pectin O-acetylation in plant cell wall: Structure, synthesis, and modification. *Cell Surf.* **2023**, *9*, 100099. [CrossRef]
62. Celus, M.; Kyomugasho, C.; Van Loey, A.M.; Grauwet, T.; Hendrickx, M.E. Influence of pectin structural properties on interactions with divalent cations and its associated functionalities. *Compr. Rev. Food Sci. Food Saf.* **2018**, *17*, 1576–1594. [CrossRef]
63. Flutto, L. Pectin: Properties and determination. In *Encyclopedia of food Sciences and Nutrition*, 2nd ed.; Caballero, B., Ed.; Academic Press: San Diego, CA, USA, 2003; pp. 4440–4449. [CrossRef]

64. Vriesmann, L.C.; de Oliveira Petkowicz, C.L. Cacao pod husks as a source of low-methoxyl, highly acetylated pectins able to gel in acidic media. *Int. J. Biol. Macromol.* **2017**, *101*, 146–152. [CrossRef] [PubMed]
65. Filippov, M.; Kohn, R. Determination of the esterification degree of carboxyl groups of pectin with methanol by means of infrared spectroscopy. *Chem. Zvesti* **1975**, *29*, 88–91.
66. Bociek, S.M.; Welti, D. The quantitative analysis of uronic acid polymers by infrared spectroscopy. *Carbohydr. Res.* **1975**, *42*, 217–226. [CrossRef]
67. Gawkowska, D.; Cybulska, J.; Zdunek, A. Structure-related gelling of pectins and linking with other natural compounds: A review. *Polymers* **2018**, *10*, 762. [CrossRef] [PubMed]
68. Grant, G.T.; Morris, E.R.; Rees, D.A.; Smith, P.J.; Thom, D. Biological interactions between polysaccharides and divalent cations: The egg-box model. *FEBS Lett.* **1973**, *32*, 195–198. [CrossRef]
69. Yang, X.; Nisar, T.; Liang, D.; Hou, Y.; Sun, L.; Guo, Y. Low methoxyl pectin gelation under alkaline conditions and its rheological properties: Using NaOH as a pH regulator. *Food Hydrocoll.* **2018**, *79*, 560–571. [CrossRef]
70. Zheng, J.; Chen, J.; Zhang, H.; Wu, D.; Ye, X.; Linardt, R.J.; Chen, S. Gelling mechanism of RG-I enriched citrus pectin: Role of arabinose side-chains in cation-and acid-induced gelation. *Food Hydrocoll.* **2020**, *101*, 105536. [CrossRef]
71. Sousa, A.G.; Nielsen, H.L.; Armagan, I.; Larsen, J.; Sørensen, S.O. The impact of rhamnogalacturonan-I side chain monosaccharides on the rheological properties of citrus pectin. *Food Hydrocoll.* **2015**, *47*, 130–139.
72. Abid, M.; Cheikhrouhou, S.; Renard, C.M.; Bureau, S.; Cuvelier, G.; Attia, H.; Ayadi, M. Characterization of pectins extracted from pomegranate peel and their gelling properties. *Food Chem.* **2017**, *215*, 318–325. [CrossRef]
73. Basak, S.; Annapure, U.S. Trends in "green" and novel methods of pectin modification-A review. *Carbohydr. Polym.* **2022**, *278*, 118967.
74. Lara-Espinoza, C.; Carvajal-Millán, E.; Balandrán-Quintana, R.; López-Franco, Y.; Rascón-Chu, A. Pectin and pectin-based composite materials: Beyond food texture. *Molecules* **2018**, *23*, 942.
75. Zhang, L.; Zheng, J.; Wang, Y.; Ye, X.; Chen, S.; Pan, H.; Chen, J. Fabrication of rhamnogalacturonan-I enriched pectin-based emulsion gels for protection and sustained release of curcumin. *Food Hydrocoll.* **2022**, *128*, 107592.
76. Bu, K.; Wu, S.; Zhu, C.; Wei, M. Comparative study of HG-type low-ester hawthorn pectin as a promising material for the preparation of hydrogel. *Carbohydr. Polym.* **2022**, *296*, 119941. [PubMed]
77. Ochoa-Villarreal, M.; Aispuro-Hernández, E.; Vargas-Arispuro, I.; Martínez-Téllez, M.Á. Plant cell wall polymers: Function, structure and biological activity of their derivatives. *Polymerization* **2012**, *4*, 63–86.
78. Schols, H.A.; Voragen, A. The chemical structure of pectins. In *Pectins and Their Manipulation*; Blackwell: Oxford, UK, 2003; pp. 1–29.
79. Mohnen, D. Pectin structure and biosynthesis. *Curr. Opin. Plant Biol.* **2008**, *11*, 266–277. [CrossRef]
80. Oakenfull, D.; Scott, A. Hydrophobic interaction in the gelation of high methoxyl pectins. *J. Food Sci.* **1984**, *49*, 1093–1098. [CrossRef]
81. Löfgren, C.; Hermansson, A.-M. Synergistic rheological behaviour of mixed HM/LM pectin gels. *Food Hydrocoll.* **2007**, *21*, 480–486. [CrossRef]
82. Kastner, H.; Einhorn-Stoll, U.; Drusch, S. Structure formation in sugar containing pectin gels-Influence of gel composition and cooling rate on the gelation of non-amidated and amidated low-methoxylated pectin. *Food Hydrocoll.* **2017**, *73*, 13–20. [CrossRef]
83. BeMiller, J.N. An introduction to pectins: Structure and properties. In *Chemistry and Function of Pectins*; Fishman, M.L., Jen, J.J., Eds.; American Chemical Society: Washington, DC, USA, 1986; pp. 2–21.
84. Willats, W.G.; McCartney, L.; Mackie, W.; Knox, J.P. Pectin: Cell biology and prospects for functional analysis. *Plant Mol. Biol.* **2001**, *47*, 9–27. [CrossRef]
85. Williamson, G.; Faulds, C.B.; Matthew, J.A.; Archer, D.B.; Morris, V.J.; Brownsey, G.J.; Ridout, M.J. Gelation of sugarbeet and citrus pectins using enzymes extracted from orange peel. *Carbohydr. Polym.* **1990**, *13*, 387–397. [CrossRef]
86. Tibbits, C.W.; MacDougall, A.J.; Ring, S.G. Calcium binding and swelling behaviour of a high methoxyl pectin gel. *Carbohydr. Res.* **1998**, *310*, 101–107. [CrossRef]
87. Genovese, D.B.; Ye, A.; Singh, H. High methoxyl pectin/apple particles composite gels: Effect of particle size and particle concentration on mechanical properties and gel structure. *J. Texture Stud.* **2010**, *41*, 171–189. [CrossRef]
88. Liners, F.; Thibault, J.-F.; Van Cutsem, P. Influence of the degree of polymerization of oligogalacturonates and of esterification pattern of pectin on their recognition by monoclonal antibodies. *Plant Physiol.* **1992**, *99*, 1099–1104. [CrossRef]
89. Luzio, G.A.; Cameron, R.G. Demethylation of a model homogalacturonan with the salt-independent pectin methylesterase from citrus: Part II. Structure–function analysis. *Carbohydr. Polym.* **2008**, *71*, 300–309. [CrossRef]
90. Powell, D.; Morris, E.; Gidley, M.; Rees, D. Conformations and interactions of pectins: II. Influence of residue sequence on chain association in calcium pectate gels. *J. Mol. Biol.* **1982**, *155*, 517–531. [CrossRef]
91. Braccini, I.; Pérez, S. Molecular basis of Ca2+-induced gelation in alginates and pectins: The egg-box model revisited. *Biomacromolecules* **2001**, *2*, 1089–1096. [CrossRef]
92. Fraeye, I.; Colle, I.; Vandevenne, E.; Duvetter, T.; Van Buggenhout, S.; Moldenaers, P.; Van Loey, A.; Hendrickx, M. Influence of pectin structure on texture of pectin–calcium gels. *Innov. Food Sci. Emerg. Technol.* **2010**, *11*, 401–409. [CrossRef]

93. Ngouémazong, D.E.; Tengweh, F.F.; Fraeye, I.; Duvetter, T.; Cardinaels, R.; Van Loey, A.; Moldenaers, P.; Hendrickx, M. Effect of de-methylesterification on network development and nature of Ca2+-pectin gels: Towards understanding structure–function relations of pectin. *Food Hydrocoll.* **2012**, *26*, 89–98. [CrossRef]
94. Sharma, B.; Naresh, L.; Dhuldhoya, N.; Merchant, S.; Merchant, U. An overview on pectins. *Times Food Process. J.* **2006**, *23*, 44–51.
95. Da Silva, J.; Rao, M. 11 pectins: Structure, functionality, and uses. In *Food Polysaccharides and Their Applications*; Taylor & Francis: Boca Raton, FL, USA, 2006.
96. Begum, R.; Aziz, M.G.; Yusof, Y.A.; Saifullah, M.; Uddin, M.B. Evaluation of gelation properties of jackfruit (*Artocarpus heterophyllus*) waste pectin. *Carbohydr. Polym. Technol. Appl.* **2021**, *2*, 100160. [CrossRef]
97. Lootens, D.; Capel, F.; Durand, D.; Nicolai, T.; Boulenguer, P.; Langendorff, V. Influence of pH, Ca concentration, temperature and amidation on the gelation of low methoxyl pectin. *Food Hydrocoll.* **2003**, *17*, 237–244. [CrossRef]
98. Cardoso, S.M.; Coimbra, M.A.; Da Silva, J.L. Temperature dependence of the formation and melting of pectin–Ca^{2+} networks: A rheological study. *Food Hydrocoll.* **2003**, *17*, 801–807. [CrossRef]
99. Gilsenan, P.; Richardson, R.; Morris, E. Thermally reversible acid-induced gelation of low-methoxy pectin. *Carbohydr. Polym.* **2000**, *41*, 339–349. [CrossRef]
100. Chai, Q.; Jiao, Y.; Yu, X. Hydrogels for biomedical applications: Their characteristics and the mechanisms behind them. *Gels* **2017**, *3*, 6. [CrossRef]
101. Mohite, P.B.; Adhav, S. A hydrogels: Methods of preparation and applications. *Int. J. Adv. Pharm* **2017**, *6*, 79–85.
102. Mishra, R.; Majeed, A.; Banthia, A. Development and characterization of pectin/gelatin hydrogel membranes for wound dressing. *Int. J. Plast. Technol.* **2011**, *15*, 82–95. [CrossRef]
103. Amirian, J.; Zeng, Y.; Shekh, M.I.; Sharma, G.; Stadler, F.J.; Song, J.; Du, B.; Zhu, Y. In-situ crosslinked hydrogel based on amidated pectin/oxidized chitosan as potential wound dressing for skin repairing. *Carbohydr. Polym.* **2021**, *251*, 117005. [CrossRef]
104. Giri, T.K.; Thakur, D.; Alexander, A.; Ajazuddin; Badwaik, H.; Tripathy, M.; Tripathi, D.K. Biodegradable IPN hydrogel beads of pectin and grafted alginate for controlled delivery of diclofenac sodium. *J. Mater. Sci. Mater. Med.* **2013**, *24*, 1179–1190. [CrossRef] [PubMed]
105. Sutar, P.B.; Mishra, R.K.; Pal, K.; Banthia, A.K. Development of pH sensitive polyacrylamide grafted pectin hydrogel for controlled drug delivery system. *J. Mater. Sci. Mater. Med.* **2008**, *19*, 2247–2253. [CrossRef]
106. Hoare, T.R.; Kohane, D.S. Hydrogels in drug delivery: Progress and challenges. *Polymer* **2008**, *49*, 1993–2007. [CrossRef]
107. Dai, S.; Wang, S.; Yan, H.; Xu, J.; Hu, H.; Ding, J.; Yuan, N. Stretchable and self-healable hydrogel-based capacitance pressure and strain sensor for electronic skin systems. *Mater. Res. Express* **2019**, *6*, 0850b0859. [CrossRef]
108. Xu, L.; Cui, L.; Jia, M.; Li, Y.; Gao, J.; Jin, X. Self-assembly of flexible graphene hydrogel electrode based on crosslinked pectin-cations. *Carbohydr. Polym.* **2018**, *195*, 593–600. [CrossRef]
109. Torpol, K.; Sriwattana, S.; Sangsuwan, J.; Wiriyacharee, P.; Prinyawiwatkul, W. Optimising chitosan–pectin hydrogel beads containing combined garlic and holy basil essential oils and their application as antimicrobial inhibitor. *Int. J. Food Sci. Technol.* **2019**, *54*, 2064–2074. [CrossRef]
110. Song, K.; Hao, Y.; Liu, Y.; Cao, R.; Zhang, X.; He, S.; Wen, J.; Zheng, W.; Wang, L.; Zhang, Y. Preparation of pectin-chitosan hydrogels based on bioadhesive-design micelle to prompt bacterial infection wound healing. *Carbohydr. Polym.* **2023**, *300*, 120272. [CrossRef] [PubMed]
111. Popov, S.; Smirnov, V.; Khramova, D.; Paderin, N.; Chistiakova, E.; Ptashkin, D.; Vityazev, F. Effect of Hogweed Pectin on Rheological, Mechanical, and Sensory Properties of Apple Pectin Hydrogel. *Gels* **2023**, *9*, 225. [CrossRef] [PubMed]
112. Feng, S.; Yi, J.; Ma, Y.; Bi, J. The role of amide groups in the mechanism of acid-induced pectin gelation: A potential pH-sensitive hydrogel based on hydrogen bond interactions. *Food Hydrocoll.* **2023**, *141*, 108741. [CrossRef]
113. Elma, M.; Saraswati, N.K.D.A.; Simatupang, P.F.A.; Febriyanti, R.; Rahma, A.; Mustalifah, F.R. Hydrogel derived from water hyacinth and pectin from banana peel as a membrane layer. *Mater. Today Proc.* **2023**, *87*, 13–17. [CrossRef]
114. Lee, Y.-e.; Kang, Y.-R.; Chang, Y.H. Effect of pectic oligosaccharide on probiotic survival and physicochemical properties of hydrogel beads for synbiotic encapsulation of Lactobacillus bulgaricus. *Food Biosci.* **2023**, *51*, 102260. [CrossRef]
115. Wu, W.; Wu, Y.; Lin, Y.; Shao, P. Facile fabrication of multifunctional citrus pectin aerogel fortified with cellulose nanofiber as controlled packaging of edible fungi. *Food Chem.* **2022**, *374*, 131763. [CrossRef] [PubMed]
116. Chen, K.; Zhang, H. Alginate/pectin aerogel microspheres for controlled release of proanthocyanidins. *Int. J. Biol. Macromol.* **2019**, *136*, 936–943. [CrossRef]
117. Méndez, D.A.; Schroeter, B.; Martínez-Abad, A.; Fabra, M.J.; Gurikov, P.; López-Rubio, A. Pectin-based aerogel particles for drug delivery: Effect of pectin composition on aerogel structure and release properties. *Carbohydr. Polym.* **2023**, *306*, 120604. [CrossRef]
118. Horvat, G.; Žvab, K.; Knez, Ž.; Novak, Z. Hybrid Polylactic-Acid–Pectin Aerogels: Synthesis, Structural Properties, and Drug Release. *Polymers* **2023**, *15*, 407. [CrossRef]
119. Nešić, A.; Gordić, M.; Davidović, S.; Radovanović, Ž.; Nedeljković, J.; Smirnova, I.; Gurikov, P. Pectin-based nanocomposite aerogels for potential insulated food packaging application. *Carbohydr. Polym.* **2018**, *195*, 128–135.
120. Groult, S.; Buwalda, S.; Budtova, T. Pectin hydrogels, aerogels, cryogels and xerogels: Influence of drying on structural and release properties. *Eur. Polym. J.* **2021**, *149*, 110386. [CrossRef]

121. Luo, S.-Z.; Hu, X.-F.; Jia, Y.-J.; Pan, L.-H.; Zheng, Z.; Zhao, Y.-Y.; Mu, D.-D.; Zhong, X.-Y.; Jiang, S.-T. Camellia oil-based oleogels structuring with tea polyphenol-palmitate particles and citrus pectin by emulsion-templated method: Preparation, characterization and potential application. *Food Hydrocoll.* **2019**, *95*, 76–87. [CrossRef]
122. Dong, Y.; Wei, Z.; Xue, C. Effect of interaction between ovotransferrin fibrils and pectin on properties of oleogel-based Pickering emulsions. *Food Hydrocoll.* **2023**, *140*, 108620. [CrossRef]
123. Pan, L.H.; Wu, X.l.; Luo, S.Z.; He, H.Y.; Luo, J.P. Effects of tea polyphenol ester with different fatty acid chain length on camellia oil-based oleogels preparation and its effects on cookies properties. *J. Food Sci.* **2020**, *85*, 2461–2469. [CrossRef]
124. Konovalova, M.V.; Kurek, D.V.; Litvinets, S.G.; Martinson, E.A.; Varlamov, V.P. Preparation and characterization of cryogels based on pectin and chitosan. *Prog. Chem. Appl. Chitin Its Deriv.* **2016**, *XXI*, 114–121. [CrossRef]
125. Ma, Y.; Bi, J.; Yi, J.; Feng, S.; Peng, J.; Zou, S.; Guo, S.; Wu, Z. Modulation of ice crystal formation behavior in pectin-sucrose hydrogel by freezing temperature: Effect on ice crystal morphology and drying properties. *Dry. Technol.* **2023**, *41*, 1771–1782. [CrossRef]
126. Martinez, Y.N.; Piñuel, L.; Castro, G.R.; Breccia, J.D. Polyvinyl alcohol–pectin cryogel films for controlled release of enrofloxacin. *Appl. Biochem. Biotechnol.* **2012**, *167*, 1421–1429. [CrossRef]
127. Mata, Y.; Blázquez, M.; Ballester, A.; González, F.; Muñoz, J. Studies on sorption, desorption, regeneration and reuse of sugar-beet pectin gels for heavy metal removal. *J. Hazard. Mater.* **2010**, *178*, 243–248. [CrossRef]
128. Mata, Y.; Blázquez, M.; Ballester, A.; González, F.; Munoz, J. Optimization of the continuous biosorption of copper with sugar-beet pectin gels. *J. Environ. Manag.* **2009**, *90*, 1737–1743. [CrossRef]
129. Slavutsky, A.M.; Bertuzzi, M.A. Formulation and characterization of hydrogel based on pectin and brea gum. *Int. J. Biol. Macromol.* **2019**, *123*, 784–791. [CrossRef]
130. Kim, J.U.; Kim, B.; Shahbaz, H.M.; Lee, S.H.; Park, D.; Park, J. Encapsulation of probiotic Lactobacillus acidophilus by ionic gelation with electrostatic extrusion for enhancement of survival under simulated gastric conditions and during refrigerated storage. *Int. J. Food Sci. Technol.* **2017**, *52*, 519–530.
131. de Moura, S.C.; Berling, C.L.; Germer, S.P.; Alvim, I.D.; Hubinger, M.D. Encapsulating anthocyanins from Hibiscus sabdariffa L. calyces by ionic gelation: Pigment stability during storage of microparticles. *Food Chem.* **2018**, *241*, 317–327. [CrossRef] [PubMed]
132. Sharma, M.; Dash, K.K.; Badwaik, L.S. Physicochemical and release behaviour of phytochemical compounds based on black jamun pulp extracts-filled alginate hydrogel beads through vibration dripping extrusion. *Int. J. Biol. Macromol.* **2022**, *194*, 715–725. [CrossRef] [PubMed]
133. Patil, J.; Kamalapur, M.; Marapur, S.; Kadam, D. Ionotropic gelation and polyelectrolyte complexation: The novel techniques to design hydrogel particulate sustained, modulated drug delivery system: A review. *Dig. J. Nanomater. Biostructures* **2010**, *5*, 241–248.
134. Winkleman, A.; Bracher, P.J.; Gitlin, I.; Whitesides, G.M. Fabrication and manipulation of ionotropic hydrogels cross-linked by paramagnetic ions. *Chem. Mater.* **2007**, *19*, 1362–1368. [CrossRef]
135. Mishra, R.K.; Datt, M.; Banthia, A.K. Synthesis and characterization of pectin/PVP hydrogel membranes for drug delivery system. *Aaps Pharmscitech* **2008**, *9*, 395–403. [CrossRef]
136. Bušić, A.; Belščak-Cvitanović, A.; Cebin, A.V.; Karlović, S.; Kovač, V.; Špoljarić, I.; Mršić, G.; Komes, D. Structuring new alginate network aimed for delivery of dandelion (*Taraxacum officinale* L.) polyphenols using ionic gelation and new filler materials. *Food Res. Int.* **2018**, *111*, 244–255.
137. Konovalova, M.V.; Markov, P.A.; Durnev, E.A.; Kurek, D.V.; Popov, S.V.; Varlamov, V.P. Preparation and biocompatibility evaluation of pectin and chitosan cryogels for biomedical application. *J. Biomed. Mater. Res. Part A* **2017**, *105*, 547–556. [CrossRef]
138. Lozinsky, V.I.; Galaev, I.Y.; Plieva, F.M.; Savina, I.N.; Jungvid, H.; Mattiasson, B. Polymeric cryogels as promising materials of biotechnological interest. *Trends Biotechnol.* **2003**, *21*, 445–451.
139. Erkey, C.; Türk, M. Support materials. In *Supercritical Fluid Science and Technology*; Elsevier: Amsterdam, The Netherlands, 2021; Volume 8, pp. 19–30.
140. Jiang, L.; Wen, Y.; Zhu, Z.; Liu, X.; Shao, W. A Double cross-linked strategy to construct graphene aerogels with highly efficient methylene blue adsorption performance. *Chemosphere* **2021**, *265*, 129169. [CrossRef]
141. Erkey, C.; Turk, M. *Synthesis of Nanostructured Materials in Near and/or Supercritical Fluids: Methods, Fundamentals and Modeling*; Elsevier: Amsterdam, The Netherlands, 2021.
142. Czarnobaj, K. Preparation and characterization of silica xerogels as carriers for drugs. *Drug Deliv.* **2008**, *15*, 485–492. [CrossRef]
143. Nayak, A.K.; Das, B. Introduction to polymeric gels. In *Polymeric Gels*; Elsevier: Amsterdam, The Netherlands, 2018; pp. 3–27.
144. Buchtová, N.; Budtova, T. Cellulose aero-, cryo-and xerogels: Towards understanding of morphology control. *Cellulose* **2016**, *23*, 2585–2595. [CrossRef]
145. Pinto, T.C.; Martins, A.J.; Pastrana, L.; Pereira, M.C.; Cerqueira, M.A. Oleogel-based systems for the delivery of bioactive compounds in foods. *Gels* **2021**, *7*, 86. [CrossRef]
146. Manzoor, S.; Masoodi, F.; Naqash, F.; Rashid, R. Oleogels: Promising alternatives to solid fats for food applications. *Food Hydrocoll. Health* **2022**, *2*, 100058. [CrossRef]
147. Farris, S.; Schaich, K.M.; Liu, L.; Piergiovanni, L.; Yam, K.L. Development of polyion-complex hydrogels as an alternative approach for the production of bio-based polymers for food packaging applications: A review. *Trends Food Sci. Technol.* **2009**, *20*, 316–332. [CrossRef]

148. Moghaddam, R.H.; Dadfarnia, S.; Shabani, A.M.H.; Moghaddam, Z.H.; Tavakol, M. Electron beam irradiation synthesis of porous and non-porous pectin based hydrogels for a tetracycline drug delivery system. *Mater. Sci. Eng. C* **2019**, *102*, 391–404. [CrossRef] [PubMed]
149. Farris, S.; Schaich, K.M.; Liu, L.; Cooke, P.H.; Piergiovanni, L.; Yam, K.L. Gelatin–pectin composite films from polyion-complex hydrogels. *Food Hydrocoll.* **2011**, *25*, 61–70. [CrossRef]
150. Miran, M.; Salami, M.; Emam-Djomeh, Z.; Moreno, F.J.; Montilla, A. Isolation and structural evaluation of pectin, pectin-based polymer blends, composites, IPNs and gels. *Handb. Nat. Polym.* **2023**, *1*, 369–398.
151. Maciel, V.B.V.; Yoshida, C.M.; Franco, T.T. Chitosan/pectin polyelectrolyte complex as a pH indicator. *Carbohydr. Polym.* **2015**, *132*, 537–545. [CrossRef] [PubMed]
152. Shishir, M.R.I.; Karim, N.; Gowd, V.; Xie, J.; Zheng, X.; Chen, W. Pectin-chitosan conjugated nanoliposome as a promising delivery system for neohesperidin: Characterization, release behavior, cellular uptake, and antioxidant property. *Food Hydrocoll.* **2019**, *95*, 432–444. [CrossRef]
153. Chen, W.; Yuan, S.; Shen, J.; Chen, Y.; Xiao, Y. A composite hydrogel based on pectin/cellulose via chemical cross-linking for hemorrhage. *Front. Bioeng. Biotechnol.* **2021**, *8*, 627351. [CrossRef]
154. Taokaew, S. Recent Advances in Cellulose-Based Hydrogels Prepared by Ionic Liquid-Based Processes. *Gels* **2023**, *9*, 546. [CrossRef] [PubMed]
155. Yan, W.; Jia, X.; Zhang, Q.; Chen, H.; Zhu, Q.; Yin, L. Interpenetrating polymer network hydrogels of soy protein isolate and sugar beet pectin as a potential carrier for probiotics. *Food Hydrocoll.* **2021**, *113*, 106453. [CrossRef]
156. Peng, H.; Chen, S.; Luo, M.; Ning, F.; Zhu, X.; Xiong, H. Preparation and self-assembly mechanism of bovine serum albumin–citrus peel pectin conjugated hydrogel: A potential delivery system for vitamin C. *J. Agric. Food Chem.* **2016**, *64*, 7377–7384. [CrossRef]
157. Zhou, M.; Wang, T.; Hu, Q.; Luo, Y. Low density lipoprotein/pectin complex nanogels as potential oral delivery vehicles for curcumin. *Food Hydrocoll.* **2016**, *57*, 20–29. [CrossRef]
158. Jung, J.; Arnold, R.D.; Wicker, L. Pectin and charge modified pectin hydrogel beads as a colon-targeted drug delivery carrier. *Colloids Surf. B Biointerfaces* **2013**, *104*, 116–121. [CrossRef] [PubMed]
159. Guo, J.; Kaletunç, G. Dissolution kinetics of pH responsive alginate-pectin hydrogel particles. *Food Res. Int.* **2016**, *88*, 129–139. [CrossRef] [PubMed]
160. Du, Q.; Zhou, L.; Lyu, F.; Liu, J.; Ding, Y. The complex of whey protein and pectin: Interactions, functional properties and applications in food colloidal systems–A review. *Colloids Surf. B Biointerfaces* **2022**, *210*, 112253. [CrossRef]
161. Xu, X.; Zhang, H.; Li, L.; Sun, L.; Jia, B.; Yang, H.; Zuo, F. Preparation of fat substitute based on the high-methoxyl pectin of citrus and application in moon-cake skin. *Food Sci. Technol.* **2021**, *42*, e92121. [CrossRef]
162. Kavya, M.; Jacob, A.R.; Nisha, P. Pectin emulsions and emulgels: Bridging the correlation between rheology and microstructure. *Food Hydrocoll.* **2023**, *143*, 108868. [CrossRef]
163. Lu, Y.; Rai, R.; Nitin, N. Image-based assessment and machine learning-enabled prediction of printability of polysaccharides-based food ink for 3D printing. *Food Res. Int.* **2023**, *173*, 113384. [CrossRef]
164. Vancauwenberghe, V.; Katalagarianakis, L.; Wang, Z.; Meerts, M.; Hertog, M.; Verboven, P.; Moldenaers, P.; Hendrickx, M.E.; Lammertyn, J.; Nicolaï, B. Pectin based food-ink formulations for 3-D printing of customizable porous food simulants. *Innov. Food Sci. Emerg. Technol.* **2017**, *42*, 138–150. [CrossRef]
165. Nisar, T.; Wang, Z.-C.; Yang, X.; Tian, Y.; Iqbal, M.; Guo, Y. Characterization of citrus pectin films integrated with clove bud essential oil: Physical, thermal, barrier, antioxidant and antibacterial properties. *Int. J. Biol. Macromol.* **2018**, *106*, 670–680. [CrossRef]
166. Norcino, L.; Mendes, J.; Natarelli, C.; Manrich, A.; Oliveira, J.; Mattoso, L. Pectin films loaded with copaiba oil nanoemulsions for potential use as bio-based active packaging. *Food Hydrocoll.* **2020**, *106*, 105862. [CrossRef]
167. Almasi, H.; Azizi, S.; Amjadi, S. Development and characterization of pectin films activated by nanoemulsion and Pickering emulsion stabilized marjoram (*Origanum majorana* L.) essential oil. *Food Hydrocoll.* **2020**, *99*, 105338. [CrossRef]
168. Lei, Y.; Wu, H.; Jiao, C.; Jiang, Y.; Liu, R.; Xiao, D.; Lu, J.; Zhang, Z.; Shen, G.; Li, S. Investigation of the structural and physical properties, antioxidant and antimicrobial activity of pectin-konjac glucomannan composite edible films incorporated with tea polyphenol. *Food Hydrocoll.* **2019**, *94*, 128–135. [CrossRef]
169. Otálora González, C.M.; De'Nobili, M.D.; Rojas, A.M.; Basanta, M.F.; Gerschenson, L.N. Development of functional pectin edible films with fillers obtained from red cabbage and beetroot. *Int. J. Food Sci. Technol.* **2021**, *56*, 3662–3669. [CrossRef]
170. Dudnyk, I.; Janeček, E.-R.; Vaucher-Joset, J.; Stellacci, F. Edible sensors for meat and seafood freshness. *Sens. Actuators B Chem.* **2018**, *259*, 1108–1112. [CrossRef]
171. Muñoz-Labrador, A.; Moreno, R.; Villamiel, M.; Montilla, A. Preparation of citrus pectin gels by power ultrasound and its application as an edible coating in strawberries. *J. Sci. Food Agric.* **2018**, *98*, 4866–4875. [CrossRef]
172. Pholsin, R.; Shiekh, K.A.; Jafari, S.; Kijpatanasilp, I.; Nan, T.N.; Suppavorasatit, I.; Assatarakul, K. Impact of pectin edible coating extracted from cocoa shell powder on postharvest quality attributes of tomato (*Lycopersicon esculentum* Mill.) fruit during storage. *Food Control* **2023**, *155*, 110023. [CrossRef]
173. Tabatabaei Moradi, L.; Sharifan, A.; Larijani, K. The effect of multilayered chitosan–pectin–Mentha piperita and lemon essential oil on oxidation effects and quality of rainbow trout fillet (*Oncorhynchus mykiss*) during refrigeration at 4 ± 1 °C storage. *Iran. J. Fish. Sci.* **2020**, *19*, 2544–2559.

174. Alvarez, M.V.; Ortega-Ramirez, L.A.; Silva-Espinoza, B.A.; Gonzalez-Aguilar, G.A.; Ayala-Zavala, J.F. Antimicrobial, antioxidant, and sensorial impacts of oregano and rosemary essential oils over broccoli florets. *J. Food Process. Preserv.* **2019**, *43*, e13889.
175. Alvarez, M.V.; Ortega-Ramirez, L.A.; Gutierrez-Pacheco, M.M.; Bernal-Mercado, A.T.; Rodriguez-Garcia, I.; Gonzalez-Aguilar, G.A.; Ponce, A.; Moreira, M.d.R.; Roura, S.I.; Ayala-Zavala, J.F. Oregano essential oil-pectin edible films as anti-quorum sensing and food antimicrobial agents. *Front. Microbiol.* **2014**, *5*, 699. [CrossRef] [PubMed]
176. Nisar, T.; Yang, X.; Alim, A.; Iqbal, M.; Wang, Z.-C.; Guo, Y. Physicochemical responses and microbiological changes of bream (*Megalobrama ambycephala*) to pectin based coatings enriched with clove essential oil during refrigeration. *Int. J. Biol. Macromol.* **2019**, *124*, 1156–1166. [CrossRef]
177. Ranjitha, K.; Rao, D.S.; Shivashankara, K.; Oberoi, H.S.; Roy, T.K.; Bharathamma, H. Shelf-life extension and quality retention in fresh-cut carrots coated with pectin. *Innov. Food Sci. Emerg. Technol.* **2017**, *42*, 91–100. [CrossRef]

MDPI
St. Alban-Anlage 66
4052 Basel
Switzerland
www.mdpi.com

Gels Editorial Office
E-mail: gels@mdpi.com
www.mdpi.com/journal/gels

www.ingramcontent.com/pod-product-compliance
Lightning Source LLC
Chambersburg PA
CBHW040311270926
51850CB00007BC/1011
* 9 7 8 3 0 3 6 5 9 6 9 8 3 *